SPRINGER
LABORATORY

R. Kuhn S. Hoffstetter-Kuhn

Capillary Electrophoresis: Principles and Practice

With 90 Figures

Springer-Verlag
Berlin Heidelberg New York London Paris
Tokyo Hong Kong Barcelona Budapest

Professor Dr. REINHARD KUHN
FB Chemie
Fachhochschule für Technik und Wirtschaft
7410 Reutlingen, Germany

Dr. SABRINA HOFFSTETTER-KUHN
Breitensteinstr. 29
7432 Bad Urach, Germany

ISBN-13:978-3-642-78060-8 e-ISBN-13:978-3-642-78058-5
DOI: 10.1007/978-3-642-78058-5

Library of Congress Cataloging-in-Publication Data. Kuhn, R. (Reinhard), 1956– Capillary electrophoresis: principles and practice / R. Kuhn, S. Hoffstetter-Kuhn. p. cm. Includes bibliographical references and index. ISBN-13:978-3-642-78060-8
1. Capillary electrophoresis. I. Hofsstetter-Kuhn, S.
(Sabrina), 1961– II. Title. QP519.9.C36K84 1993 543'.0871–dc20 93-19399

Softcover reprint of the hardcover 1st edition 1993

Typesetting: Camera ready by author
52/3145-5 4 3 2 1 0 – Printed on acid-free paper

Preface

Electrophoresis is one of the most widely used separation technique and still a fruitful field of innovative research although the theoretical principles are known for almost 100 years by Kohlrausch's pioneering work. Among numerous other electrophoretic modes such as isoelectric focusing or two dimensional electrophoresis in slab gels that are integral part of nearly all biochemical work, capillary electrophoresis (CE) is the latest development in this series.

Though the first publications on CE appeared just 10 years ago, a big number of commercial equipment is currently available. Today capillary electrophoresis has left the stage of evaluation and is becoming a routine separation technique. Due to its separation principle which is orthogonal to chromatography, CE ideally supplements HPLC. Moreover, it combines the advantages of electrophoresis such as the broad application range covering small ions up to whole living cells or particles with those of HPLC like automated operation and quantitation of separated bands. Thus, it is no wonder that CE gains more and more importance among biochemists and analysts working in pharmaceutical industry.

This book is intended to be a practical guide for beginners in CE as well as for those researchers with some experience in this field. First, the reader will be guided through two chapters summarizing the basic principles and the most important factors influencing the performance of CE. Equipped with these "theoretical" qualifications he will find a survey of current instrumentation and detailed descriptions of the different techniques of CE. Frequently she/he will find practical hints, tables of solubilities, etc., which supplies useful information for working at the laboratory bench. Many applications are presented in tabular form hoping that this kind of presentation provides the best survey and stimulates the reader to develop his own method based on the given informations.

Because errors are never completely eliminated (only those who do nothing make no mistakes), we would like to ask the readers to find these errors and to receive a "thank you" in a (possible) next edition.

We want to thank many colleagues for their valuable advises. Especially we would like to thank Roman Frei, Claude Morin, Celine Steinmetz and Francois Vogel for their technical assistance. Our friend Dr. Terry Christen-Olefirowicz reviewed parts of the manuscript. Dr. Fritz Erni, Dr. Vreni Steiner and Peter Enders supported our work. Thanks to all of them. Last but not least we would like to thank all our friends for their patience during the "genesis" of this book.

Bad Urach
April 1993

Reinhard Kuhn
Sabrina Hoffstetter-Kuhn

Contents

1 Introduction

1.1 Safety Considerations

A number of hazardous steps are involved in performing capillary electrophoresis. Since high voltages are generally applied, operating instructions of the equipment manufacturers have to be followed precisely. One has to be aware that on occasion highly toxic chemicals are used. Particular attention has to be paid to the local and national safety regulations.

1.2 History

It is the goal of any analytical technique to provide information about the composition of a material system. While in the past analysis was focused on the determination of single compounds, which means identification and quantification, today additional requirements are demanded which have to be met by modern analytical systems. Separation procedures are becoming more and more important. In many cases information about sample composition is desired rather than the analysis of one single compound. Purity control requires the separation and analysis of by-products in addition to the main component. Pattern recognition techniques, like peptide or nucleotide mapping, are firmly established in biochemical research. Last but not least, sequencing analysis provides information about the structure of biomolecules like proteins, nucleotides and carbohydrates. Ultimately, more detailed information about more complex samples have to be provided in shorter times by using highly sophisticated equipment, allowing real-time data processing and automation.

Today the most powerful separation techniques are based on the principle of electrophoresis, which can be described in general as the migration of charged substances in solutions under the influence of an applied electrical field. The description of the principle of electrophoresis goes back to the last century when Kohlrausch derived his basic equations for ionic migration in an electrolyte solution in 1897 [1]. Since paper and later gels of polyacrylamide and agarose were introduced in electrophoresis to suppress convection due to Joule heating, the different techniques of gel electrophoresis, such as zone electrophoresis (ZE) and isoelectric focusing (IEF), have become indispensable in biochemistry.

tions, identification of species and individuals in forensic medicine and serum and lipoprotein analysis in clinical chemistry are some of the most important areas of gel electrophoresis. As a result of this very broad application range, gel electrophoresis is synonymous with electrophoresis for almost everybody, even today.

Nevertheless, these techniques, which include gel preparation, sample application, staining and eventually quantification of the zones by densitometry are very time consuming and labour-intensive. Additionally, interactions between the analytes and the gel matrix affect the separation. Even though this is often desirable, e.g. the molecular sieving effect of polyacrylamide gels in zone electrophoresis, electrophoretic behavior of the separated compounds is overlaid by chromatography. Hence, many attempts were made to perform electrophoresis in free solution without any stabilizing media to overcome convection.

Zone electrophoresis in free solution was described by Hjertén in 1967 [2]. He performed zone electrophoresis in tubes of quartz glass, having inner diameters of 1-3 mm and coatings of methylcellulose to prevent electroosmosis. Convection was reduced by rotating the separation chamber about its longitudinal axis. Zone detection was accomplished with a UV detector, which scanned the length of the tube. This free zone electrophoresis technique was applied to the separation of a great variety of samples including proteins and nucleic acids as well as viruses.

Another approach is the use of narrow-bore capillary tubes of sub-millimeter diameter. Due to the high ratio of the cross-section of the separation compartment to its surface area, heat dissipation is enhanced in these systems. Based on this so-called anticonvective wall effect, Everaerts and coworkers developed capillary isotachophoresis (ITP) in narrow-bore Teflon tubes in the mid 1970s [3]. The use of Teflon instead of glass tubes has the advantage that electroosmosis, which would distort the isotachophoretic separation, is minimized. Although commercial equipment for this technique has been available since that time, the interest in capillary ITP among the scientific world was rather low in comparison to other techniques.

In 1974, Virtanen reported zone electrophoresis in glass tubes of 200-500 μm bores [4]. The separated compounds were detected by potentiometry. Several years later, Mikkers, Everaerts and Verheggen performed zone electrophoresis in narrow-bore Teflon tubes with an internal diameter of 200 μm [5]. The separation within 10 minutes of 16 small anions, ranging from chloride to benzylaspartarte, was demonstrated employing conductometric detection. Plate heights of less than 10 μm were achieved. Nevertheless, this detection mode was relatively insensitive and required sample overloading.

This shows that the general principle of capillary zone electrophoresis (CZE) has been known for a long time. Two major problems, however, were not completely solved at that time, namely the low sensitivity of the detection systems for narrow-bore tubes and electroosmosis. It was James Jorgenson and coworkers who helped to achieve the rapid development of this method in the last decade [6-8]. "Much of the interest in applying HPLC to separation and analysis of biopolymers stems from the fact that HPLC is a highly instrumental technique with autosamplers and on-line detectors connected to computers for data acquisition and analysis." With this statement Jorgenson not only pointed out the problems that have to be taken into account when using gel electrophoretic techniques, but also initiated a new way to the development of electrophoresis with a totally different approach to all electrophoreticians so far. Instead of suppressing electroosmosis by using electrically inert capillaries, he took advantage of the unique plug flow profile of the electroosmotic flow, which is generated in fused silica capillaries of very narrow internal diameters, to move the analytes through a capillary with much less dispersion than observed in HPLC. His first two publications appeared in 1981, where separations of dansyl and fluorescamine derivatives of amine-containing compounds with plate heights of only a few micrometers are shown [6,7]. In this work, on-column fluorescence detection was used to increase sensitivity.

In the following years many improvements with respect to detection systems as well as performance were made. Additionally many subtechniques related to CZE were developed to meet the requests for powerful separation techniques, especially for biological and pharmacological compounds. Today, only 12 years after its introduction, capillary electrophoresis (CE) is a well established and indispensible technique combining many of the advantages of HPLC with those of electrophoresis, i.e. high efficiency, analysis speed, low sample volumes and applicability to polar and non-polar substances.

1.3 Nomenclature

The rapid progress of capillary electrophoresis and the fact that workers developing and researching this technique came from different areas (namely electrophoresis and chromatography) led to a confusing variety of nomenclature and abbreviations used for capillary electrophoresis and its related subtechniques. A short list of the most common names and abbreviations is given below:

ACE	affinity capillary electrophoresis
CAE	***capillary affinity electrophoresis***
CE	***capillary electrophoresis***
CEC	capillary electrochromatography
CES	capillary electroseparation
CGE	***capillary gel electrophoresis***
CIEF	***capillary isoelectric focusing***
CITP	***capillary isotachophoresis***
CMEC	capillary micellar electrokinetic chromatography
CZE	***capillary zone electrophoresis***
EC	***electrochromatography***
EKC	electrokinetic chromatography
FSCE	free solution capillary electrophoresis

FZE	free zone electrophoresis
MEC	micellar electrokinetic chromatography
MECC	micellar electrokinetic capillary chromatography
MEKC	***micellar electrokinetic chromatography***
HPCE	high performance capillary electrophoresis
HPE	high performance electrophoresis
HPZE	high performance zone electrophoresis

Obviously several expressions exist for one and the same object. There is, for example, no difference between CE, HPE and HPCE. The last abbreviation was an attempt to show the similarity to HPLC, which has itself changed its meaning from high pressure to high performance liquid chromatography. As another example, MEC, MECC and MEKC describe one and the same technique. Additionally, in some cases electrophoretic techniques (CGE, MEKC) and in other cases electrophoretic separation principles (CZE, CIEF, CITP) are used to define the method. Finally, CE sometimes means capillary electrophoresis but can also mean capillary zone electrophoresis. CE should be used as the general term for CZE, MEKC and CGE. These three techniques go back to the principle of zone electrophoresis, thus making the mix-up of the terms more understandable.

To bring more order into this jungle of abbreviations at least for this book, we will try to use only those abbreviations that are written in bold italic letters in the list above. Capillary electrophoresis includes CZE, CGE and MEKC. Although other authors sometimes incorporate also capillary ITP and IEF under this term, we think that it is more convenient to treat them separately because the principles and practice of both techniques are different to CE.

2 Basic Principles

"Separation is the art and science of maximizing separative relative to dispersive transport". With this remark Calvin Giddings obviously wanted to emphasize that separation science needs fantasy (art) and knowledge (science) in order to be applied successfully. In general, separation is the result of different chemical potentials of the analytes due to the physico-chemical interaction with their environment. The counteracting process of separation is dispersion. A profound understanding of the basic concepts of the separation process in capillary electrophoresis is mandatory to realize the capabilities and limitations of the technique. In this chapter, we will present briefly the four basic electrophoretic separation modes. Afterwards, we will go into the theory of electrophoretic migration and electroosmosis and will discuss performance criteria such as efficiency and resolution. This discourse will focus essentially on aspects of capillary zone electrophoresis in open capillary tubes, which means capillaries open at the ends and containing only buffer without any stabilizer. In many respects the deduced conclusions also apply to related techniques discussed in chapter 5.

2.1 Basic Electrophoretic Separation Modes

Electrophoresis is defined as the transport of electrically charged compounds in solution under the influence of an electric field. It includes a great number of systems involving either differential or equilibrium gradient methods. They are all based on one of the four electrophoretic modes, namely moving-boundary electrophoresis (MBE), zone electrophoresis (ZE), isotachophoresis (ITP) and isoelectric focusing (IEF). All electrophoretic processes can be described with only one set of equations showing their fundamental unity [9]. Additionally, they all can be carried out in the same electrophoretic equipment. Moving-boundary electrophoresis is in fact of great historical importance [10], because it was the first electrophoretic method successfully applied to the separation of charged compounds in free solution. In contemporary electrophoresis, however, MBE has lost its previous importance. Therefore the presentation of the principle of MBE can be disregarded here. In the following description of the principles of ZE, ITP and IEF a narrow-bore tube, connected with anode and cathode compartments, is considered as the separation system.

2.1.1 Zone Electrophoresis

For zone electrophoresis (Fig. 2.1.), the column and the electrode reservoirs have to be filled with the so-called carrier or background electrolyte, which conducts the electric current and provides the buffering capacity. The sample, consisting of a mixture of anions and cations, is introduced into this continuous buffer system at one end of the tube as a sharp initial zone. This zone represents the only discontinuity of the system. Under the influence of the electric field, the ionic species of the carrier electrolyte and of the sample migrate to the corresponding electrode, cations towards the cathode (A, B, C) and anions (D) towards the anode, respectively. Due to its high concentration the carrier electrolyte determines the physical properties such as conductance and pH throughout the capillary. The influence of the sample can be neglected. Therefore the sample components migrate independently from the carrier electrolyte with their specific velocities. After some time, they will separate into distinct zones if their differences in net mobilities are high enough. Their relative position and their shapes continuously change with time. Thus, no steady-state is reached in ZE.

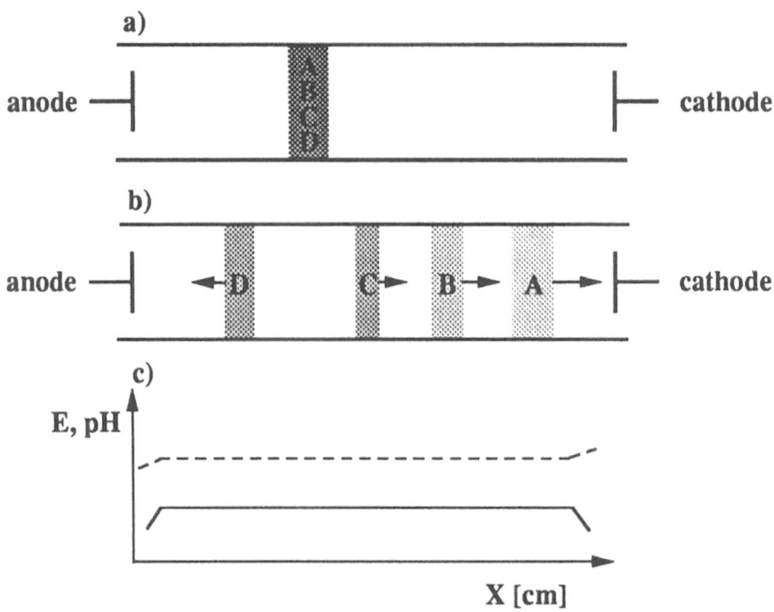

Fig. 2.1. Principle of zone electrophoresis.(**a**) initial state, (**b**) differential migration of the distinct sample zones, (**c**) profile of field strength (*plain line*) and pH (*dashed line*) across the separation chamber

ZE can be carried out either as a one phase process in free solution or in combination with a solid support medium or a second liquid phase. In the latter cases, separation is not only governed by electrophoretic migration, but also by molecular sieving or partitioning between two phases. For further details concerning the theory of ZE, the reader is referred to chapter 3 and Sects. 5.1 - 5.3.

2.1.2 Isotachophoresis

In isotachophoresis a small quantity of sample is introduced at the interface of a discontinuous buffer system, consisting of a leading (L) and a terminating (T) electrolyte. In Fig. 2.2., the separation of 3 cationic species (A, B, C) in a narrow-bore tube is simulated. The tube and the cathode reservoir are filled with the leading electrolyte whose cations must have a higher mobility than the cations that have to be separated. Additionally, the anions of the leader should possess a buffer capacity at the pH at which the sample is separated. The anode reservoir

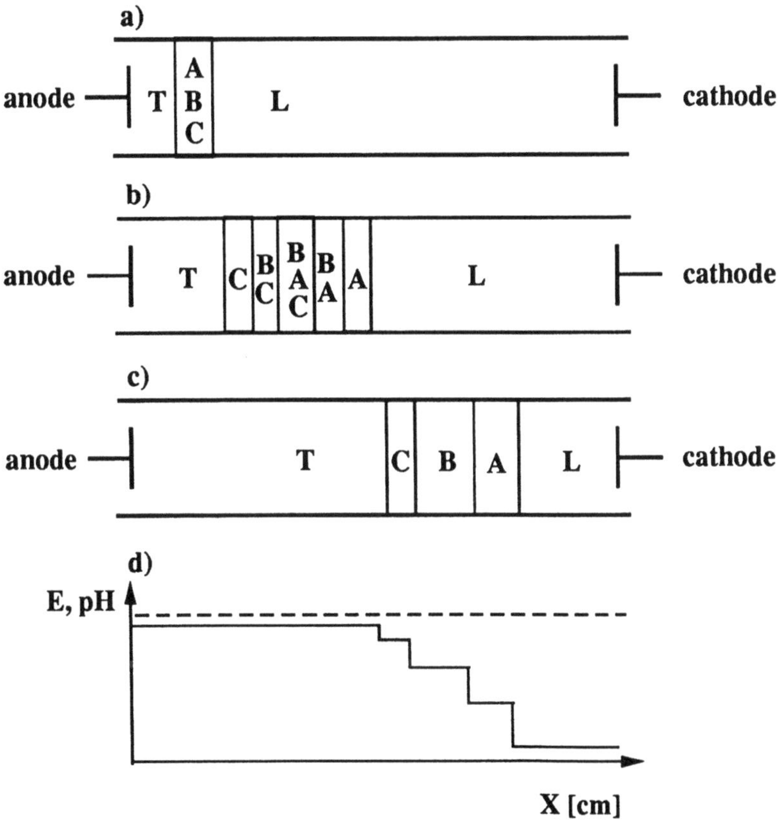

Fig. 2.2. Principle of isotachophoresis.(**a**) initial state, (**b**) intermediate state, (**c**) steady-state, (**d**) profile of field strength (*plain line*) and pH (*dashed line*) across the separation chamber

is filled with the terminating electrolyte which must possess a lower mobility than the cationic species of the sample (Fig. 2.2.a). If voltage is applied, the cationic compound with the highest mobility (A) will migrate faster, leaving behind those with lower mobilities (B and C). This results in two mixed zones in front of and behind the original sample zone (Fig. 2.2.b). Due to their higher mobilities the cations of the leading electrolyte can never be passed by sample cations. The terminating cations, however, are not able to pass the cationic compounds of the sample. Hence, the sample zones are sandwiched between leader and terminator. In order to maintain the transport of the current through the system, the mixed sample zones are separated further, until each zone contains only one cationic species (Fig. 2.2.c). No further changes occur and a steady state has been reached. All zones must migrate connected to each other like a moving train with the same velocity as the leading cation, because no background electrolyte is present that could transport the electric current, if they were released. The electric field strength shows a stepwise profile along the capillary (Fig. 2.2.d).

Unlike in ZE, the different zones will not be broadened further because of the "self-correction" of the zone boundaries: if a cation remains behind in a zone with a higher field strength, its migration velocity increases, until it reaches its own zone again. If the cation diffuses into a preceding zone, where the electric field is lower, its velocity will decrease, until it is caught up by its proper zone.

For the separation of anionic components the leading electrolyte possessing the highest mobility has to be filled into the anode reservoir and the terminating electrolyte, which must possess a lower mobility than the sample, into the cathode reservoir. The separation takes place according to the same principle, but the migration is directed towards the anode. For further details of ITP see Sect. 5.4 and also Ref. 3.

2.1.3 Isoelectric Focusing

Isoelectric focusing is limited to the separation of amphoteric substances, because the sample components are not separated due to their differences in net mobility as in the other electrophoretic modes, but due to their different isoelectric points (pI), where the pI is that pH at which an ampholyte has zero net charge. At that pH they are present mainly in the zwitterionic form and do not migrate when they are exposed to an electric field. At lower pH, they will be positively charged and migrate to the cathode. If the pH is higher than their pI, they will be negatively charged thus migrate towards the anode. In general, separation takes place in a linear pH gradient across the separation chamber (Fig. 2.3.). The pH gradient is generated by a mixture of carrier ampholytes which have isoelectric points ranging from the acidic to the basic range in close proximity to each other and possess a good buffering capacity. The ampholyte solution, commonly a mixture of polyamino polycarboxylic acids, is placed in

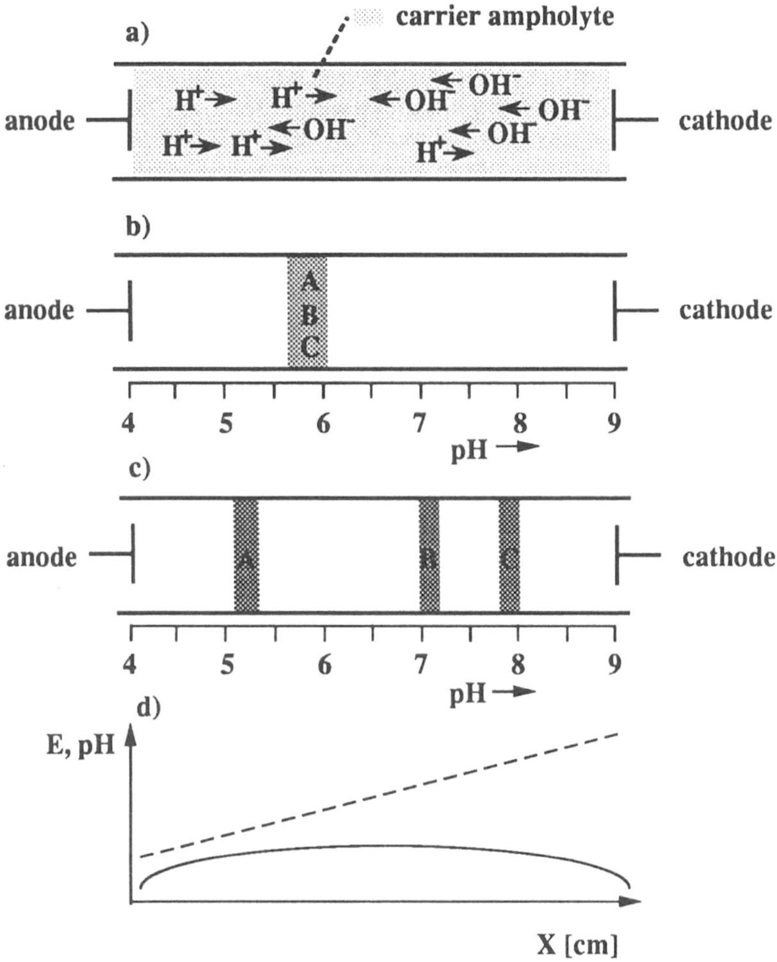

Fig. 2.3. Principle of isoelectric focusing. (**a**) generation of the pH gradient, (**b**) sample introduction, (**c**) steady-state, (**d**) profile of field strength (*plain line*) and pH (*dashed line*) across the separation chamber.

the separation chamber. The anode compartment is filled with an acidic solution, whereas the cathode compartment contains a base. If voltage is applied, H_30^+ ions migrate to the cathode and OH^- ions to the anode. The ampholytes migrate according to their charge and pI towards the corresponding elec-trodes and buffer the migrating H_30^+ and OH^- ions (Fig. 2.3.a). The local pH is given by the particular carrier ampholyte whose charge is balanced by the H_30^+ and OH^- concentration, respectively. A sample consisting of 3 amphoteric substances is introduced to this prebuilt pH gradient (Fig. 2.3.b). It should be mentioned that the sample can also be dissolved directly in the solution of carrier ampholytes, before the pH gradient is generated. In both cases, the separation mechanism is the same. Each ampholyte migrates towards the position where the pH is equal to its pI. At this position its velocity becomes zero, and the component will be concentrated into a narrow zone (Fig. 2.3.c). As in ITP, the system is able to correct zone broadening effects: if the compound changes its position by diffusion, it will be charged again, thus migrating back to the point where its velocity is zero. For a detailed description of the principles of IEF see Ref. 11, for further details of capillary IEF see Sect. 5.5.

2.2 Set-up for Capillary Electrophoresis

As already mentioned above, all electrophoretic modes can be carried out, in principle, using the same equipment, which consists of 5 units: the anode and cathode reservoirs with the corresponding electrodes, the separation chamber, the injection system and the detector. The basic instrumental set-up to accomplish capillary electrophoresis is depicted in Fig. 2.4. A capillary tube filled with the buffer solution is placed between two buffer reservoirs. The electric field is applied by means of a high voltage power supply which can generate voltages up to 30 kV. Injection of the analytes is performed by replacing one buffer reservoir by the sample vial. A defined sample volume is introduced into the capillary by either hydrodynamic flow or electromigration. An on-column detector is located at the end of the capillary which is opposite to the injection site.

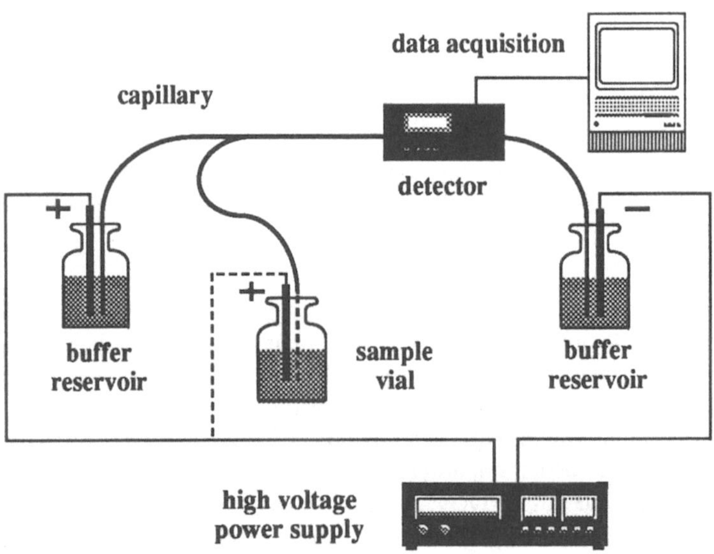

Fig. 2.4. Instrumental set-up of a capillary electrophoresis system

If an uncoated open-tube fused silica capillary is used as the separation chamber, as is mostly the case in CZE, two electrokinetic actions occur under the influence of the electric field. First, electrophoresis of the ions takes place, secondly, electroosmosis, which takes place due to the immovable charge of the capillary walls being effective from the basic to weak acid pH range. Separation, however, is based solely on electrophoresis while electroosmosis causes a liquid transport analogous to a mechanical pump. Because the electroosmotic flow in aqueous solution is mostly directed toward the cathode, the sample is injected at the anode. The sample components migrate with different migration velocities, depending on their charge densities, towards the corresponding electrodes. They are all carried through the detection system by the electroosmotic flow, which is higher than the migration velocities of the ions.

If electrophoresis is carried out in the absence of electroosmotic flow, injection is accomplished at the electrode with the same sign as the charged compounds to be separated. In practice this can easily be achieved by changing the polarities of the electrodes.

Most commercial instruments use on-column UV-VIS absorbance or fluorescence detection. The detector response is recorded versus the migration time. The output which is analogous to a chromatogram obtained in HPLC is called an electropherogram. A computer connected to the detector allows data acquisition and interpretation. Despite the similarities in instrumental set-up and data output, CZE and HPLC are orthogonal thus providing information based on independent separation principles.

2.3 Theory of Electrophoretic Migration

Electrophoretic migration of ions or charged particles is obtained by harnessing electrical forces along the axis of an electrical field gradient. The electrophoretic migration shows itself macroscopically as a conduction of electric current in a solution under the influence of an applied voltage following Ohm's law.

$$U = R \cdot I \tag{2-1}$$

U electric potential [V]
R electric resistance of the electrolyte [$V \cdot A^{-1} = \Omega$]
I electric current [A]

The resistance R of the solution is the reciprocal to the conductance L which can be measured by a conductometer. The conductance depends on the geometry of the measuring device, the ionic species and the electrolyte concentration. In practice the measured conductance L is related to the specific conductance κ by dividing by the cell constant according to

$$L = \kappa \cdot K^{-1} \tag{2-2}$$

L measured conductivity [$\Omega^{-1} = S$]
κ specific conductance [$S \cdot cm^{-1}$]
K cell constant [cm^{-1}]

If we divide the specific conductance by the concentration c we obtain the equivalent or molar conductance Λ:

$$\Lambda = \frac{\kappa}{c} \; [cm^2 \cdot \Omega^{-1} \cdot mol^{-1}] \tag{2-3}$$

In practice, equivalent weights instead of molecular weights of the ions are usually taken into account to allow a comparison of the conductive properties of electrolyte solutions. For this purpose the molarity of an ion has to be divided by the stoichiometric number.

According to the first Kohlrausch law, anions and cations contribute independently to the conductance, which means that the equivalent conductance Λ is the sum of the ionic equivalent conductances of the cations (λ^+) and the anions (λ^-). For strong (completely dissociated) 1:1 electrolytes we obtain:

$$\Lambda = \lambda^+ + \lambda^- \tag{2-4}$$

λ^+ ionic equivalent conductance of the cation [$cm^2 \cdot \Omega^{-1} \cdot mol^{-1}$]
λ^- ionic equivalent conductance of the anion [$cm^2 \cdot \Omega^{-1} \cdot mol^{-1}$]

If we look at Eq. 2-3 the equivalent conductance should be independent of the concentration. This is, however, only true at infinite dilution. For strong electrolytes, the second Kohlrausch law describes empirically the influence of the concentration on the equivalent conductance:

$$\Lambda_c = \Lambda_0 - k \cdot \sqrt{c} \tag{2-5}$$

Λ_c equivalent conductance at a concentration c [$cm^2 \cdot \Omega^{-1} \cdot mol^{-1}$]
Λ_0 limiting equivalent conductance [$cm^2 \cdot \Omega^{-1} \cdot mol^{-1}$]
c electrolyte concentration [M]
k constant [$M^{-1/2}$]

The equivalent conductance at zero concentration Λ_0 is called the limiting equivalent conductance. Ionic interactions are responsible for the fact that the equivalent conductance decreases with increasing electrolyte concentration. Their interpretation on the molecular level will be shortly discussed later in this chapter. Limiting equivalent conductances for both electrolytes and for separate ions can be found in tabular form in corresponding handbooks. To give an impression of their quantity some limiting equivalent conductances are listed in Table 2.1.

Table 2.1. Limiting equivalent conductances of some selected ions. Data are taken from Ref. 12

Ion	λ_0^+ [$cm^2\Omega^{-1}mol^{-1}$]	Ion	λ_0^- [$cm^2\Omega^{-1}mol^{-1}$]
H^+	349.8	OH^-	198.6
Li^+	38.7		
Na^+	50.1	F^-	55.4
K^+	73.5	Cl^-	76.4
Rb^+	77.8	Br^-	78.1
Cs^+	77.2	I^-	76.8
NH_4^+	73.6	NO_3^-	71.5
$N(CH_3)_4^+$	44.9	HCO_3^-	44.5
$N(C_2H_5)_4^+$	32.7	CH_3COO^-	40.9
1/2 Mg^{2+}	53.1	1/2 SO_4^{2-}	80.0
1/2 Ca^{2+}	59.5	1/2 CO_3^{2-}	69.3
1/2 Ba^{2+}	63.6		

In the case of weak electrolytes the equivalent conductance Λ_c is strongly affected by the dissociation of the ions. The quotient of the equivalent conductance measured at a concentration c and the limiting equivalent conductance is equal to the dissociation degree α:

$$\frac{\Lambda_c}{\Lambda_0} = \alpha \qquad (2\text{-}6)$$

α describes the degree of dissociation and is defined as the ratio of the concentration of the dissociated ion to the total concentration of the analyte (see also Sect. 3.3.2). α depends both on the pH and on the concentration of the electrolyte solution. Ostwald's law describes the relationship between the dissociation constant K_c, the concentration of the electrolyte solution c and the equivalent conductance:

$$K_c = \frac{\Lambda_c^2}{(\Lambda_0 - \Lambda_c)\Lambda_0} \cdot c \qquad (2\text{-}7)$$

Since K_c and Λ_0 are constant, this relationship shows the influence of the electrolyte concentration on Λ_c. The equivalent conductance decreases with increasing concentration.

For weak electrolytes ionic interactions do not play such an important role as for strong electrolytes, because the concentration dependence of Λ_c is more affected by the degree of dissociation. Nevertheless, to complete the picture, it has to be mentioned that, for an exact description of all factors, the right side of Eq. 2-6 has to be multiplied by the so-called coefficient of the conductance f_Λ.

If a charged compound is dissolved in an electrolyte solution, the macroscopically measured conductance of the solution does not provide any information about its migration behavior. In electrophoretic separations, however, we are more interested in the electrophoretic behavior of the charged analyte rather than that of the whole solution. Let us therefore consider the migration of a charged compound in an electrolyte solution at infinite dilution, where no ionic interactions occur. In a homogenous electric field the charged component i is accelerated by the electric force F_e:

$$F_e = z_i \cdot e_0 \cdot E \tag{2-8}$$

z_i charge number of component i
e_0 elemental charge $[1.602 \cdot 10^{-19}\ A \cdot s = C]$
E electric field strength $[V \cdot cm^{-1}]$

In a viscous hydrodynamic medium the drag force F_d which acts on the moving species i is proportional to its migration velocity v_i^0 and to the Newtonian viscosity η of the medium:

$$F_d = k \cdot \eta \cdot v_i^0 \tag{2-9}$$

k constant [cm]
η Newtonian viscosity of the solution [Pa·s]
v_i^0 migration velocity of component i at infinite dilution $[cm \cdot s^{-1}]$

According to Stokes' law the constant k can be substituted by $6\pi r$ for a spherical particle. For non-spherical species and small ions, the numerical value is lower than 6.

If the acceleration caused by the electric force F_e is counterbalanced by the drag force F_d, the charged species i moves with a constant migration velocity which is given by

$$v_i^0 = \frac{z_i \cdot e_0}{6\pi\eta r_i} \cdot E \tag{2-10}$$

The hydrodynamic or Stokes radius r_i of the ion i represents the radius of the solvated or, in aqueous solution, hydrated form of the ion. It differs from the crystallographic radius, which is easily accessible by X-ray analysis. If we look at the equivalent conductances of, for instance, the alkali ions (see Table 2.1.), it is obvious that the values do not decrease from Li^+ to K^+ as it would be the case for unsolvated ions. Because of its higher charge density Li^+ is more hydrated than K^+ and so on. Thus, the conductances increase in the opposite direction to that expected when there is no hydration.

According to Eq. 2-10, the migration velocity is proportional to the electric field strength. The proportionality factor is called the absolute or limiting electrophoretic mobility μ_i^0. The migration velocity as well as the electrophoretic mobility can have negative or positive values depending on the sign of the charge number z_i. μ_i^0 is related to the migration velocity by division by the electric field strength:

$$\mu_i^0 = \frac{v_i^0}{E} = \frac{z_i \cdot e_0}{6\pi\eta r_1} \ [\text{cm}^2 \cdot \text{V}^{-1} \cdot \text{s}^{-1}] \qquad (2\text{-}11)$$

The absolute electrophoretic mobility μ_i^0 represents the average velocity of a charged species per unit of electric field strength at zero concentration. For a given ion, μ_i^0 is only dependent on the viscosity of the medium. The influence of temperature on the electrophoretic mobility has its origin almost completely in the change in the solvent viscosity with temperature which can be expressed by:

$$\eta = C \cdot e^{\frac{E_A}{R \cdot T}} \qquad (2\text{-}12)$$

C constant [Pa·s]
E_A activity energy for the viscous flow [$\text{J} \cdot \text{mol}^{-1}$]
R molar gas constant [8.314 $\text{J} \cdot \text{mol}^{-1} \cdot \text{K}^{-1}$]
T temperature [K]

Since the viscosity drops exponentially with increasing temperature, the electrophoretic mobility is increased exponentially with the temperature. In a first approximation for small ΔT values, however, the change in mobility with temperature can be calculated by:

$$\mu_{T_2} = \mu_{T_1} \ (1 + \alpha \cdot \Delta T) \qquad (2\text{-}13)$$

ΔT temperature difference T_2-T_1 [K]
α $\approx 0.02\ \text{K}^{-1}$ for aqueous solutions

Thus, as a rule of thumb, the mobility increases with rising temperature approximately 2% per one degree Kelvin. For further details about the influence of temperature on the electrophoretic mobility see Sect. 3.2.3.

μ_i^0 is a characteristic constant for a given species in a certain solvent at constant temperature and is proportional to the equivalent conductance at infinite dilution:

$$\Lambda_0 = \lambda_0^+ + \lambda_0^- = (\mu_i^{0+} + \mu_i^{0-}) \cdot F \qquad (2\text{-}14)$$

λ_0^+ limiting ionic equivalent conductance of the cation [$cm^2 \cdot \Omega^{-1} \cdot mol^{-1}$]
λ_0^- limiting ionic equivalent conductance of the anion [$cm^2 \cdot \Omega^{-1} \cdot mol^{-1}$]
μ_i^{0+} absolute electrophoretic mobility of the cation [$cm^2 \cdot V^{-1} \cdot s^{-1}$]
μ_i^{0-} absolute electrophoretic mobility of the anion [$cm^2 \cdot V^{-1} \cdot s^{-1}$]
F Faraday constant [96 485 $C \cdot mol^{-1}$]

The equivalent conductances in Table 2.1. can easily be transformed into the absolute electrophoretic mobilities by dividing by the Faraday constant. In practice, we do not work at infinite dilution and other ionic species are present in the electrolyte solution. They strongly affect the mobility in exactly the same way as already mentioned in the case of the equivalent conductance. Additionally, the net charge of a weak electrolyte is smaller than its theoretical charge $z_i \cdot e_0$ because of incomplete dissociation (see Sect. 3.3.2). Let us here consider the ionic interactions influencing the mobility of a charged species in a real solution.

The phenomenon of electrostatic interactions in electrolyte solutions has been treated extensively by Debye, Hückel and Onsager and is based on the fact that an ion is always surrounded by oppositely charged counterions. These counterions forming the so-called ionic atmosphere are responsible for the action of two additional forces slowing down the ionic species, namely the electrophoretic retardation F_{ret} and the relaxation effect F_{rel}. This is schematically shown in Fig. 2.5. The electrophoretic retardation is caused by the fact that the central ion does

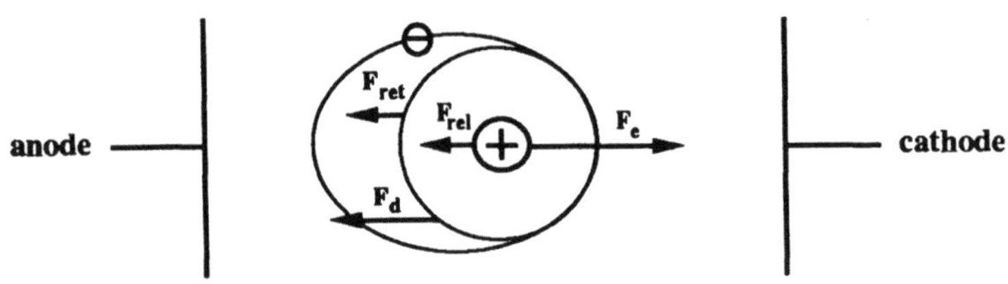

Fig. 2.5. Forces acting on a charged species in an electrolyte solution

not migrate in a stationary environment as is assumed by Stokes law. Since the counterions move in the opposite direction to the central ion, the friction force is higher resulting in a decrease of the mobility. Additionally the directed movement of the central ion permanently deforms the ionic atmosphere. The charge density of the ionic atmosphere in front of the central ion is always somewhat lower than behind it. Coulomb forces between the ions tend to rebuild it in its proper arrangement, which takes a finite time. During this relaxation time the central ion is slowed down by the electrical force F_{rel} acting in the opposite direction to its migration.

The higher the electrolyte concentration, the stronger are the electrostatic interactions not only between the ions of the electrolyte solution, but also between the charged analyte and its counterions. The easiest way to consider the influence of the ionic atmosphere on the mobility is to exchange the theoretical charge by the smaller effective charge and the hydrodynamic radius by the effective radius of the ion including its atmosphere of counterions:

$$\mu_i = \frac{Q_{eff}}{6\pi\eta R} \tag{2-15}$$

μ_i effective electrophoretic mobility [$cm^2 \cdot s^{-1} \cdot V^{-1}$]
Q_{eff} effective charge of the ion [C]
R total radius of the ion [cm]

As a consequence, if ionic interactions occur, the effective mobility will always be lower than the absolute mobility. A more quantitative investigation of the impact of ionic interactions on the mobility will follow in Sect. 3.3.4.

According to Eqs. 2-3 and 2-14 the effective mobilities of all components i are related to the specific conductance of the solution κ as follows:

$$\kappa = F \cdot \sum_{i=1}^{n} c_i \cdot \mu_i \tag{2-16}$$

Especially for colloids and particles where electrophoretic migration is induced as a result of the surface charge caused by adsorbed ions, another definition of the electrophoretic mobility has been developed by von Smoluchowski based on the Debye-Hückel theory of the diffuse double layer at the surface of charged particles:

$$\mu_i = \frac{\zeta \cdot \varepsilon}{4\pi\eta} \tag{2-17}$$

ζ zeta potential [V]
ε permittivity (dielectric constant) of the medium (= $4\pi\varepsilon_0\varepsilon_r$) [$F \cdot m^{-1}$]

The theory of the diffuse double layer, which is equivalent to the ionic atmosphere described above, and the explanation of the zeta potential will follow in Sect. 2.5 dealing with electroosmosis. It should only be mentioned here that, in the ideal case, the mobility defined by Eq. 2-17 is independent from the size and shape of the particle. In the literature, you can also find a similar equation to Eq. 2-17, where the factor $4\pi\eta$ is exchanged by $6\pi\eta$. In general, Eq. 2-17 should be used for the electrophoretic mobility of particles that are large compared to the thickness of their double layer, whereas the other equation should be used for particles which are relatively small compared to their double layer.

The mobilities μ_i^0 and μ_i defined so far have been concerned only with individual components. For a substance S composed of several components S_1, S_2,..., S_n, it can be useful to use a further characteristic, that we will call the average electrophoretic mobility $\overline{\mu_s}$. The following facts require its introduction:

- the substance is composed of several components that are in rapid dynamic equilibrium with one another,
- the components exhibit different absolute mobilities, and
- the individual components cannot be separated electrophoretically and the substance moves as a whole in the electric field.

If the components are present in the solution with the molar fractions x_1, x_2, ..., x_n and if their effective mobilities are μ_1, μ_2, ..., μ_n, $\overline{\mu_s}$ can be expressed as follows:

$$\overline{\mu_s} = \frac{1}{c_s}\sum_{i=1}^{n} c_i \cdot \mu_i = \sum_{i=1}^{n} x_i \cdot \mu_i \qquad (2\text{-}18)$$

c_s total or analytical concentration of S [M]
c_i concentration of the components i [M]

To give an impression of the usefulness of $\overline{\mu_s}$ the mobility of oxalic acid in solution is brought up as a practical example:

$$\begin{matrix}COOH\\|\\COOH\end{matrix} + H_2O \rightleftharpoons \begin{matrix}COOH\\|\\COO^-\end{matrix} + H_3O^+$$

$$\begin{matrix}COOH\\|\\COO^-\end{matrix} + H_2O \rightleftharpoons \begin{matrix}COO^-\\|\\COO^-\end{matrix} + H_3O^+$$

The undissociated part of oxalic acid does not contribute to the average mobility, which is then given by:

$$\overline{\mu_{[COOH]_2}} = x_{COOHCOO^-} \cdot \mu_{COOHCOO^-} + x_{(COO_2)^{2-}} \cdot \mu_{(COO_2)^{2-}}$$

The actual magnitude of the molar fractions of the oxalate ions and thus the apparent mobility can be affected by a change of the pH of the solution.

To summarize, the electrophoretic mobility depends either directly or indirectly on a number of factors such as radius, shape and charge of the ion, solvation, viscosity and dielectric constant of the medium, degree of dissociation and temperature. Their different influences on the electrophoretic mobility will be treated in more detail in chapter 3.

2.4 Determination of Effective Mobility

Calculation of the ionic mobilities using Eqs. 2-11, 2-15 or 2-17 is very difficult or even impossible. However, μ_i can be calculated from an electropherogram as illustrated in Fig. 2.6. which shows a typical separation pattern achieved by CZE in open fused silica tubes in the presence of electroosmotic

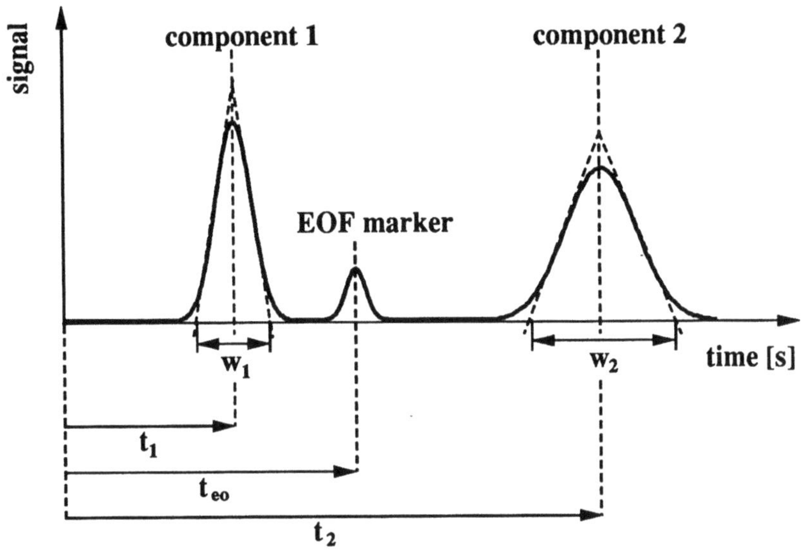

Fig. 2.6. Schematic diagram of a capillary electrophoretic separation. t_1 migration time of component 1, t_2 migration time of component 2, t_{eo} migration time of EOF marker, w_1 temporal peak width of component 1 and w_2 temporal peak width of component 2

flow (EOF). The sample has been injected at the anodic end of the capillary. The migration of the sample components through the detector cell is recorded versus time. The sample consists of a cationic component (1), an anionic component (2) and a neutral substance which moves with the velocity of the electroosmotic flow and serves as an electroosmotic flow marker (EOF marker). If the electroosmotic flow velocity is higher than the velocity of the anionic compound, a simultaneous detection of cations, anions and neutral species is possible as illustrated in Fig. 2.7.

The net velocity $v_{i(net)}$ of component i is to be calculated by dividing the length of the capillary from the injection point to the detector L_D by the migration time t_i. The electrophoretic velocity v_i can be calculated from the net velocity and the electroosmotic flow velocity v_{eo} as follows:

$$v_i = v_{i(net)} - v_{eo} = \frac{L_D}{t_i} - \frac{L_D}{t_{eo}} \text{ [cm·s}^{-1}\text{]} \qquad (2\text{-}19)$$

L_D capillary length to detector or effective capillary length [cm]
t_i migration time of component i [s]
t_{eo} migration time of EOF marker [s]
$v_{i(net)}$ net velocity of component i [cm·s^{-1}]
v_{eo} electroosmotic flow velocity [cm·s^{-1}]

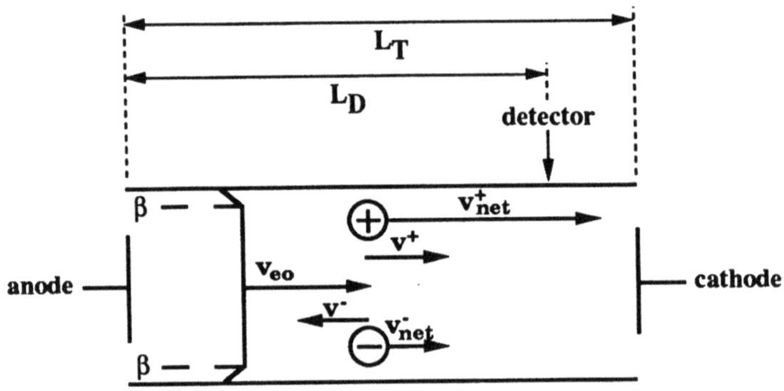

Fig. 2.7. Schematic representation of the migration of cations and anions in the presence of electroosmosis. v^+_{net} net velocity of the positively charged compound, v^-_{net} net velocity of the negatively charged compound, v_{eo} electroosmotic flow velocity, v^+ electrophoretic velocity of the positively charged compound, v^- electrophoretic velocity of the negatively charged compound, L_D effective capillary length, L_T total capillary length and β thickness of the diffuse double layer

It should be noted that v_i will take negative values in the case of anions since v_{eo} is higher than $v_{i(net)}$. Thus, the signs indicate the relative direction of the flow velocity and the electrophoretic or electroosmotic mobility. In practice, however, the negative sign is often neglected. The effective electrophoretic mobility of component i is then given by:

$$\mu_i = \frac{v_i}{E} = \frac{v_i \cdot L_T}{V} \text{ [cm}^2\text{·V}^{-1}\text{·s}^{-1}\text{]} \qquad (2\text{-}20)$$

L_T distance between the electrodes or total capillary length [cm]
V applied voltage [V]

In practice, a typical electropherogram looks like that in Fig. 2.8. which shows the separation of 5 benzoyl derivatives by CZE. At the chosen conditions benzyltrimethylammonium (1) migrates as a cation and elutes first, followed by

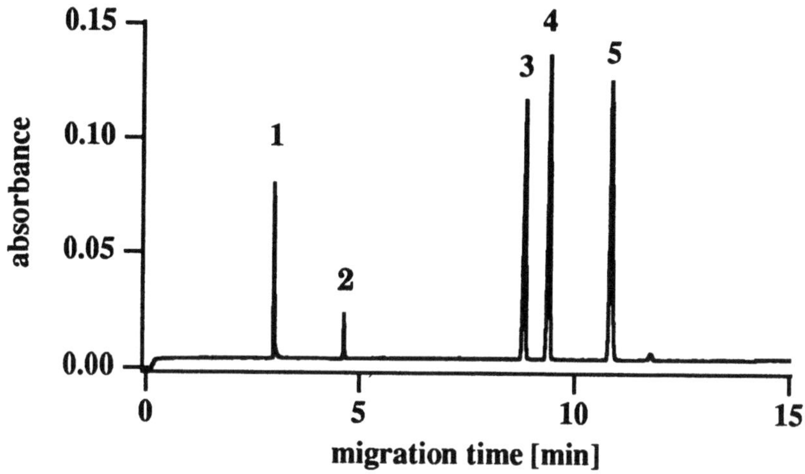

Fig. 2.8. Electropherogram of benzyltrimethylammonium chloride (**1**), benzyl alcohol (**2**), acetylsalicylate (**3**), 4-hydroxybenzoate (**4**) and benzoate (**5**). Instrument: Beckman P/ACE 2000; experimental conditions: fused silica capillary, 57 cm x 75 µm i.d., hydrodynamic injection for 1 s, field strength 263 $V \cdot cm^{-1}$, temperature 25 °C, UV detection at 200 nm, electrolyte system 50 mM sodium phosphate buffer, pH 7.0

benzyl alcohol (2) which is neutral and can therefore be used as EOF marker. The later eluting components (3-5) are negatively charged and migrate in the opposite direction to the electroosmotic flow. The relevant electrophoretic data calculated with the aid of the deduced equations are summarized in Table 2.2.

Table 2.2. Calculated data of the electrophoretic separation of benzoyl derivatives shown in Fig. 2.8.

Compound	t [min]	v_{net} [$cm \cdot s^{-1}$]	v [$cm \cdot s^{-1}$]	µ [$cm^2 \cdot V^{-1} \cdot s^{-1}$]
benzyltrimethyl-ammonium chloride	3.00	0.278	0.098	$3.7 \cdot 10^{-4}$
benzyl alcohol	4.62	0.180		$6.8 \cdot 10^{-4}$
acetylsalicylate	8.84	0.094	- 0.086	$- 3.3 \cdot 10^{-4}$
4-hydroxybenzoate	9.42	0.088	- 0.092	$- 3.5 \cdot 10^{-4}$
benzoate	10.84	0.077	- 0.103	$- 3.9 \cdot 10^{-4}$

If CE is performed in the absence of the electroosmotic flow, where v_i is equal to $v_{i(net)}$, the following simplified equation can be used instead of Eqs. 2-19 and 2-20:

$$\mu_i = \frac{v_i}{E} = \frac{L_D \cdot L_T}{t_i \cdot V} \ [cm^2 \cdot V^{-1} \cdot s^{-1}] \qquad (2\text{-}21)$$

2.5 Electroosmosis

Electroosmosis or also known as "electroendosmosis" is one of the oldest electrokinetic effects discovered. "It is probably the presence of electroendosmosis that has consistently frustrated previous attempts to develop a practical method for the performance of electrophoresis on a micro scale." This statement of Stellan Hjertén [2], one of the pioneers of electrophoresis, indicates one of the main difficulties in the development of electrophoresis which had to be solved in the past. While at the beginning of electrophoresis many attempts were made to suppress electroosmosis, it was James Jorgenson [6-8] who first demonstrated the power of CZE in narrow bore capillaries utilizing the effect of electroosmosis.

Electroosmosis is a basic phenomenon in all electrophoretic separation processes. It can be described as the relative motion of a liquid to a fixed charged surface caused by an electric field. This motion is also called electroosmotic flow (EOF). You can find it whenever you apply a voltage to a liquid system that is in close contact with a charged surface, as is the case if you run CE in a fused silica capillary. The magnitude and the direction of the resulting electroosmotic flow depend on the composition of the capillary and the nature of the solution within the tube. Empirically it was found that the phase with the higher dielectric constant is positively polarized versus the other. Because of its extremely high dielectric constant, water is usually positively polarized in comparison to the fused silica surface. Hence, if an electric field is applied across the fused silica capillary, the mobile ions of the solution migrate with their hydrate water towards the cathode resulting in a flow of the whole solution.

In a pressure-driven flow system such as HPLC, frictional forces at the liquid-solid boundaries cause a strong pressure drop across the capillary. These forces result in a laminar or parabolic flow profile. As a consequence, a cross-sectional velocity gradient occurs within the capillary resulting in a velocity profile such that the velocity vector is highest in the middle of the tube and goes toward zero approaching the walls (see Fig. 2.9.a). In electrically-driven systems the liquid flow caused by electroosmosis shows a plug profile because the driving force is uniformly distributed along the capillary. Consequently, a uniform flow velocity vector across the tube occurs. Only in the double layer region, very close to the capillary surface, the flow velocity approaches zero (see Fig. 2.9.b). Zone broadening caused by the laminar flow profile in HPLC is therefore negligible in CZE.

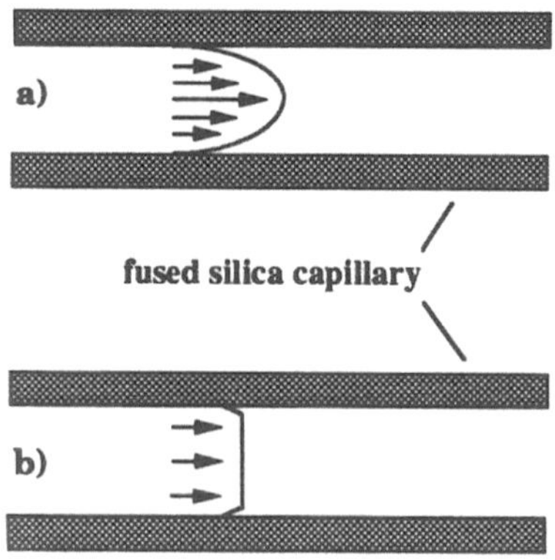

Fig. 2.9. Flow profiles of pressure-driven systems (**a**) and electrically-driven systems (**b**). Arrows indicate flow velocity vectors

The reason for the relative motion of the buffer solution can be found in the nature of the capillary surface which can be described by Stern's model of the double layer (Fig. 2.10.). When silica is in contact with an aqueous solution, its surface hydrolyzes to form silanol surface groups. These groups may be positively charged as $SiOH_2^+$, neutral as SiOH or negatively charged as SiO^-, depending on the pH value of the surrounding electrolyte solution. The surface group density of silica is in the order of $5 \cdot 10^{14}$ cm^{-2}. Counterions tend to adsorb onto the silica wall by electrostatic attractions to balance the surface charge. According to Stern's model, a rigid double layer of adsorbed ions (Helmholtz layer) is superposed by a diffuse double layer (Gouy-Chapman or Debye-Hückel layer) allowing diffusion of ions by thermal motion. This double layer system causes an electric potential Ψ at the interface between silica surface and electrolyte solution. Within the rigid double layer the potential φ decreases linearly with the distance x from the surface. The potential of the diffuse double layer, known as zeta potential ζ, drops exponentially with the distance x from the surface. The thickness d of the rigid layer lies in the molecular range, assuming that a monomolecular layer of counterions is adsorbed. The thickness β of the diffuse double layer which is equivalent to the radius of the ionic atmosphere is defined as:

$$\beta = \sqrt{\frac{\varepsilon_r \varepsilon_0 kT}{2N_A e_0^2 I}} \ [\mathrm{m}] \tag{2-22}$$

ε_r relative dielectric constant
ε_0 permittivity (dielectric constant) of free space [$8.854 \cdot 10^{-12}$ $F \cdot m^{-1}$]
k Boltzman's constant [$1.38 \cdot 10^{-23}$ $J \cdot K^{-1}$]
N_A Avogadro's constant [$6.022 \cdot 10^{23}$ mol^{-1}]

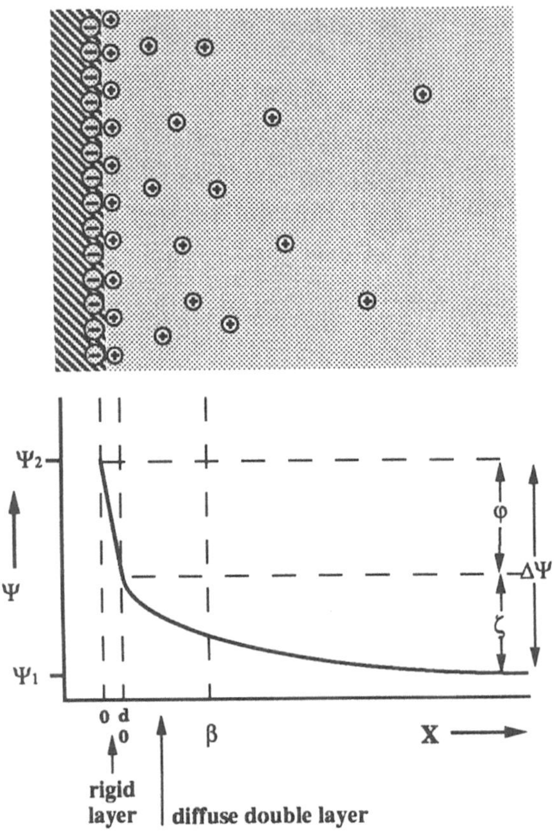

Fig. 2.10. Sterns model of the double layer occurring at the interface between an electrolyte solution and the surrounding surface

The ionic strength I of the electrolyte solution is defined according to Lewis and Randall as follows:

$$I = \frac{1}{2}\sum_i z_i^2 \cdot c_i \ [M] \tag{2-23}$$

c_i ionic concentration of component i [M]

As a consequence, the electrophoretic mobilities of analytes decrease with increasing ionic strength of the solution (see Sect. 3.3.4).

In aqueous solution at 298 K with $\varepsilon_r = 78.30$, β can be approximately calculated as follows:

$$\beta = \frac{3 \cdot 10^{-8}}{\sqrt{I}} \text{ [cm]} \tag{2-24}$$

The thickness β of the diffuse double layer in an aqueous solution is in the range of molecular dimensions for a 0.1 M electrolyte solution (5 - 10 Å). It is increased with decreasing ionic strength and reaches 50 - 100 nm for 1 mM solutions.

Figure 2.11. demonstrates the influence of the two most important factors on the zeta potential of a glass surface: (i) the pH value and (ii) the ionic strength of the surrounding electrolyte solution [13]. For a given salt concentration the

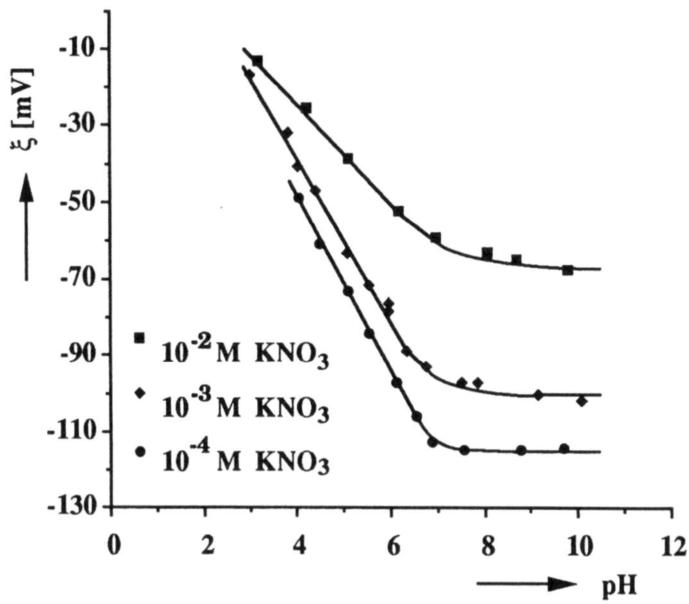

Fig. 2.11. The variation of the zeta potential of vitreous silica as a function of pH in aqueous solutions of potassium nitrate. (With permission from Ref. 13)

surface silanols of the silica behave like a weak acid and dissociate with increasing pH of the solution to SiO^- ions. As a consequence, the overall charge of the surface increases, which leads to a growth of the zeta potential. After the titration of the silanols is completed at pH 7 - 8, the zeta potential no longer changes and the curve levels off. The ionic strength effects the zeta potential such that more negative charges on the surface are balanced by counterions with increasing ionic strength, thus leading to a general decrease of the surface charge and consequently of the zeta potential. In addition, as mentioned above, the double layer thickness decreases with increasing ionic strength, thus leading to a decrease of the zeta potential. By extrapolation of the curves to a zeta potential of zero, a pH of 2.5 could be approximated. This means that, for this type of glass, the surface charge is nearly zero at pH 2.5. If the pH is decreased further, the silanol groups will behave like a weak base and the surface will be positively charged by forming $SiOH_2^+$ ions.

The velocity of the EOF is proportional to the applied electric field as given in the following equation:

$$v_{eo} = \mu_{eo} \cdot E \tag{2-25}$$

μ_{eo} electroosmotic mobility [$cm^2 \cdot V^{-1} \cdot s^{-1}$]

The electroosmotic mobility depends on the zeta potential, the permittivity of the medium and the viscosity of the solution as following:

$$\mu_{eo} = \frac{\zeta \cdot \varepsilon}{4\pi\eta} \tag{2-26}$$

ζ zeta potential [V]

One can see, that the equation for the electroosmotic mobility is identical to Eq. 2-17 which describes the electrophoretic migration, showing that both are based on the same physical phenomenon. Since the zeta potential is influenced by the pH and the ionic strength of the solution as shown above, μ_{eo} varies also with these factors. If the pH of an electrolyte solution is plotted versus the electroosmotic flow, a sigmoidal curve shape similar to a titration curve of a weak acid is found as illustrated in Fig. 2.12. For a 50 mM phosphate buffer solution the inflection point of the curve is at approx. pH 5.2. In the pH range of 3 - 7 small changes of the buffer pH have an immense effect on the electroosmotic mobility. The EOF does not change significantly for pH values higher than pH 7 and lower than pH 3.

Several attempts have been made to investigate the influence of the ionic strength of the carrier electrolyte on the electroosmotic mobility in CZE. Salomon et al. [14] have developed a model that accounts for the decrease of EOF with increasing buffer concentration. The dependence of the electroosmotic mobility on the buffer concentration is caused by two terms as given by the following equation, where a monovalent buffer is assumed:

$$\mu_{eo} \approx \frac{Q \cdot x}{\eta} = \frac{Q_0}{\eta(1 + K_W \cdot [M^+])} \left(d + \frac{3 \cdot 10^{-8}}{\sqrt{[M^+]}} \right) \tag{2-27}$$

Q charge per unit area at the silica surface ($\approx c_{SiO^-}$) [$sites \cdot cm^{-2}$]

x double layer thickness ($= d + \beta$) [cm]

Q_0 total number of ionized silanol groups at the surface before adsorption ($\approx c_{SiO^-} + c_{SiO^-M^+}$) [$sites \cdot cm^{-2}$]

K_W equilibrium constant between the cations in the buffer and at the adsorption sites on the capillary wall [mM^{-1}]
[M^+] concentration of the cation M^+ [mM]
d thickness of the rigid layer [cm]
η Newtonian viscosity [Pa·s]

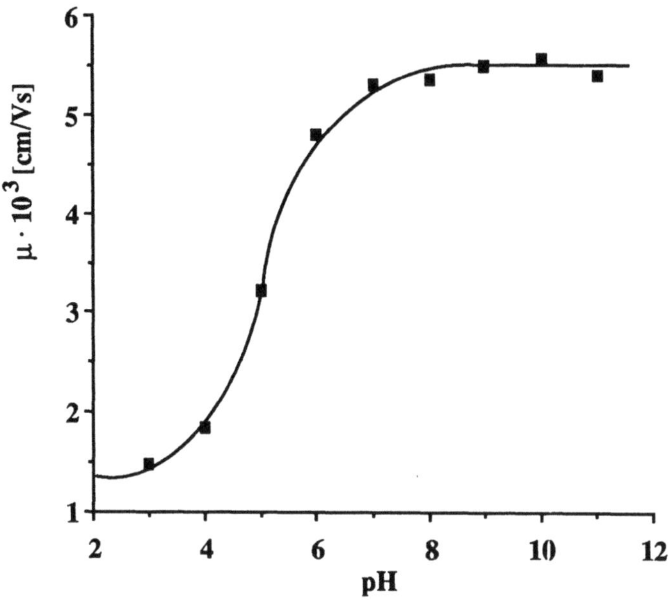

Fig. 2.12. Plot of pH versus electroosmotic mobility. Instrument: Beckman P/ACE 2000; experimental conditions: fused silica capillary, 57 cm x 75 μm i.d., hydrodynamic injection for 1 s, field strength 263 V·cm^{-1}, temperature 25 °C, electrolyte system 50 mM sodium phosphate buffer. Benzyl alcohol is used as EOF marker and detected at 200 nm

Whereas the first term on the right side of Eq. 2-27 is related to the dependence of the surface charge on the adsorption of cations on the capillary surface, the second term describes the increase of the diffuse double layer thickness β with increasing number of cations (see Eq. 2-24). This means that the electroosmotic mobility is decreased with increasing buffer concentration, because the surface charge and the double layer thickness are decreased. More investigations of the impact of buffer composition and concentration on the electroosmotic flow are made in Sects. 3.3.4 and 3.3.5.

Experimentally, the electroosmotic mobility can be measured analogous to the procedure described in Fig. 2.6. using the following equation:

$$\mu_{eo} = \frac{v_{eo}}{E} = \frac{L_D \cdot L_T}{t_{eo} \cdot V} \qquad (2\text{-}28)$$

For the calculation of electrophoretic mobilities from the migration time and the electroosmotic flow, the velocity of the EOF has to be precisely known. The simplest way is to choose the "water" dip indicating the migration of the former injection plug and being equal to the EOF displacement. The "water" dip appears in the electropherogram because the sample solution often has a lower UV absorbance than the electrolyte system resulting in a negative UV signal. On the other hand, if the UV absorbance of the sample is higher than that of the buffer solution, a positive system peak can be observed at the site of the former injection plug. A more precise approach is to inject a neutral marker into the capillary that migrates with the electroosmotic velocity. From the migration time to the detector and the capillary length, the EOF can be calculated. The neutral marker has to fulfill some requirements. The compound must be water soluble, neutral over a wide pH range and no adsorption on the capillary walls must occur. Additionally it should show a high UV absorbance in order to allow small amounts to be injected. Benzyl alcohol, riboflavin and with restrictions mesityl oxide, acetone and benzene have been described as well suited for this purpose. One has to take care, however, that the EOF marker indicates the real EOF displacement. If the marker is attracted by the capillary surface or partially charged by complexation with the carrier electrolyte, it will be slower or faster than the real flow.

Another approach was reported by Wanders et al. [15]. Determination of EOF was carried out on-line using an analytical balance to weigh the volume flow continuously. Variations of the flow velocity with the time can be detected with this procedure whereas the first method solely results in a mean value of the EOF for a given time interval.

There have been several approaches to the proper control of the electroosmotic flow with particular emphasis on its reduction or elimination. To some extent buffer additives can be used to modify dynamically the capillary surface. Cationic surfactants like cetyltrimethylammonium bromide (CTAB) and related homologous compounds are claimed to be reagents which are able to change the sign of the surface charge by adsorption [16]. Depending on the concentration of e.g. CTAB, the EOF slows down and even reverses in the opposite direction at concentrations higher than approx. $3.5 \cdot 10^{-4}$ M. The impact of the concentration of CTAB and related compounds on the electroosmotic mobility is given in Fig. 2.13. While decyltrimethylammonium bromide (DeTAB) decreases the EOF but does not invert the migration direction, reversed migration can be accomplished using dodecyltrimethylammonium bromide (DoTAB), tetradecyltrimethylammonium bromide (TTAB) or cetyltrimethylammonium bromide (CTAB). As shown in Fig. 2.13.a and b, extreme changes of the EOF take place with even low concentrations of TTAB and CTAB. For practical use

DoTAB is better suited to adjust the electroosmotic flow, since it changes approximately linearly with the modifier concentration.

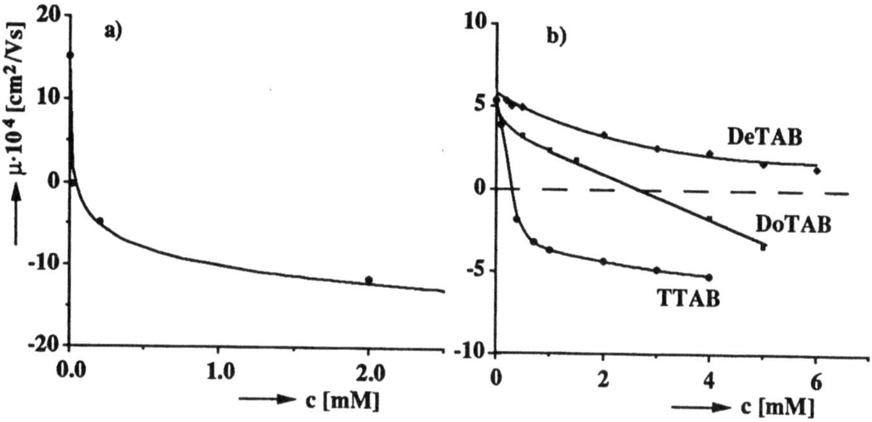

Fig. 2.13. Effect of cationic surfactants like CTAB (**a**) and DeTAB, DoTAB and TTAB (**b**) on the electroosmotic mobility. Experimental conditions in (**a**): fused silica capillary, 100 cm x 75 μm i.d., applied voltage 30 kV; in (**b**): fused silica capillary, 70 cm x 100 μm i.d., applied voltage ca. 10 kV, electrolyte system 100 mM boric acid, adjusted to pH 9.1 with KOH. With permission from Ref. 17 (**a**) and 18 (**b**)

The addition of organic modifiers such as alcohols or hydrophilic linear polymers also leads to a reduction of EOF (see Sect. 3.3.7). As an alternative to the dynamic coating, polymers like polyacrylamide are covalently bound to the silica to eliminate entirely the EOF, with the additional benefit of minimizing the adsorption of proteins to the capillary wall (see Sect. 3.1.2). Finally, gel-filled capillaries are another approach to minimize electroosmosis. A detailed description of the different procedures for the control or elimination of EOF by capillary coating can be found in Sect. 5.1.

2.6 Performance Criteria

Analytical separations are performed to obtain, in a reasonable short time, information, both qualitative and quantitative, about the composition of a sample. This information is only provided, if the separation technique exhibits satisfactory performance. The exceptional achievements of field-driven separation processes like capillary electrophoresis is put down directly to the extremely high electric field forces which can be applied. By extending the nomenclature of chromatography and its basic concepts to electrophoresis, factors are derived which allow the performance to be quantified using parameters like efficiency and resolution.

2.6.1 Efficiency

The efficiency of the electrophoretic system is gauged by the number of theoretical plates N achieved by the capillary. Experimentally, N can be calculated from an electropherogram, e.g. as given in Fig. 2.6., using one of the following equations:

$$N = 16 \cdot \left(\frac{t}{w}\right)^2 \tag{2-29}$$

$$N = 5.54 \cdot \left(\frac{t}{w/2}\right)^2 \tag{2-30}$$

t migration time of the component [s]
w temporal peak width at the base line [s]
$w/2$ temporal peak width at half of the peak height [s]

Eq. 2-30 is to be preferred, if the peak of interest shows a tailing which makes the determination of the peak width more difficult. To compare different separation systems with respect to the separation efficiency, the plate number is often referred to a capillary length of one meter. Plate numbers as high as $10^6 \cdot m^{-1}$ for open tubes and $3 \cdot 10^7 \cdot m^{-1}$ for gel-filled capillaries have been obtained. Although the plate number in CE does not have the same importance as in HPLC, it is suited to estimate the peak capacity n of the system as

$$n \approx 0.5\sqrt{N_{max}} \tag{2-31}$$

N_{max} maximal plate number

The peak capacity as defined in Eq. 2-31 indicates, how many components can be theoretically separated as peaks with the resolution 1 (see below) in between the EOF marker and the last eluting component. Table 2.3. shows the calculated plate numbers and the peak capacity for the separation of the 5 benzoyl derivatives depicted in Fig. 2.8. using Eqs. 2-30 and 2-31.

In 1969 Giddings [19] derived the following fundamental equations for the mathematical description of the plate height H and the corresponding plate number N in field-driven separation processes. Twenty years later, he was able to show that in particular CE closely approaches these models [20]. The distance x that a charged species covers at a migration time t is given by

$$x = v \cdot t \text{ [cm]} \qquad (2\text{-}32)$$

v velocity [$cm \cdot s^{-1}$]

Table 2.3. Calculated plate numbers and peak capacity of the separation of 5 benzoyl derivatives shown in Fig. 2.8.

Compound	t [s]	w [s]	N	n
benzyltrimethyl-ammonium chloride	180.0	1.3	106 000	
benzyl alcohol	277.2			
benzoate	650.5	4.4	120 000	
4-hydroxybenzoate	565.3	3.8	122 000	
acetylsalicylate	530.5	3.5	127 000	178

Figure 2.14. illustrates the zone broadening caused by axial diffusion. Ideally a sample is injected as a sharp zone into the capillary. During migration

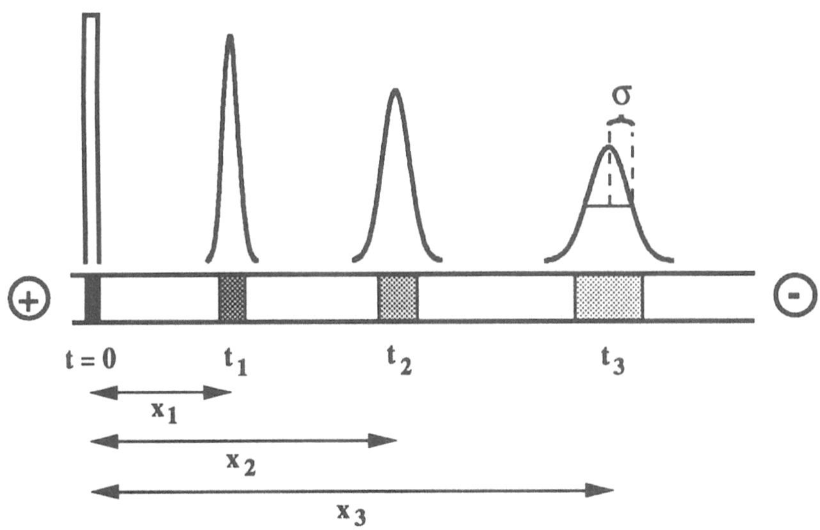

Fig. 2.14. Change of the concentration distribution with time; $t_1 < t_2 < t_3$. s represents the time-related standard deviation of the Gaussian curve. x_1, x_2, x_3 represent the migration distance from the point of injection

through the tube the zone spreads as a function of time forming a Gaussian curve. With the spreading of the band the sample components diffuse more and more into the electrolyte solution. The half width of the peak at half height is called the standard deviation σ of the curve. The time-related standard deviation obtained from an electropherogram can be transformed into the spatial standard deviation of the peak by multiplying by the migration velocity.

Zone broadening by diffusion processes can be described by Einstein's equation:

$$\sigma^2 = 2 \cdot D \cdot t \ [cm^2] \tag{2-33}$$

D diffusion coefficient [$cm^2 \cdot s^{-1}$]
σ^2 spatial variance [cm^2]

The plate height H, which is equal to σ^2/x, can be expressed as

$$H = \frac{2 \cdot D \cdot t}{x} = \frac{2 \cdot D}{v} \ [cm] \tag{2-34}$$

Because v is equal to $\mu \cdot E$ we obtain

$$H = \frac{2 \cdot D}{\mu \cdot E} \ [cm] \tag{2-35}$$

Since the theoretical plate number N is related to H by $N = x/H$, we have

$$N = \frac{x \cdot \mu \cdot E}{2 \cdot D} \tag{2-36}$$

If we assume that the migration takes place across the whole length of the capillary, we can replace the field strength by V/x and obtain

$$N = \frac{\mu \cdot V}{2 \cdot D} \tag{2-37}$$

or taking the electroosmotic mobility into account

$$N = \frac{(\mu + \mu_{eo}) \cdot V}{2 \cdot D} \tag{2-38}$$

Equations 2-37 and 2-38 are interesting in several respects as discussed by Jorgenson and Lukacs [7]. For a given analyte, μ and D are constant parameters. Thus, the theoretical plate number can only be influenced by the applied voltage. Since N is directly proportional to the applied voltage, the highest voltage possible is suggested to obtain high plate numbers. Unlike in chromatography, N is independent of the capillary length and the migration time.

Knox [21] discussed the ultimate possible performance of capillary electrophoresis and was able to show that up to a million theoretical plates per meter are in principle achievable. In practice, additional band broadening occurs due to effects like Joule heating and convection, adsorption and sample overloading thus decreasing the theoretically expected plate number. This will be treated in detail in chapter 3. As stated already above, the electroosmotic flow shows a plug profile which contributes only to a minor extent to band broadening.

As given in Eq. 2-38, the electroosmotic mobility is directly related to the plate number suggesting that efficiency can be improved by increasing μ_{eo}. This approach, however, is misleading because the resolution of two components will decrease with the electroosmotic flow as explained in the following section. Thus, an increase of EOF will result in shorter analysis times and sharper but more poorly resolved peaks.

2.6.2 Resolution

The resolution R of two peaks is defined as the quotient of the distance between the peak centers Δx and 4σ, the mean of the two standard deviations of the peaks [19].

$$R = \frac{\Delta x}{4\sigma} \tag{2-39}$$

Δx is proportional to the incremental migration velocity Δv. So we obtain

$$\Delta x / \bar{x} = \Delta v / \bar{v} \quad \text{or} \quad \Delta x = \bar{x} \cdot \frac{\Delta v}{\bar{v}} \tag{2-40}$$

$\bar{x}$ mean of migration distance of two components [cm]
$\bar{v}$ mean of migration velocity of two components [$cm \cdot s^{-1}$]

The resolution can therefore be written as:

$$R = \frac{\bar{x}}{4\sigma} \cdot \frac{\Delta v}{\bar{v}} \tag{2-41}$$

Because $H = \sigma^2 / \bar{x}$ and $N = \bar{x} / H$ the resolution is related to the plate number N as follows:

$$R = \frac{\sqrt{N}}{4} \cdot \frac{\Delta v}{\bar{v}} \tag{2-42}$$

The ratio of the second term in Eq. 2-42 can be transformed as

$$\frac{\Delta v}{\bar{v}} = \frac{\mu_1 - \mu_2}{\bar{\mu}} \tag{2-43}$$

$\bar{\mu}$ mean of mobilities of components 1 and 2

In the presence of electroosmosis, Eq. 2-43 must be rewritten as

$$\frac{\Delta v}{\bar{v}} = \frac{\mu_1 - \mu_2}{\bar{\mu} + \mu_{eo}} \tag{2-44}$$

By substituting Eq. 2-41 and 2-33 into Eq. 2-44 one obtains

$$R = \frac{1}{4} \cdot \sqrt{\frac{(\bar{\mu} + \mu_{eo}) \cdot V}{2 \cdot D}} \cdot \frac{\mu_1 - \mu_2}{\bar{\mu} + \mu_{eo}} \tag{2-45}$$

Rearranging leads to

$$R = 0.177 \cdot (\mu_1 - \mu_2) \cdot \sqrt{\frac{V}{D \cdot (\bar{\mu} + \mu_{eo})}} \tag{2-46}$$

From Eq. 2-46 it becomes obvious that the resolution of two peaks decreases with the magnitude of EOF, if the EOF has the same direction as the electrophoretic migration. In fact, from the theoretical point of view, the best resolution is obtained when the electrophoretic mobility is just balanced by the EOF as

$$\mu_{eo} \approx - \bar{\mu} \tag{2-47}$$

The calculation of resolution from the difference in electrophoretic mobility of two components according to Eq. 2-46 is shown in Fig. 2.15. For the calculation it is supposed that the electrophoretic mobility of one component (μ_2) is kept constant at $5 \cdot 10^{-5}$ $cm^2 \cdot V^{-1} \cdot s^{-1}$ whereas μ_1 is variable between 0 to 30% higher values. Simulation is carried out in a 50 cm capillary at an applied field strength of 300 $V \cdot cm^{-1}$. The diffusion coefficient is 10^{-5} $cm^2 \cdot s^{-1}$. The graph shows that resolution is low for positive electroosmotic mobilities and increases with decreasing mobility. The highest resolution is obtained if the electroosmotic flow moves in the opposite direction to the components. In the range studied, a linear relationship between resolution and difference in mobility was calculated for positive μ_{eo}. As the electroosmotic mobility approaches $-\bar{\mu}$ the curve shape becomes more and more convex. Eq. 2-46 is not defined for μ_{eo} being equal to $-\bar{\mu}$ or lower.

Although highest resolutions are obtained if electroosmotic and electrophoretic mobilities are adjusted such that Eq. 2-47 is fulfilled, these separations need to run for a long time. In practice, a resolution of 1.25 is sufficient even for quantitative analysis and resolutions higher than 1.5 are not desired because of the time wasted.

The resolution between two peaks in an electropherogram can be calculated as follows:

$$R = \frac{2(t_2 - t_1)}{w_1 + w_2}, \text{ where } t_2 \geq t_1 \tag{2-48}$$

t_1 migration time of component 1 [s]
t_2 migration time of component 2 [s]
w_1 temporal peak width of component 1 [s] (see Fig. 2.6.)
w_2 temporal peak width of component 2 [s]

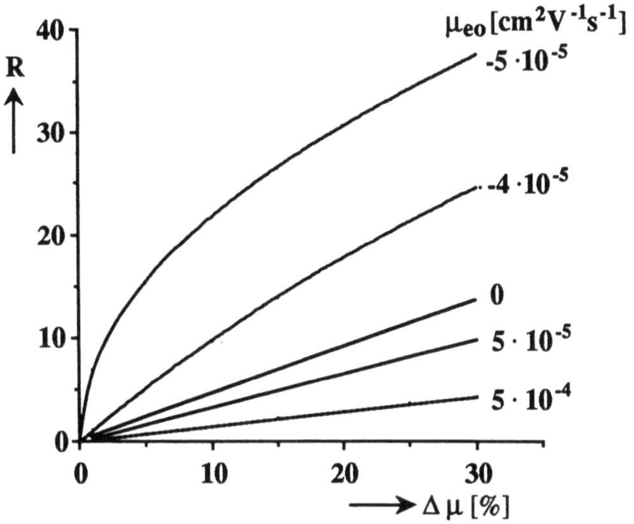

Fig. 2.15. Resolution dependence on the difference in electrophoretic mobility of two components calculated from Eq. 2-46. Values indicate electroosmotic mobilities. Further details are given in the text

Resolution as defined in Eq. 2-42 is the product of two terms. While the first term containing the plate number expresses the efficiency, the second term represents the selectivity of the separation system. In chromatography, a measure for the selectivity is given by the separation factor α, which is defined by the quotient of the capacity factors of the two analytes to be separated:

$$\alpha = \frac{k_2}{k_1} = \frac{t_2 - t_0}{t_1 - t_0}, \text{ where } t_2 \geq t_1 \tag{2-49}$$

α separation factor
k_1' capacity factor of component 1
k_2' capacity factor of component 2
t_0 dead time of the column [s]
t_1 retention time of component 1 [s]
t_2 retention time of component 2 [s]

In analogy to this definition, the separation factor was also introduced in CE, especially for the characterization of chiral separations. Because no dead volume exists in CE, α can be defined as follows:

$$\alpha = \frac{t_2}{t_1} \tag{2-50}$$

t_1 migration time of component 1 [s]
t_2 migration time of component 2 [s]

Because $t_2 \geq t_1$, α is always ≥ 1.

Selectivity seems to be the most critical factor for the optimization in CE. In chromatography a variety of stationary and liquid phases are available which cover nearly all purposes, whereas in CE, method optimization in terms of selectivity is limited to modifications of the electrolyte system.

3 Factors Influencing Performance

Electrophoresis in capillaries has already demonstrated its great value as an analytical tool in pharmaceutical analysis, biochemistry, etc., as witnessed by an increasing stream of publications. The resolution even of complicated mixtures obtained by CE often excels that of chromatographic techniques. The extremely high resolving power of CE certainly originates from the fact that most dispersive effects can be kept well under control in modern instrumentation. Nevertheless: "Felix qui potuit rerum cognoscere causas" (Vergil), or in our words: it is eminently important to understand these effects and to estimate their influences on the separation. This is the objective of the following chapter, where we discuss the most important dispersive effects and demonstrate possibilities to minimize them. In addition, influences of operational conditions like field strength, capillary dimensions and temperature are pointed out. Furthermore, we show how selectivity can be manipulated by the choice of buffer electrolyte. Here, the most effective parameters such as pH, ionic strength, buffer composition, complex formation and organic modifiers are discussed.

3.1 Fundamental Dispersive Effects

Band broadening in capillary electrophoresis is the result of a number of effects which all contribute additively to the spreading of a zone. Supposing Gaussian peak shapes, the band broadness can be expressed by the variance σ^2. Thus, the total variance σ_t^2 is the sum of variances due to the particular sources of dispersion:

$$\sigma_t^2 = \sigma_D^2 + \sigma_A^2 + \sigma_J^2 + \sigma_E^2 + \sigma_I^2 + \sigma_W^2 + \sigma_O^2 \qquad (3\text{-}1)$$

The terms on the right side of the equation represent the variances due to diffusion, adsorption, Joule heating, electrophoretic dispersion, injection, the width of the detection zone and other effects, respectively. The different effects can never be totally eliminated, but their impact on the separation efficiency can be controlled by appropriate design of the instrumental equipment and by careful selection of the working conditions. In the following sections the contributions of the most important band broadening effects to the separation efficiency are treated in detail. The impact of the width of the detection zone will be treated in Sect. 4.2.2.

In CE separated molecules pass the detector cell with different velocities. Molecules whose electrophoretic motion is in the same direction as the electroosmotic flow move more rapidly than neutral molecules, which in turn move more rapidly than molecules whose electrophoretic motion is in the opposite direction to the EOF. As a consequence, the time that sample compounds with different electrophoretic mobilities require to pass the detector is not the same for all compounds. This causes the temporal widths w_t for the various zones to be unequal or, in other words, variations in the temporal width may arise solely from differences in electrophoretic mobilities. The peak width obtained from the electropherogram has to be converted into the spatial peak width w_s by dividing it by the zone velocity, which is equal to L_D/t. L_D is the length of the capillary from the injection point to the detector and t is the migration time of a particular zone.

3.1.1 Diffusion

Formally two different diffusion processes can be observed in fluids. First, the so-called self diffusion or Brownian molecular motion, which represents an irregular movement of mass caused by local fluctuations in thermal energy. Secondly, if a concentration gradient exists within the fluid, a movement takes place in such a way as to equalize the concentration. This latter motion is directed along the concentration gradient and is driven by the difference in the chemical potential between high and low concentration. While the first phenomenon contributes only to a minor extend to band broadening, the second one represents an important factor in the performance of all differential separation techniques. Macroscopically the directed diffusion along a concentration gradient can be observed as a flux of mass J_i, which is described mathematically by

$$J_i = -D \cdot \left(\frac{dc_i}{dx} \right) \tag{3-2}$$

J_i flux of mass $[mol \cdot cm^{-2} \cdot s^{-1}]$
D diffusion constant $[cm^2 \cdot s^{-1}]$

The proportionality constant is the diffusion coefficient D which is defined by Fick's law as:

$$D = \frac{k \cdot T}{6 \pi \eta r} \tag{3-3}$$

D is directly proportional to the temperature T but reciprocal to the apparent hydrodynamic radius r. This latter relationship is illustrated in Table 3.1. summarizing diffusion coefficients and corresponding molecular masses for several compounds.

Table 3.1. Relative molecular masses MW and diffusion coefficients D of several compounds in aqueous solutions at 298 K [22]

Compound	MW [$g \cdot mol^{-1}$]	D [$cm^2 \cdot s^{-1} \cdot 10^5$]
Na^+	22.99	1.25
ethanol	46	1.08
valine	117.2	0.74 (293 K)
tryptophan	204.2	0.61 (293 K)
glucose	180	0.56
sucrose	342	0.42
cytochrome C	13,370	0.11
human serum albumin	68,500	0.069
human fibrinogen	339,700	0.019
tobacco mosaic virus	40,590,000	0.0046

In capillary electrophoresis a sample is supposed to be injected as a sharp zone into the tube. During migration through the capillary the profile of the sample zone is broadened due to directed diffusion. Because it takes place along the axis of the capillary, it is also called axial diffusion in contrast to lateral diffusion which occurs across the capillary. The latter type of diffusion is important in regard to dispersion caused by temperature gradients (see Sect. 3.1.3). The zone profile of the analyte is shaped like a Gaussian bell curve and can be quantified by the spatial variance σ_D^2, which is related directly to the diffusion coefficient at the time t as described in Eq. 3-4 (see also Sect. 2.6).

$$\sigma_D^2 = 2 \cdot D \cdot t \text{ [cm}^2\text{]} \tag{3-4}$$

Simultaneously to the broadening of the band the maximal absorbance of the zone decreases with the migration time.

As a consequence of Eq. 3-3, band broadening caused by diffusion is less pronounced for compounds with high molecular masses than for small ions. Therefore CE should be especially suited for the separation of large molecules possessing small diffusion coefficients. Because of the temperature dependence of D, lowering the temperature for the separation reduces diffusion of the bands on account of longer migration times.

With the knowledge of both the diffusion coefficient and the electrophoretic mobility of a species, the molecule's effective charge can be calculated as demonstrated by Walbroehl and Jorgenson [23]. By combining Eq. 3-3 with Eq. 2-15 a relationship between the diffusion coefficient and the electrophoretic mobility is derived:

$$Q_{eff} = \mu_i \cdot \frac{k \cdot T}{D} \tag{3-5}$$

Using Eq. 3-5 the charge of any molecule can be calculated if the electrophoretic mobility and the diffusion coefficient are known. Whereas the electrophoretic mobility can be derived directly from an electropherogram, the diffusion coefficient is obtained indirectly by CE by using the "stopped flow" approach. For this technique, the sample is permitted to migrate halfway into the capillary. Then the electrical field is turned off and the analytes are allowed to diffuse for a given period of time. Depending on the diffusion coefficient of the molecule, it will take up to several hours until diffusion has progressed in such a way that a measurable band broadening occurs. After electrophoresis is started again the band broadening is recorded by the detector and compared to the non-interrupted run. The difference of both variances is the result of diffusion. The time-related variance obtained by this procedure has to be converted into a spatial variance by multiplying with the square of the migration velocity to allow the calculation of the diffusion coefficient according to Eq. 3-3. Finally, using Eq. 3-5 the charge of the molecule under the given conditions can be calculated.

3.1.2 Adsorption

Besides a number of factors like diffusion and thermal effects, interferences of the electrophoretic migration based on adsorption phenomena occur. These disturbing effects are caused by the fact that solid surfaces are never completely indifferent with regard to a sample constituent. As a result of this chemical and structural behavior of the surface, analytes can be adsorbed either reversibly or irreversibly. In extreme cases practically all of the analyte is irreversibly bound to the surface, whereas in minor severe cases reversible adsorption results in a band broadening or tailing. It is well known from liquid chromatography that silica exhibits strong adsorption of many compounds, in particular of macromolecules like proteins. Although many attempts are made to reduce adsorption by derivatizing silanol groups to reduce active adsorption sites, complete deactivation has not succeeded so far. The adsorption tendency to the fused silica tubes used in CE is also well known.

The distribution coefficient K may be constant or vary with sample concentration. Adsorption processes where K is independent of concentration are based on linear adsorption isotherms. Linearity means that the distribution coefficient is constant and the sample elutes as a symmetric band. Besides linear isotherms two particular isotherm types may be distinguished at low sample concentrations: convex and concave. Both non-linear adsorption isotherms affect the shape of the analyte peak depending on the sample concentration. While concave isotherms are of minor importance, convex isotherms are typical for adsorption processes where K decreases with increasing sample concentration. In such a system the normally symmetric zones of the analytes are disturbed because the sorption-desorption reaction is not sufficiently fast and the steady-state equilibrium of the analytes is distorted in such a way that adsorbed solute molecules will lag behind the others. Hence, the zone center tends to catch up the leading edge of the zone resulting in a "tailing" of the eluted zone. For extremely convex isotherms zone tailing can be so pronounced that entire elution of the sample from the silica wall is impossible. Incomplete sample recovery is a consequence in these cases. Often non-linear adsorption isotherms become linear at sufficiently low sample concentrations.

Liu et al. [24] have found that, as the diameter of the capillary decreases, the tendency for adsorptive interactions between the solutes and the fused silica surface is increased, especially at higher field strength. According to Ref. 24, the contribution of adsorptive interactions to band broadening in an open tubular column can be expressed as the variance σ_A^2:

$$\sigma_A^2 = \frac{2c_f(1-c_f)^2(\mu_{eo}+\mu_i)\cdot r\cdot L_D\cdot n}{Z\cdot\alpha} \tag{3-6}$$

c_f	fractional concentration of a free solute
r	inner radius of the capillary [cm]
Z	number of molecules striking a unit surface area per second [$cm^{-2}s^{-1}$]
n	number of molecules per unit volume inside the tube [cm^{-3}]
α	fraction of molecules which adhere on collision
L_D	effective capillary length [cm]

As one can see from Eq. 3-6, σ_A^2 is proportional to the applied field strength and the inner diameter of the tube. This fact can be an explanation of the phenomenon that band broadening in tubes of small diameters and elevated voltages is higher than theoretically expected, even if thermal effects can be neglected.

Many approaches are described in literature dealing with reducing adsorption of analytes. In the simplest way, wall adsorption can be minimized by choosing the buffer pH such that analyte and silica have the same charge sign. Hence wall adsorption tendencies are suppressed by Coulombic repulsion. In the case of ampholytes, e.g. proteins, the buffer pH should be higher than the isoelectric point of the proteins to make sure that the analyte and the surface are negatively charged. Biological buffers have shown themselves to be well suited for this purpose [25]. On the one hand, their buffer capacity is the highest in the basic pH range, on the other hand their amine functionalities compete with the proteins for the remaining active sites of the silica. In Appendix 8.1 a list of suitable biological buffers is presented. Although a million theoretical plates are provided by using buffers of high pH, this approach suffers from one major disadvantage. Owing to the restriction by the basic pH, method optimization in terms of pH is limited to a narrow range. Additionally many proteins and peptides are sensitive to basic pH which causes degradation and denaturation.

In contrast to the analysis in the basic pH range, buffers with low pH are also claimed to work well [26]. Since dissociation of the acidic silanol groups is suppressed at low pH values, adsorption by Coulombic interactions is reduced. This approach, however, suffers from the same problems as discussed above. In addition, mobility differences of ampholytes are often too small to ensure separation at this low pH value.

Dynamic capillary coatings involve the use of buffer additives to reduce adsorption. Buffers containing high salt concentrations minimize adsorption of proteins to the silica by an ion exchange mechanism. Potassium sulfate (0.25 M) has proven to be superior to alkaline halogenides with respect to the prevention of adsorption and UV absorbance interferences [27]. Instead of potassium sulfate, highly concentrated phosphate buffers (0.5 M) in combination with small i.d. capillaries (25 µm) can also be employed. Using zwitterions instead of high concentrations of potassium sulfate is a similar approach to prevent adsorption. Zwitterions do not contribute significantly to conductivity even at high concentrations, but compete strongly for the active surface sites. Trimethylammonium propylsulfonate and betaine [28] have been suggested for this purpose. Competition for the active sites of the silica by addition of cationic, divalent amines such as 1,4-diaminobutane (DAB), 1,5-diaminopentane (DAP) 1,3-diaminopropane and morpholine has also been proposed for reducing adsorption. Whereas in the case of peptides, amine concentrations of 0.1 - 5 mM are sufficient to increase separation efficiency, much higher concentrations must be employed for protein separations (30 - 60 mM). Since cationic surfactants like CTAB reverse the charge on the wall (Sect. 2.5) they are especially useful to reduce adsorption of cationic proteins. Finally, non-ionic surfactants can be used to dynamically coat deactivated capillaries (Sect. 5.1.2.3).

A more sophisticated possibility for reducing adsorption is represented by the surface treatment of the silica by static coating. Hjertén [29] described static coating with a polymer layer. Bifunctional compounds are bound via one functional group to the silica surface whereas the other group takes part in the polymerization process. γ-Methacryloxypropyl trimethylsiloxane or vinyltrichlorsilane are used as bifunctional compounds. The methoxy and chloro functionalities bind to the silanol groups whereas the methacryloxy or vinyl group reacts in a second step with monomers like acrylamide, vinyl alcohol or vinylpyrrolidone to form a polymeric film on the surface. In another method [30] the silica is silylated with 3-aminopropyltriethoxysilane followed by a reaction of the amine functionality with glutardialdehyde. After this step alpha-lactoglobulin or any other protein can be covalently bound to the surface. As anticipated the pI value of the surface is approximately the pI of the protein. Thus, not only can the electroosmotic flow be adjusted with the buffer pH, but also adsorption of proteins is suppressed. A more detailed description of the different static coating procedures that are used to overcome adsorption is provided in Sect. 5.1.2.

Hjertén [31] recommends a procedure which makes it possible to estimate whether adsorption has a major effect on band broadening. If in a plot of the plate height H (σ^2/L) versus the inverse field strength 1/E at low values of E a straight line results, adsorption is negligible and the starting zone width ΔX_0 can be estimated from the plot by using the following equation.

$$H = \frac{\Delta X_0^2}{12 L_D} + \frac{2D}{\mu_i} \cdot \frac{1}{E} \tag{3-7}$$

If, however, a convex curve shape results from the plot, then significant adsorption occurs. A deeper discussion of the procedure and the evaluation is given in Ref. 31.

Adsorption can also be recognized very easily by the following procedure. The analyte is injected at the detector side of the column and is allowed to move to the opposite end of the tube. As it passes the detector cell the peak shape of the component is recorded. Before the analyte can leave the tube, the polarity of the electrodes is changed, the analyte moves back through the capillary and thus is recorded a second time by moving through the detector. Increased distortion of the peak indicates adsorption. Irreversible adsorption of the analyte on the silica can easily be measured by comparing the peak areas.

3.1.3 Joule Heating

The change of energy per time unit - the power of the electric system - is defined by

$$P = U \cdot I = R \cdot I^2 \tag{3-8}$$

P power [W]
U potential [V]
I current [A]
R resistance [Ω]

When a current passes along a capillary, electrical energy is partially converted into Joule heating. As a consequence of heat generation the temperature increases within the capillary. The internal temperature can be calculated by measuring the differential electroosmotic mobility at two different voltages [32]. One voltage should be low assuming there is no heating (< 5 kV). The other voltage should be the running voltage, where one needs to know the temperature. The temperature is related to the viscosity of the medium according to Eq. 2-12. Because $\mu_{eo} \approx 1/\eta$ it follows:

$$\mu_{eo} = C \cdot e^{-\frac{E_A}{RT}} \tag{3-9}$$

C is the proportionality constant. For the quotient of two EOF mobilities at two different temperatures, it follows:

$$\frac{\mu_{eo_1}}{\mu_{eo_2}} = e^{-\frac{E_A}{RT_1}+\frac{E_A}{RT_2}} \tag{3-10}$$

Index 1 refers to the measurement at low voltage and 2 to the measurement at running voltage. In aqueous solution E_A/R is 1820 K. Assuming that T_1 is equal to room temperature (298 K), the temperature at the running voltage can be calculated by

$$T_2 = \frac{1820}{\ln(\mu_{eo_1}) - \ln(\mu_{eo_2}) + 6.11} \tag{3-11}$$

Empirically, it has been shown that below a power of 3 W the temperature is related to the power by:

$$T = 11.5 \cdot P[W] + 298\ [K] \tag{3-12}$$

A rule of thumb for a quick calculation of the internal temperature is, that a power of 0.1 W increases the temperature by 1.1 K for natural convection and 0.6 K for forced air convection (see below). If, for instance, the voltage is 20 kV and the current 50 μA, the temperature increase in the capillary is about 11 K.

Along a temperature gradient from the center of the capillary through the wall the heat is transferred out of the system. There are three fundamental types of heat transfer: conduction, convection and radiation.

Conduction is the transfer of heat from one part of a body to another part of the same body, or from one body to another one being in physical contact with it but without motion of the particles of the body. Convection is the transfer of heat from one point to another within a fluid, gas or liquid by the mixing of one portion of the fluid with another. In natural convection the movement of the fluid is completely the result of differences in density as a result of temperature differences. In forced convection, the motion is produced by mechanical means, e.g. a fan or a liquid cooling circuit. Finally, radiation is the transfer of heat from one body to another one which is not in close contact by means of wave motion through space.

Heat generation in a capillary by electric current occurs homogeneously across the bore. The rate of heat per unit volume Q is given by [21]:

$$Q = E^2 \cdot \Lambda \cdot c \cdot \varphi \tag{3-13}$$

Q heat output per unit volume [$W \cdot cm^{-3}$]
E field strength [$V \cdot cm^{-1}$]

Λ equivalent conductance of the electrolyte solution [$cm^2 \cdot \Omega^{-1} \cdot mol^{-1}$]
c electrolyte concentration [M]
φ total porosity of the medium (=1 for an open tube)

Using typical values for CZE with E = 300 $V \cdot cm^{-1}$, Λ = 150 $cm^2 \cdot \Omega^{-1} \cdot mol^{-1}$ and c = 50 mM, we obtain 675 $W \cdot cm^{-3}$. For a capillary of 57 cm · 75 μm i.d. with a total volume of 2.5 μL the total heat generation is 1.7 W.

The transportation of heat out of the tube takes place through the capillary walls and causes a parabolic temperature gradient directed radially from the center to the wall. The difference in temperature by heat dissipation in capillary electrophoresis is schematically shown in Fig. 3.1. Whereas the temperature profile in the liquid inside the capillary is parabolic with the maximum temperature being at the center, the temperature drops logarithmically in the capillary wall

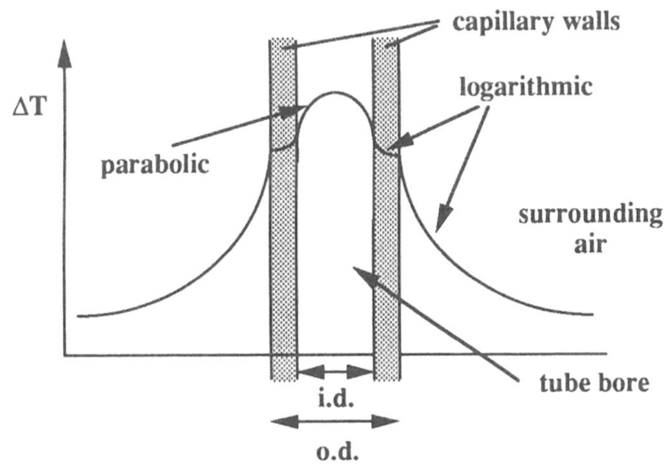

Fig. 3.1. Schematic representation of the temperature profile as a result of current flow across a capillary which contains an electrolyte solution. (With permission from Ref. 21)

and in the surrounding medium. While heat dissipation through the wall takes place mainly by conduction, heat is transferred through the surrounding air by convection and radiation. The temperature difference ΔT between the center of the capillary and its wall is proportional to the heat output Q and the square of the inner diameter of the tube according to:

$$\Delta T = \frac{Q \cdot (\text{i.d.})^2}{16K} \qquad (3\text{-}14)$$

i.d. inner diameter of the capillary [cm]

K thermal conductance of the medium [$W \cdot cm^{-1} \cdot K^{-1}$]

The thermal conductance depends on the medium and is approximately $6 \cdot 10^{-3}$ $W \cdot cm^{-1} \cdot K^{-1}$ for water. Table 3.2. summarizes typical thermal conductances of some media which are important in CE. In our example discussed above the temperature difference between the center of the tube and the wall is approx. 0.4 K.

Medium	K [$W \cdot cm^{-1} \cdot K^{-1}$]
air	$2.5 \cdot 10^{-4}$
polyimide	$1.6 \cdot 10^{-3}$
water	$6.0 \cdot 10^{-3}$
methanol	$2.0 \cdot 10^{-3}$
fused silica	$1.5 \cdot 10^{-2}$
quartz	$1.4 \cdot 10^{-2}$
borosilicate glass	$1.0 \cdot 10^{-2}$

Table 3.2. Characteristic thermal conductances K according to Ref. 21 and 33.

The heat transfer across the capillary wall ΔT_W is given by

$$\Delta T_W = \frac{Q \cdot (i.d.)^2}{8 K_W} \cdot \ln\left(\frac{o.d.}{i.d.}\right) \qquad (3\text{-}15)$$

o.d. outer diameter of the capillary [cm]
K_W thermal conductance of the wall [$W \cdot cm^{-1} \cdot K^{-1}$]

The thermal conductance of the fused silica is double the conductance of water which means that heat is more effectively conducted across the capillary than through the liquid. Because of this reason the temperature difference per volume will be smaller than within the bore. Again, using typical values for o.d./i.d. = 4 and $K_W = 1.4 \cdot 10^{-2}$ $W \cdot cm^{-1} \cdot K^{-1}$, we obtain a temperature difference between the inner and outer capillary wall of $\Delta T_W = 0.47$ K. Thus, the temperature difference per unit volume is about 0.1 K, approximately 1/4 of the temperature difference of the solution.

From the thermal conductances given in Table 3.2. it becomes evident that the efficiency of heat removal is governed by the heat transfer between the outer wall of the tube and the surrounding air, which is the limiting factor, and not by the inside diameter of the capillary. As already mentioned heat transportation in air is mainly based on natural convection rather than conduction. Alternatively, enhanced convection can be forced by using a fan to produce an air stream with high velocity. The temperature rise ΔT for different capillary diameters under the conditions of natural and forced convection for a heat output of 300 $W \cdot cm^{-3}$ is shown in Fig. 3.2. It is obvious from Fig. 3.2. that forced convection is highly

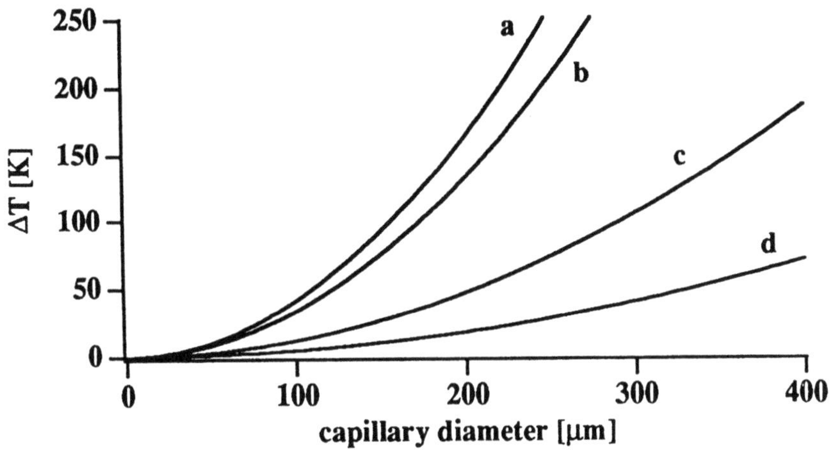

Fig. 3.2. Differences in temperature, ΔT, for various tube diameters under natural and forced convection according to Ref. 21. Air velocities: natural convection (**a**), 0.1 $m \cdot s^{-1}$ (**b**), 1 $m \cdot s^{-1}$ (**c**) and 10 $m \cdot s^{-1}$ (**d**)

recommended for capillary electrophoresis especially if higher field strengths are desired or buffer solutions of high conductivities have to be used. For this purpose forced air cooling with a fan of at least 10 $m \cdot s^{-1}$ or better liquid cooling are particularly recommended. A simple method to illustrate the efficiency of heat removal by a cooling system in capillary electrophoresis is the measurement of the electroosmotic mobility at several field strengths. Remember that, although the electroosmotic mobility is not directly dependent on the temperature, it changes with the viscosity of the solution. The effect of Joule heating on the electroosmotic mobility is illustrated by plotting μ_{eo} versus the field strength. The deviation of the curve from a straight line parallel to the x-axis is due to the change in viscosity by the temperature increase of the electrolyte. This is shown in Fig. 3.3. for electrophoretic instruments without cooling and with a fan and a liquid cooling system, respectively. With higher field strengths a significant increase of μ_{eo} is observed in all cases. Nevertheless, heat removal is more efficient with liquid cooling than with forced air cooling. The highest increase is found for the instrument without any cooling device.

Now, what are the effects of Joule heating and temperature gradients on the separation? The variation of temperature within a tube can contribute significantly to the plate height and can seriously reduce the efficiency of the system. The temperature gradient may cause alterations in migration because of several effects:

- change of solution viscosity and density
- change in partition ratios between two phases
- change in the rates of kinetic processes

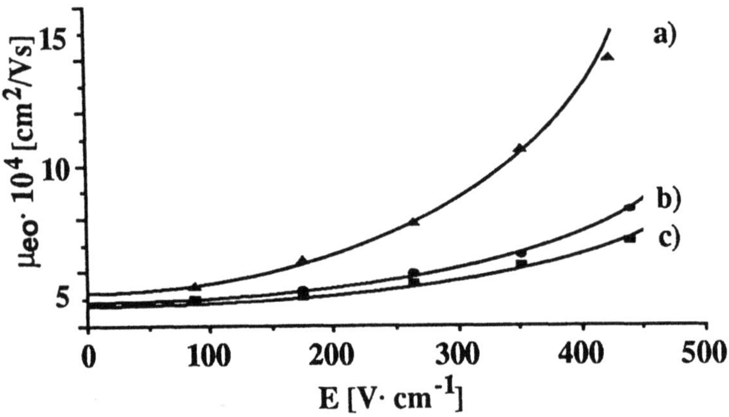

Fig. 3.3. Plot of electroosmotic mobility versus field strength for a capillary electrophoresis system without cooling (**a**), with a fan (**b**) and with a liquid (c) cooling system. Experimental conditions: fused silica capillary, 57 cm x 75 µm i.d., hydrodynamic injection for 1 s, electrolyte system 50 mM sodium phosphate buffer, pH 7.0. In both systems with forced cooling the temperature has been set to 30 °C. Benzyl alcohol is used as EOF marker and detected at 200 nm

The parabolic temperature profile across the tube (see Fig. 3.1.) results in a parabolic variation of the migration velocity, which is highest in the center of the capillary and lowest at the walls. The rate of the migration velocity in the temperature gradient is determined almost completely by the temperature dependence of the viscosity. Remember that a temperature rise of 1 °C increases the migration velocity for approximately 2%. The most effective ways to keep heat generation low during electrophoresis can be deduced from Eq. 3-13. Heat production is proportional to the square of the field strength, the equivalent conductance and the electrolyte concentration. Moreover, according to Eq. 3-14, the temperature difference between the center of the capillary and the walls is proportional to the heat production and to the square of the inner diameter of the capillary. In this regard lower field strengths commonly yield better resolution than high field strengths although this contradicts the theory. Because heat is directly related to the conductance and the concentration, highly conductive electrolytes and high concentrations increase production of heat. Ions with low conductance like TRIS, lithium, borate or phosphate ought to be preferred over those with high conductance like potassium, chloride or sulfate, unless these are especially requested. Narrow bore capillaries improve efficiency in three respects. First, while in large bore capillaries molecules essentially remain in the same radial position during their migration across the capillary, in narrow bore tubes the analyte molecules are able to diffuse across the entire cross section of the capillary. This lateral diffusion leads to a randomization of the position of the molecules. The higher the randomization is, the more uniform is the migration velocity of the zone. Secondly, heat generation is smaller in narrow bore capillaries compared to larger diameters because of the lower charge density inside the capillary. Finally, heat dissipation is more effective in narrow bores owing to the higher ratio of surface to volume.

In the above reflections, the electric field strength is thought to be constant over the whole length of the capillary. This is only true if the conductivity of the sample zone is equal to that of the surrounding buffer. Especially if the sample is dissolved or diluted with distilled water as is the case in sample stacking (see Sect. 5.8.1), the field strength in the sample zone and therefore the temperature can get so high that even boiling may occur. This has to be taken into account when heat-labile analytes are employed.

Using representative values for diffusion coefficient (10^{-5} $cm^2 \cdot s^{-1}$), viscosity (10^{-3} Pa·s), thermal conductance (see Table 3.2.), etc., Knox [21] derived the following boundary condition under which plate height contribution from thermal effects is considered negligible compared to plate height contribution from axial diffusion.

$$(\text{i.d.})^3\ [\mu m]^3 \cdot E\ [kV \cdot m^{-1}] \cdot c\ [M] < 3.3 \cdot 10^9 \qquad (3\text{-}16)$$

The effect of this boundary condition on maximal allowed capillary diameters under several operating conditions is shown in Table 3.3. It can be deduced from Table 3.3. that, for typical tube diameters of 10 - 100 µm, these boundary conditions are not violated even for high field strengths, as long as the electrolyte

Table 3.3. Boundary condition for maximal allowed tube diameter (in µm) where the contribution from thermal effects on plate height is smaller than 0.1 times the contribution from axial diffusion. Data are taken from Ref. 21.

c [M]	10 $kV \cdot m^{-1}$	20 $kV \cdot m^{-1}$	50 $kV \cdot m^{-1}$	100 $kV \cdot m^{-1}$
0.001	1500	750	300	150
0.01	700	350	140	70
0.1	320	160	60	30

concentration does not exceed 0.01 M. Nevertheless, one has to keep in mind that the relation 3-16 can only be considered as a rough guideline. Grushka et al. [34] have also calculated maximal allowable capillary diameters under several conditions. Their results will be presented in Sect. 3.2.2.

For further information about thermal effects, temperature control and calculation of internal temperatures see Refs. 21 and 32 - 40. The influence of column temperature on the selectivity of the separation is treated in Sect. 3.2.3.

3.1.4 Electrophoretic Dispersion

Changes in concentrations during electrophoresis are restricted by Kohlrausch's "beharrliche Funktion", regulating function or omega function [1] which results from the fact that all electrophoretic processes are essentially charge-transport processes obeying Ohm's law. The omega function is established by the initial electrolyte distribution along the axis of the separation chamber and is not changed by the applied current. It is only a function of the location x of the species i in the separation chamber and is independent of time. The regulating function in its simplest form, as it has been derived by Kohlrausch for strong electrolytes, can be written as follows:

$$\omega(x) = \sum_{i=1}^{n} \frac{c_i \cdot z_i}{\mu_i} = \text{constant} \qquad (3\text{-}17)$$

It is assumed that electrophoretic mobilities are constant and that no diffusion takes place. The regulating function says that any of the migrating zones follow the concentration profile which has been formed along the separation compartment before electromigration starts. This means that an electric current does not induce any concentration changes in a system with only one uniform electrolyte distribution described by one single omega function. On the other hand, when the separation chamber is filled with a discontinuous electrolyte system before the start of the experiment, there are as many different regulating functions as there are different electrolytes. When electromigration begins, the migrating zones copy all discontinuities and concentration distributions at locations where they were created before the electric field was applied across the separation chamber. In the case of zone electrophoresis where a sample solution is introduced into the capillary which is filled with a homogeneous background solution, two zones are present separated by two sharp boundaries: the leading or frontal boundary and the trailing or rear boundary. Thus, two omega functions describe the system. It is assumed that the background electrolyte consists of the coion B^+, which has the same charge as the sample component, and the counterion C^-. The sample consists of the component S^+ and the counterion C^- (Fig. 3.4.). Now the corresponding omega functions are:

$$\omega_1 = \frac{c_B}{\mu_B} + \frac{c_C}{\mu_C} \qquad (3\text{-}18)$$

$$\omega_2 = \frac{c_S}{\mu_S} + \frac{c_C}{\mu_C} \qquad (3\text{-}19)$$

where ω_1 describes the situation at the initial sampling compartment and ω_2 that of the separation compartment (Fig. 3.4.a). The sample is introduced into the capillary as a narrow rectangular zone which is called the sampling compartment. As soon as the sample zone begins to migrate out of the sampling

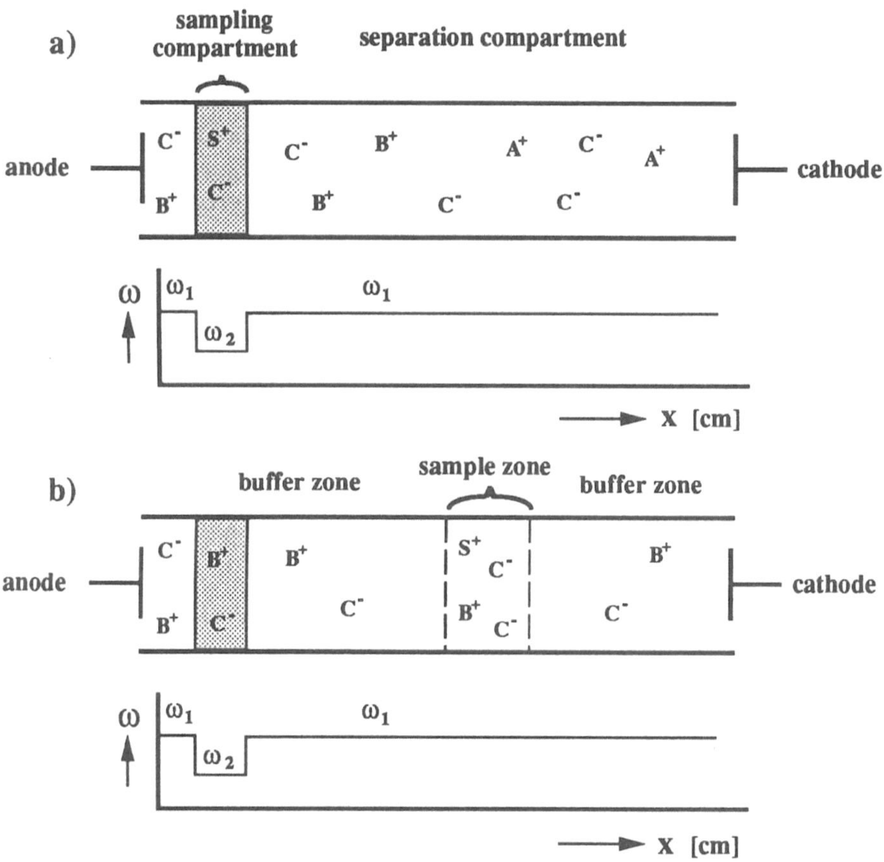

Fig. 3.4. Distribution of the omega functions over sampling and separation compartment in zone electrophoresis. (**a**) Initial situation before the electric field is applied; (**b**) situation during electrophoretic migration. B^+ represents the carrier constituent, having the same charge as the sample component S^+, and C^- the counterion. x reflects the migration coordinate

compartment into the separation compartment, its concentration distribution will be ruled by the omega function ω_2 (Fig. 3.4.b).The flux of ions out of the sampling compartment is exactly equal to the flux into it. Consequently, the boundaries between sampling and separation compartment are stationary, neglecting the effect of diffusion. In the given example it is assumed that no electroosmotic flow occurs. The apparent movement of the stationary boundaries of the sampling compartment in the presence of EOF comes solely from the electroosmotic movement of the entire bulk solution.

The phenomenon of electrophoretic or electromigration dispersion, which is responsible for the peak asymmetry frequently observed in capillary zone electrophoresis, was first treated by Mikkers et al. [41]. Based on Kohlrausch's regulating functions they calculated the concentration distributions in the sample and buffer zone. An even more sophisticated approach to investigate the influence of sample and buffer composition on the electrophoretic behavior of different analytes are computer simulations which are also based on the regulating functions. Recently, Mosher, Saville and Thormann have published a comprehensive monograph about computer simulations [42].

Two factors are responsible for peak broadening by electrophoretic dispersion: the difference in specific conductance κ between the sample and the buffer zone and the concentration ratio of the sample constituent to the buffer's coion c_S/c_B. Both factors have their origin in mobility differences between sample and coion. To illustrate the behavior of the sample zone in dependence on these mobility differences, let us first assume that the sample zone is not distorted by the boundary structure at its initial location (see Fig. 3.4.). The electrophoretic system has the same constituents as discussed in Fig. 3.4. If an electric field is applied, the sample zone begins to migrate out of the sampling compartment into the separation compartment and is mixed with the buffer constituents. During the migration of the sample through the capillary, local changes of electrolyte concentrations as well as conductivity changes occur which result in boundary anomalies. If the sample is a weak electrolyte there may be pH changes as well. Because the sample ions partially replace the buffer ions according to the regulating function, the buffer concentration in the sample zone is always lower than in the buffer zone. Therefore, the mobility of the sample ion determines the conductivity of the sample zone. Depending on the mobility of the sample constituent three different cases can be distinguished (Fig. 3.5.).

a) $\mu_S < \mu_B$: If the mobility of the sample constituent is lower than that of the coion of the buffer (Fig. 3.5.a), the electric field strength will always be higher in the sample zone than in the buffer, because the conductivity in the sample zone is lower than in the buffer zone. Thus, a sample molecule S^+, which enters the front phase of surrounding buffer by diffusion or convection, will be slowed down and will migrate with a lower velocity than it did in the sample zone. The front boundary will therefore become sharpened. The migration velocity of this boundary is constant. The trailing boundary, however, broadens with time and has a decreasing velocity: if the sample constituent enters the rear phase of the surrounding buffer, its velocity will also be decreased, thus resulting in a diffuse rear boundary of the zone. The peak shows a pronounced tailing and the peak height is decreased.

b) $\mu_S > \mu_B$: If the mobility of the sample constituent is higher than that of the coion, the electric field strength in the sample zone will be lower than in the buffer zone. When a sample molecule diffuses into the frontal buffer zone, its velocity will be increased by the higher field strength of the buffer zone.

Conversely, a sample molecule entering the rear buffer zone will be accelerated until it reaches its own zone again. Therefore, the leading side of the sample zone will be diffuse, whereas the rear side will be sharp (Fig. 3.5.b). The peak shows a fronting.

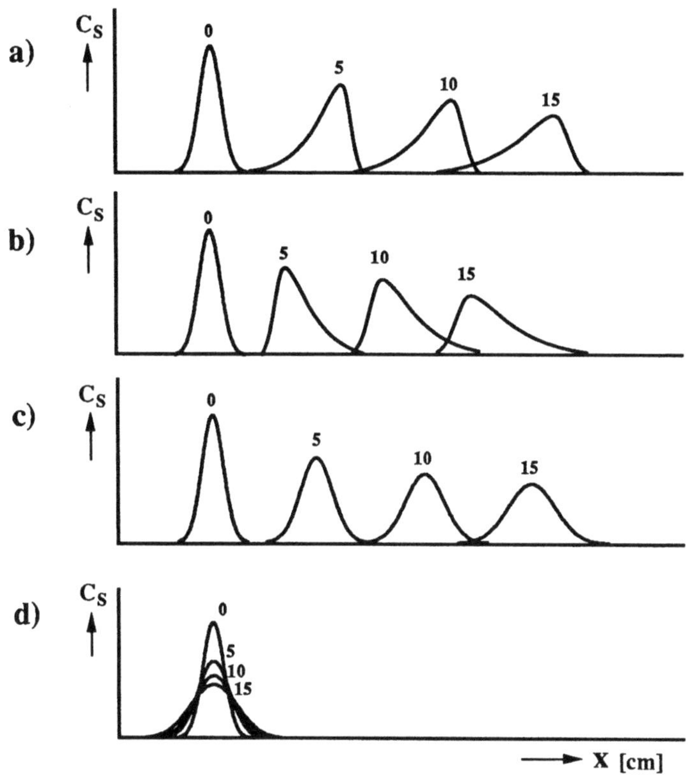

Fig. 3.5. Concentration distribution in zone electrophoresis as a function of the sample mobility: (**a** - **c**) dispersion as a result of diffusion and electromigration, if $\mu_S < \mu_B$ (**a**), $\mu_S > \mu_B$ (**b**) and $\mu_S = \mu_B$ (**c**). Concentration distributions are shown for 0, 5, 10 and 15 min of electromigration; (**d**) diffusive dispersion in the absence of electromigration

c) $\mu_S = \mu_B$: If the mobility of the sample constituent is equal to the mobility of the buffer constituent, the electric field strength will be constant across the whole capillary (Fig. 3.5.c). Only diffusional broadening is observed.

The higher the sample concentration is in comparison to the buffer concentration, the more pronounced are local changes in conductivity and field strength across the sample zone. Since the conductivity change relative to the buffer is much smaller for low sample concentrations, the field strength can be regarded as constant all over the separation compartment leading only to diffusional broadening of the sample zone. According to Mikkers et al. [41] band broadening due to diffusion and electrophoretic migration is of the same order of magnitude when

$$D \approx 0.1 \cdot \frac{c_S}{c_B} \cdot l \cdot v_S \tag{3-20}$$

c_S concentration of sample component S in the sample zone [M]
c_B concentration of buffer component B in the buffer zone [M]
l initial width of the sample plug [mm]
v_S velocity of sample component S in the buffer zone [$mm \cdot s^{-1}$]

Using typical values for CZE with $D = 10^{-5}\ cm^2 \cdot s^{-1}$, $v_S = 1\ mm \cdot s^{-1}$ and $l = 1$ mm, respectively, diffusion and electromigration will have a comparable adverse effect at a concentration ratio c_S/c_B of 10^{-2}. Below this value diffusion is mainly responsible for band broadening, whereas above this value the contribution of electrophoretic migration to dispersion is dominant. In other words, if the ratio is low enough, the conductivity change between sample and buffer zone is negligible and the carrier electrolyte determines the conductivity and the pH along the whole separation compartment. In this case, ideal ZE conditions are given. If, on the other hand, the sample concentration becomes too high, electrophoretic dispersion manifests itself in sample overloading of the electrolyte system.

In the former discussion we have presupposed that the counterion of the sample and the carrier are identical. In practice, however, the counterions often differ from each other and the migration is also influenced by the counterion of the carrier electrolyte. As long as the ratio c_S/c_B does not exceed 10^{-2} the mobility and concentration of the counterion does not affect the separation. But if, for instance, weakly absorbing solutes have to be detected by UV detection, the sample concentration cannot be reduced below the sensitivity of the detector used. On the other hand, an increase of the buffer concentration would minimize electrophoretic dispersion, but the conductivity would increase in the same way, resulting in excess Joule heating. It has been found experimentally [43] that electrophoretic dispersion can be reduced substantially by choosing a counterion of the carrier electrolyte having an effective mobility close to that of the sample ion.

The resulting peak asymmetries due to electromigration can affect tremendously the resolution between two peaks. Figure 3.6. illustrates how the resolution of three peptides is affected if the ratio of sample concentration to buffer concentration becomes too high. Sodium phosphate is used as buffer electrolyte, and the peptides are positively charged at the chosen pH. Thus, Na^+ ions represent the buffer's coion. Note that the concentration of Na^+ is not equal to the

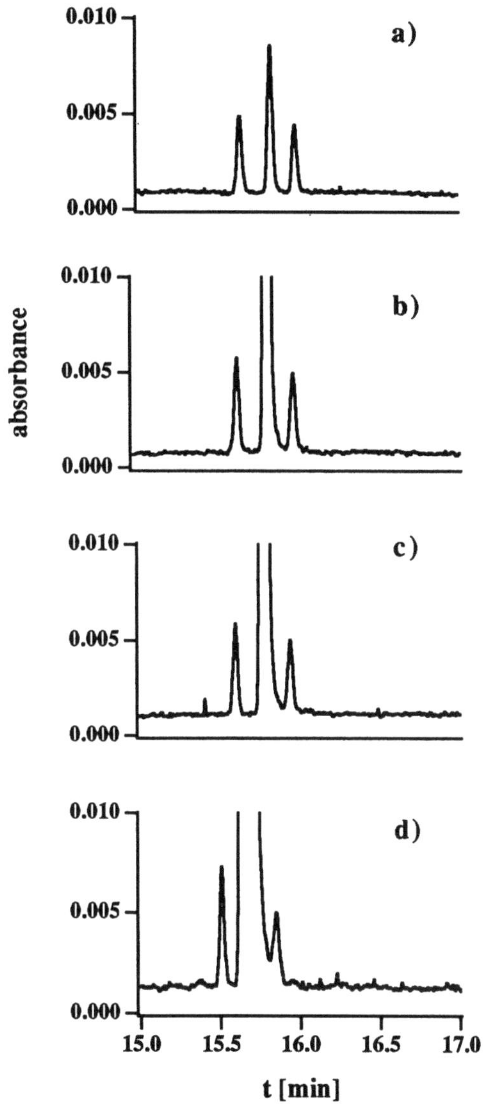

Fig. 3.6. Influence of the ratio of the peptide concentration c_S^+ to the concentration of sodium c_{Na}^+ on the resolution of a mixture of 3 peptides. Instrument: Beckman P/ACE 2000; experimental conditions: fused silica capillary, 107 cm x 75 μm i.d., hydrodynamic injection for 1 s, field strength 200 $V \cdot cm^{-1}$, temperature 25 °C, UV detection at 200 nm, electrolyte system 50 mM sodium phosphate buffer, pH 7.1. The ratio of the concentration of the second peptide to the concentration of sodium $c_S/c_{Na}+$ is (**a**) 0.0009, (**b**) 0.0019, (**c**) 0.0094 and (**d**) 0.02. The concentrations of the first and the third peptide are kept constant in all runs and are 0.036 mM

concentration of sodium phosphate at pH 7.1, but 82.5 mM. Because the electrophoretic mobility of Na^+ is higher than that of the peptides, electrophoretic dispersion manifests itself in a tailing. As long as the ratio of the concentration of the second peptide to the concentration of sodium is low enough, baseline separation is obtained (Fig. 3.6.a and b). Peak tailing caused by electromigration occurs if the ratio exceeds a certain value. While the two peaks are still separated at a ratio of ≈0.01 (Fig. 3.6.c), a ratio of ≈0.02 (Fig. 3.6.d) leads to a strong tailing of the second peptide peak. Obviously only the tailing is responsible for the low resolution if the concentration of the second peptide becomes too high. The resolution between the first and the second peak is not affected.

In the previous discussion of electrophoretic dispersion it has been assumed that no distortion occurs at the initial location of the admitted sample pulse. This holds, however, only if neither conductivity nor concentration gradients exist at the boundary between sampling and separation compartment. In practice, the sample is diluted or dissolved in either the carrier electrolyte or in water resulting in conductivity as well as concentration gradients at the boundary. Before dispersion occurs, the sample will be diluted or concentrated over the boundary between sampling and separation compartment. Nine different concentration distributions can result from the combination of these two distortion effects as illustrated in Table 3.4.

Table 3.4. Impact of the ratio of the conductivity of the sample solution κ_S to the conductivity of the buffer solution κ_B

	$\kappa_S > \kappa_B$	$\kappa_S = \kappa_B$	$\kappa_S < \kappa_B$
$\mu_S^+ < \mu_B^+$	dilution peak tailing	peak tailing	concentration peak tailing
$\mu_S^+ = \mu_B^+$	dilution		concentration
$\mu_S^+ > \mu_B^+$	dilution peak fronting	peak fronting	concentration peak fronting

If the conductivity of the sample solution is lower than that of the buffer solution, the sample will be concentrated over the stationary boundary between the sampling and the separation compartment because of the higher field strength in the latter. Figure 3.7. shows the electrophoretic development of a sample constituent from the time of injection onwards. The sample is dissolved in pure water and the mobility of the sample constituent is higher than that of the buffer constituent B^+. First of all, the sample is concentrated over the stationary boundary between the sampling and the separation compartment. The concentration leads to a decrease in the zone length of the sample. After a short time t_2, the sample still contains a homogeneous part, but the fronting region is already visible. During its migration through the separation compartment, the sample zone develops in the same way as shown in Fig. 3.5.b. If the sample constituent has the same mobility as the buffer constituent, only diffusional broadening occurs after the sample has been concentrated over the boundary between the sampling and the separation compartment. If the mobility of the sample constituent is lower than that of the buffer constituent, the migration process is more complicated than that shown in Fig. 3.7. After the concentration step, transient double peaks can occur.

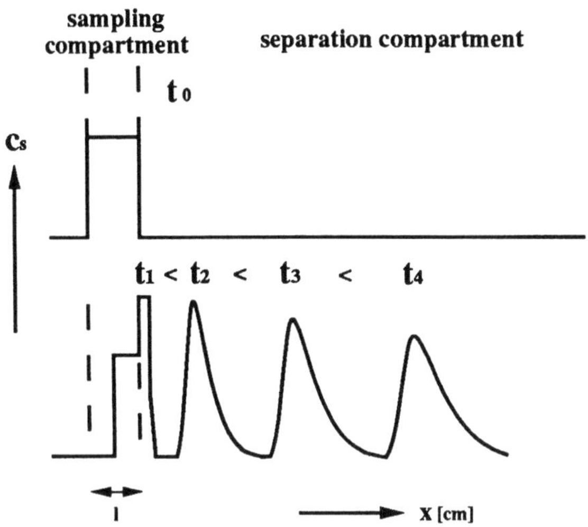

Fig. 3.7. Development of a zone electrophoretic process for a sample dissolved in pure water having a mobility which is higher than that of the buffer constituent with the same sign

In the opposite case, where the conductivity of the sampling compartment is higher than that of the separation compartment, the sample will be diluted by entering the separation compartment. This will result in an additional peak broadening, because the sample zone length will be increased.

The phenomenon of the concentration of a highly diluted sample at the boundary between sampling and separating compartments is called sample stacking. It is used in CZE to improve detection sensitivity and to narrow the length of the sample zone leading to an increase of the theoretical plate numbers. It has been shown by Chien and Burgi [44] that a factor of several hundred in signal enhancement can be achieved if the sample buffer is removed prior to separation. The technique of sample stacking will be treated in detail in Sect. 5.8.2.

It can be concluded that peak asymmetry and overloading effects as a result of mobility differences between the sample and the buffer zone can be minimized either by decreasing the concentration of the sample in such a way that the ratio c_S/c_B does not exceed 10^{-2} or by adjusting the mobilities of the sample and the buffer constituent. Additionally, the conductivity of the sampling zone should be kept lower than that of the buffer zone to prevent sample dilution at the boundary between sampling and separation compartment. For a more quantitative treatment of the phenomenon of electrophoretic dispersion the reader should see Refs. 31, 41, 42 and 45.

3.1.5 Sample Injection Width

The injection width of the sample is the most important extrinsic contribution to band broadening. If the sample is introduced into the capillary as a rectangular pulse the variance due to injection [4] is:

$$\sigma_I^2 = \frac{l^2}{12} \tag{3-21}$$

l width of the initial sample pulse [cm]

Typical injection lengths in a 75 μm i.d. capillary are in the range of 1 - 2 mm. Inside the capillary the injection plug is turned into a Gaussian zone through the action of diffusion. Huang, Coleman and Zare [46] have developed an equation to account for peak broadening in CZE. The equation applies to conditions where electrophoretic dispersion and Joule heat can be neglected. Three major contributions to the peak width are identified, namely axial diffusion, the injection length of the sample and adsorptive interactions of the analytes with the wall. They have found that the length of the injection plug is the most significant factor in respect to peak broadening, if the injection length is several times the diffusion width for the species in the zone. Therefore the number of theoretical plates N varies linearly with the applied voltage, as has been stated theoretically, only in the case of very small injection lengths. If the injection length becomes more than twice the diffusion width, the expression for N approaches

$$N = \frac{12 \cdot L_D^2}{l^2} \tag{3-22}$$

L_D effective capillary length [cm]

The contribution of diffusion to peak width can be calculated by:

$$w_s = 2\sqrt{2Dt} \tag{3-23}$$

w_s spatial peak width of the component [cm]
t migration time of the component [s]
D diffusion constant of the compound [$cm^2 \cdot s^{-1}$]

Table 3.5. gives some calculated values of w_s for different diffusion coefficients and migration times. Especially for species having low diffusion constants and short migration times, the width of the sample plug can become considerably higher than the diffusion width.

D [$cm^2 \cdot s^{-1}$]	t [min]	w_s [mm]
10^{-5}	5	1.550
	10	2.191
	15	2.683
10^{-6}	5	0.490
	10	0.693
	15	0.849
10^{-7}	5	0.155
	10	0.219
	15	0.268

Table 3.5. Calculated values of w_s for different diffusion coefficients and migration times

Vinther and Søeberg [47] have also found that peak efficiency dramatically decreases with increasing injection times and thus with increasing sample zone lengths. In one of their presented experiments, for an injection time of 0.2 s the injection term σ_I^2 contributes ca. 1% to the total dispersion whereas an injection time of 15 s leads to a contribution to the injection length of about 98%. In the latter case, a stacking process can be very effective in narrowing the sample zone length. The efficiency of a stacking run with an injection time of 15 s is 5-10 times higher than under non-stacking conditions. They suggest, therefore, that the injection time should be as short as possible depending on the detector performance, and that moderate stacking conditions and low applied potentials should be employed during the stacking period. Moderate stacking means that the conductivity of the sample solution should only be slightly lower than that of the buffer zone.

For the calculation of injection volumes and sample plug lengths see Sect. 4.1, for more details about sample stacking see Sect. 5.8.2.

3.1.6 Comparative Evaluation of the Different Dispersive Effects

In the preceding sections we have seen how the different dispersive effects influence the performance in CE and how they can be minimized or suppressed. Obviously, separation efficiency is mostly affected by a combination of several dispersive effects rather than only by a single one. Let us therefore compare these effects with respect to their relative influence on peak broadening.

Several authors have devoted themselves to the quantitative treatment of the most important dispersive effects in capillary electrophoresis [24, 31, 45-48]. Foret, Deml and Bocek [48], for instance, have studied the contribution of diffusion, Joule heating, sampling, electrophoretic dispersion and electroosmotic flow on the dispersion of zones in open and closed fused silica capillaries ranging from 125 to 400 µm in i.d. Their theoretical model predicts that an optimum field strength exists, at which the number of theoretical plates is a maximum. For open capillaries they have derived the following formula to estimate the electric field strength E_{opt} at the maximum of efficiency for a given experimental arrangement:

$$E_{opt} \approx 2.92 \cdot \sqrt[3]{\frac{D \cdot K}{\delta_i \cdot \kappa \cdot \mu_i} \cdot \frac{1}{r}} \qquad (3\text{-}24)$$

K	thermal conductance of the background electrolyte $[W \cdot cm^{-1} \cdot K^{-1}]$
κ	specific conductance of the background electrolyte $[S \cdot cm^{-1}]$
r	inner radius of the capillary [cm]
δ_i	temperature coefficient of the mobility of component i $[K^{-1}]$

In this equation, adsorptive interactions between the solute and the capillary wall as well as electrophoretic dispersion are neglected. This theory has shown to agree fairly well with the experiment, especially in the case of low conductive background electrolytes. However, the limited sensitivity of the detector used in this study prevented measurements with capillaries narrower than 125 µm.

Hjertén [31] has derived approximate equations for efficiency and resolution as a function of the width of the starting zone and of the zone broadening caused by diffusion, Joule heat, adsorption and the differences in conductivity between a solute zone and the background electrolyte. Two cases are stressed: the conductivity differences between sample and background electrolyte eliminate entirely (a) or partially (b) the diffusional broadening at one boundary of a zone. Again, when adsorption is negligible, one can derive from these equations the field strength at the maximum efficiency. At this optimal field strength, contributions to the zone broadening from diffusion, Joule heat and conductivity differences have the ratio 4:1:1 in case (a). In case (b) the ratio between diffusional dispersion and broadening due to Joule heat is 4:1.

Liu and coworkers [24] have evaluated dispersion phenomena in untreated and surface-treated open tubular as well as in gel-filled capillaries, with emphasis on inner diameters of 10 - 100 µm, as they are commonly used in CE. Because they use fluorescence detection, the sample concentration is low enough to neglect electrophoretic dispersion. In addition to diffusional dispersion in the relatively low electric field range, adsorptive interactions are believed to play a certain role in band broadening. The sorption-desorption kinetics become important with increasing field strength and manifest themselves in asymmetric tailing zones. This is especially true for small-bore capillaries. On the other hand, coated and gel-filled capillaries show slight trends towards peak distortion. Thermal effects appear to contribute only to a minor extent to dispersion in small capillaries (< 50 µm), but could become significant in larger bores (> 75 µm). They have found further that diffusion is minimized in gel-filled capillaries with diameters less than 50 µm and that thermal effects do not play a significant role at voltages up to 350 V/cm.

Finally, they have developed an equation which takes into account the influence of injection, axial diffusion, adsorption and Joule heat on the plate height in open tube CE :

$$H=\frac{l^2}{12L_D}+\frac{2D}{(\mu_i \pm \mu_{eo})E}+\frac{c_f(1-c_f)^2(\mu_i \pm \mu_{eo})rEn}{Z\alpha}+\frac{\delta_i^2\kappa^2\mu_i^2}{K^2D}\cdot \left(K_1\frac{E^5r^6}{(\mu_i \pm \mu_{eo})}+K_2\frac{E^6r^8}{LD}\right) \qquad (3\text{-}25)$$

l	initial width of the sample plug [cm]
L_D	effective capillary length [cm]
c_f	fractional concentration of a free solute
r	inner radius of the capillary [cm]
Z	number of molecules striking a unit surface area per second [$cm^{-2}\cdot s^{-1}$]
n	number of molecules per unit volume inside the tube [cm^{-3}]
α	fraction of molecules which adhere on collision
δ_i	temperature coefficient of the mobility of component i [K^{-1}]
κ	specific conductance of the background electrolyte [$S\cdot cm^{-1}$]
K	thermal conductance of the background electrolyte [$W\cdot cm^{-1}\cdot K^{-1}$]
K_1	$6.5\cdot 10^{-4}$
K_2	$4.34\cdot 10^{-5}$

Eq. 3-25 can simply be written as

$$H = A + \frac{B}{E} + CE + G_1E^5 + G_2E^6 \qquad (3\text{-}26)$$

which predicts a Van Deemter plot as in chromatography. Obviously, the initial sample plug length is the only factor which is not influenced by the applied field strength. Whereas, as theory predicts, band broadening caused by diffusion is decreased with increasing field strength, dispersion caused by adsorption and Joule heat increases at the same time.

Again, all of the treated dispersive effects can be kept well under control if the following guidelines are followed:

> Because almost every dispersive effect is temperature dependent, the system should not only be cooled by forced convection, but also be thermostated.

> Capillaries in the range of 50 to 75 µm have shown to be a good compromise with regard to the surface-to-volume ratio in respect to Joule heating and sorption-desorption kinetics as well as detection sensitivity for UV-VIS absorbance.

> The sample concentration should not exceed 1% of the electrolyte concentration (or the mobility of the sample ion should be similar to that of the coion) and

the sample plug length should be kept as short as possible. If detection problems can occur due to low sensitivity, sample stacking is preferred over longer injection times.

> The conductivity of the sample should be kept lower than that of the background electrolyte to suppress additional band broadening by dilution at the boundary between sampling and separation compartment.

> The specific conductance and/or the ionic strength of the background electrolyte should be as low as possible (as the last two guidelines allow) to avoid excess Joule heating. Buffer constituents of low mobilities are preferred over those having high mobilities.

It has been shown that operational parameters like field strength, capillary dimensions and temperature play a key role in the performance of CE. These parameters will be stressed in the next sections.

3.2 Operational Parameters

In the previous section, we have seen several times that operational parameters like field strength, capillary dimensions and temperature have to be chosen carefully to minimize band broadening effects. These parameters are discussed in the following.

3.2.1 Field Strength

The driving force behind the migration of ions in CE is the field strength applied across the capillary, which is related to the applied voltage by dividing by the total capillary length. Since both the electrophoretic migration velocity and the electroosmotic flow velocity are directly proportional to the electric field, highest field strengths will bring about the shortest analysis times. Jorgenson and Lukacs [6-8] regarded axial diffusion as the exclusive dispersion factor in CE. They obtained a linear relationship between the plate number and the applied voltage. Therefore, one could imagine that the highest efficiency would be obtained by working at highest possible field strengths. But theoretical plate numbers are proportional to the voltage only for low values of E, because heat production limits the application of high field strengths. The influence of voltage on the separation is demonstrated in Fig. 3.8.a. Four positional isomers of dihydroxybenzoic acids are separated at 260, 350 and 440 $V \cdot cm^{-1}$. While there is only a small decrease in resolution of the four components by increasing the field from 260 to 350 $V \cdot cm^{-1}$, a dramatic loss in resolution is found if field strength is further increased to 440 V·cm^{-1}. The influence of excessive heat production can be visualized by plotting the applied field strength versus the resulting current (Fig. 3.8.b). According to Ohm's law this plot should be linear. In

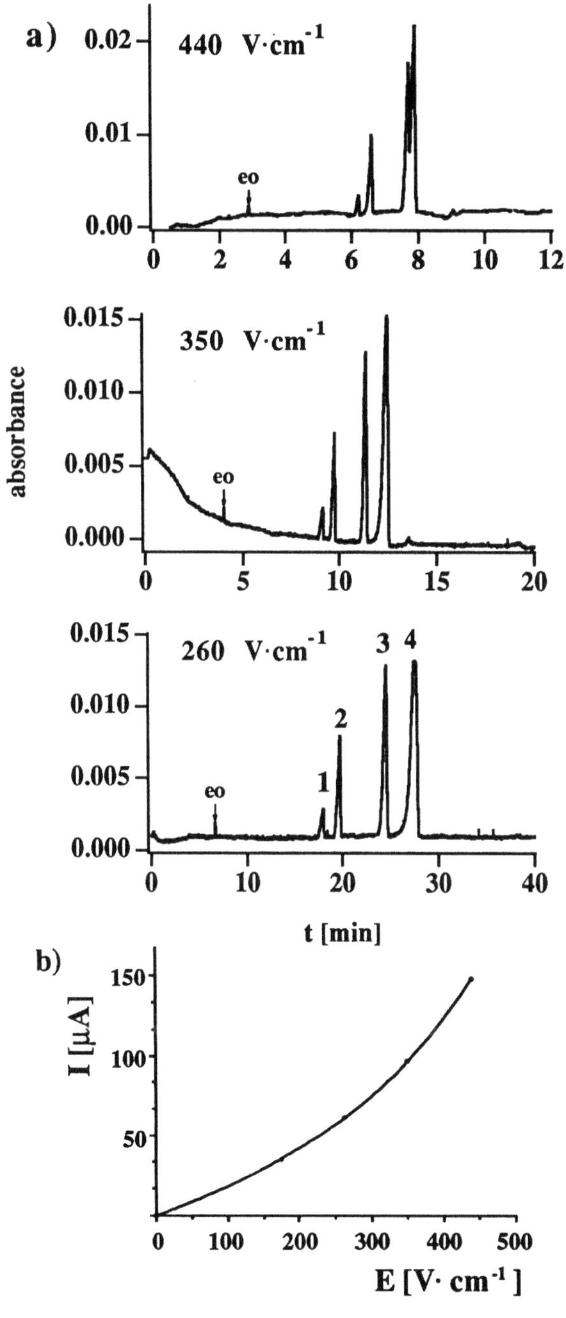

Fig. 3.8. (**a**) Influence of the field strength on the electrophoretic separation of four positional isomers of dihydroxybenzoic acid. Elution order of the isomers: **1**) 2,4-, **2**) 2,3-, **3**) 2,6- and **4**) 2,5-dihydroxybenzoic acid. Instrument: Beckman P/ACE 2000; experimental conditions: fused silica capillary 57 cm x 75 μm i.d., hydrodynamic injection for 1 s, temperature 25 °C, UV detection at 200 nm, electrolyte system 25 mM Na_2HPO_4 - 25 mM $Na_2B_4O_7$, pH 9.0. Benzyl alcohol is used as EOF marker. (**b**) Plot of field strength versus resulting current for the buffer given in (**a**)

practice, however, the plot is linear only in the lower voltage range as shown in Fig. 3.8.b. Deviations from linearity are caused by increased heat generation at higher potential differences. The optimal field strength which should be applied can be determined from the plot as the point where deviation from linearity begins.

Hint: This procedure can be carried out without the need to perform a separation. If the capillary is filled with buffer, just change the voltage in steps of 2 - 5 min. and record the resulting current. A step-like profile of the current will be measured, from which the current values can be readily determined.

Changes of the current with the voltage simultaneously impact the electroosmotic and electrophoretic velocities such that plots of i.e. the electroosmotic velocity versus field strength do not show linearity at high field strengths. However, plots of v_{eo} versus I are straight lines even at high values of I. This behavior can be explained theoretically as following: by replacing E in Eq. 2-26 by the quotient i/κ the following expression for the EOF velocity can be derived [2, 49, 50]:

$$v_{eo} = \frac{\varepsilon \cdot \zeta \cdot i}{4\pi \cdot \kappa \cdot \eta} \qquad (3\text{-}27)$$

ε permittivity of the medium $[F \cdot m^{-1}]$
i current density $[A \cdot cm^{-2}]$
κ specific conductance $[\Omega^{-1} \cdot cm^{-1}]$
η Newtonian viscosity of the solution $[Pa \cdot s]$

ε and ζ vary only slightly with temperature and their variation is such that their product is essentially independent on temperature. Therefore, Eq. 3-28 can be written in the form:

$$v_{eo} = A \cdot \frac{i}{\kappa \cdot \eta} \qquad (3\text{-}28)$$

A constant $[C \cdot cm^{-1}]$

Although both the specific conductance κ and the viscosity η exhibit rather sensitive temperature dependencies, they change in such a way that their product $\kappa \cdot \eta$ remains constant for small variations in temperature, which is known as the Walden rule. Consequently, the electroosmotic velocity depends only on the current density and not on the temperature of the system. Therefore, a constant-current mode is preferable to a constant-voltage mode of operation in CE in terms of precision of the EOF and, hence, the electrophoretic velocity of the analytes. This is demonstrated in Table 3.6. for the separation of the four dihydroxy-derivatized benzoic acids shown in Fig. 3.8. The experiments are carried out in triplicate maintaining constant either the voltage at 15 kV or the current at 67 μA. Benzyl alcohol is used as EOF marker. Standard deviations of the velocity for all benzoic acid derivatives and for the EOF are calculated. Keeping the current constant significant lower values are found compared to those obtained at constant voltage.

Table 3.6. Evaluation of three consecutive electrophoretic separations of dihydroxy-substituted derivatives of benzoic acid at either constant voltage (a) or constant current (b). All runs were performed one after the other under the same conditions

	a) 15 kV		b) 67 μA	
Component	Average Velocities · 10 [$cm \cdot s^{-1}$]	Rel. Standard Deviation [%]	Average Velocities ·10 [$cm \cdot s^{-1}$]	Rel. Standard Deviation [%]
benzyl alcohol (EOF marker)	1.343	0.56	1.407	0.32
2,4-dihydroxy-benzoic acid	0.565	1.51	0.608	0.76
2,3-dihydroxy-benzoic acid	0.525	1.62	0.567	0.76
2,6-dihydroxy-benzoic acid	0.445	1.71	0.486	1.00
2,5-dihydroxy-benzoic acid	0.380	2.45	0.421	0.71

Figure 3.9. compares a plot of the electroosmotic velocity versus the field strength and the current density. While the curve shape of Fig. 3.9.a is concave as already known from Fig. 3.3., a plot of the current density versus the electroosmotic velocity gives a straight line because the current density changes analogously to EOF with the viscosity. It should be mentioned that runs which are performed at constant voltage are called isoelectrostatic and those performed at constant current isorheic runs.

3.2.2 Capillary Dimensions

The heart of each capillary electrophoresis system is the capillary itself. Because of its intrinsic properties fused silica has become the material of choice. Tube diameters down to a few micrometers are commercially available and light transmission of fused silica is high even at 190 nm. To some extent fused silica capillaries are flexible as long as the outer polyimide coating has not been removed. Tube dimensions commonly used in CE range between 10 to 100 μm internal diameter, 375 µm outer diameter and 10 to 100 cm in length. As already mentioned the choice of the capillary dimensions has an effect on several factors such as migration time and resolution, detection sensitivity, heat dissipation and adsorption.

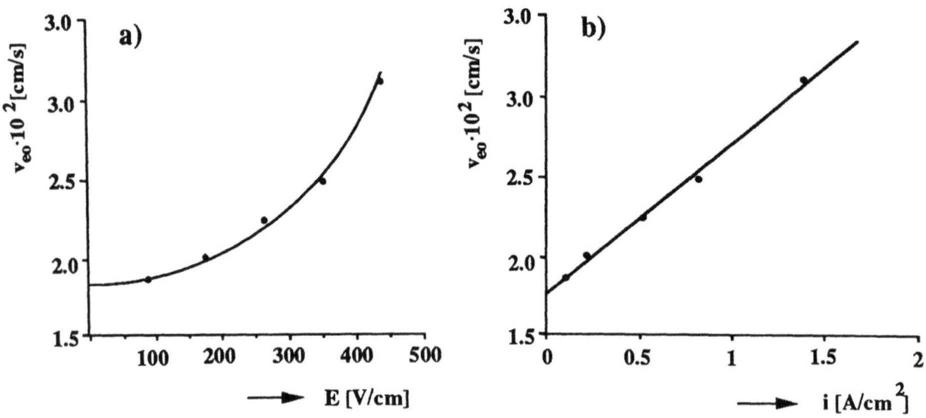

Fig. 3.9. Plots of electroosmotic mobility versus field strength (**a**) and current density (**b**) for the separation shown in Fig. 3.8

The impact of the capillary length on the migration time is shown by Colburn et al. [51] for the migration of a peptide in two capillaries with lengths of 50 and 100 cm, respectively. Their experiments are carried out at constant field strength with the detector placed 25 cm from the cathodic end of the capillary in both cases. Thus, the migration distance to the detector is 25 cm for the short capillary and 75 cm for the long one. Based on the different effective capillary lengths, the migration time increases from 18 minutes to about 48 minutes for the long capillary. With the longer migration time the separation efficiency and the peak resolution have been improved dramatically for the longer capillary. The corresponding electrophoretic data are summarized in Table 3.7.

Table 3.7. Influences of the capillary length on the performance in CE (data taken from Ref. 51)

Capillary length [cm]	50	100
Length to the detector [cm]	25	75
Resolution of peak A	1.00	1.96
Plate number of peak A	75,000	300,000

A more detailed investigation of the effect of the capillary length on the plate number has been provided by Jorgenson [52]. By measuring the plate number of dansyl-labeled isoleucine at tube lengths ranging from 50 to 150 cm, they have found (see Fig. 3.10.) that the efficiency does not significantly improve at tubes longer than 100 cm but considerably longer analysis times result. Plate numbers decrease dramatically in capillaries shorter than 100 cm. This loss is assumed to be based on thermal effects of the system. Shorter capillaries show lower electrical resistance and cause higher currents. While the surface area for dissipation of Joule heating decreases, the generation of heat is enlarged by the increased power of the system.

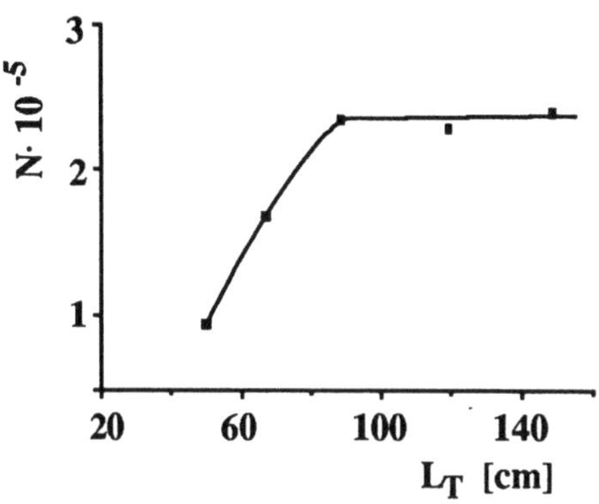

Fig. 3.10. Theoretical plate number as a function of capillary length. Experiments are carried out in 50 mM phosphate buffer, pH 6.86, at an applied voltage of 15 kV. (With permission from Ref. 52)

A more theoretical approach is derived from Eq. 2-46 to visualize the dependence of a separation problem on the capillary length, which is required to maintain resolution constant for differing mobilities of the analytes. By substituting the voltage V in Eq. 2-46 by the product of the field strength E times the capillary length L_T, resolution is directly proportional to the square root of the length. The capillary length, required to obtain a resolution of 1.25 in dependence of the difference in mobility of two analytes is shown in Fig. 3.11. The individual curves represent data at different electroosmotic mobilities. It is assumed that no dispersive effects occur except that of axial diffusion with a diffusion coefficient of 10^{-5} $cm^2 \cdot s^{-1}$. An applied field strength of 300 $V \cdot cm^{-1}$ is supposed. The electrophoretic mobility μ_1 is regarded as constant at $5 \cdot 10^{-5}$ $cm^2 \cdot V^{-1} \cdot s^{-1}$ and μ_2 is variable. As clearly demonstrated the electroosmotic flow counteracts the separation. In this respect high electroosmotic flows require long capillaries to maintain the resolution. To give an example, we suppose that the mobility difference of the two analytes is 7.5%. If electroosmosis with a mobility of $5 \cdot 10^{-4}$ $cm^2 \cdot V^{-1} \cdot s^{-1}$ takes place in the same direction as electrophoresis of the analytes, a capillary length of approx. 65 cm is needed to obtain a resolution of 1.25. If μ_{eo} has about the same value as the mobilities of the analytes ($5 \cdot 10^{-5}$ $cm^2 \cdot V^{-1} \cdot s^{-1}$), a capillary of at least 12 cm is still required. From Fig. 3.11. it becomes obvious that shorter capillaries can be used, if electroosmosis counteracts the electrophoretic migration of the analytes. It should be noted that short capillaries are not equivalent to short run times in this consideration.

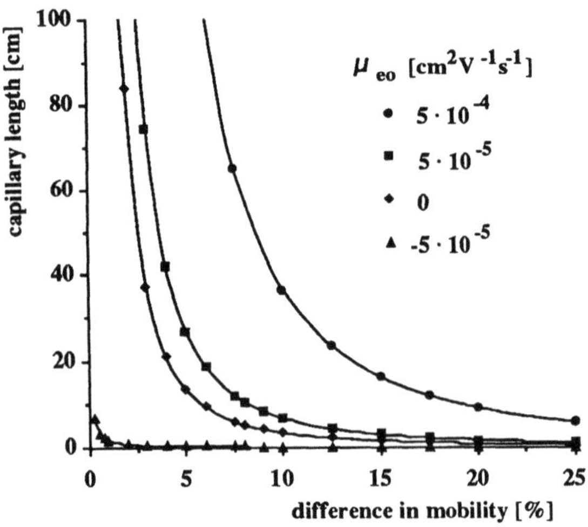

Fig. 3.11. Required capillary length as a function of the difference in mobility of two components to maintain a resolution of 1.25. A field strength of 300 $V{\cdot}cm^{-1}$ and a diffusion coefficient of 10^{-5} $cm^2{\cdot}s^{-1}$ is supposed. The simulation is based on Eq. 2-46. The values represent different electroosmotic mobilities

The inner diameter (i.d.) of the capillary tube influences the separation performance in several respects. Tubes with small diameters are advantageous because of the Joule heat produced by the electric current (see also Sect. 3.1.3). Heat is generated uniformly across a tube section, but dissipation takes place through the silica walls only. A temperature gradient results which effects a density gradient (leading to convection) and a viscosity change across the capillary. Jorgenson [52] studied the effect of the tube diameter on the efficiency. As he demonstrated, the plate numbers decrease from approx. 250 000 theoretical plates for capillaries with i.d.'s below 80 μm to approx. 100 000 plates for a 100 μm capillary. Since heat is much better dissipated in small diameter capillaries, analysis times can be shortened significantly, if high voltages are applied to short, 10 - 20 μm i.d. capillaries. If the mobility differences of the analytes are high enough, separation can be performed in less than 1 minute.

Grushka et al. [34] calculated the maximum allowable capillary radius as a result of excess Joule heat. Figure 3.12. shows the maximal allowable radii as a function of voltage and buffer concentration for limiting cases, where the plate height increase is 10%, 20% and 40% above the theoretical minimum value. In contrast to the suggestions of Knox [21] for maximal allowable capillary diameters for small molecules with D = 10^{-5} $cm^2{\cdot}s^{-1}$, which are presented in

Sect. 3.1.3, diffusion constants are chosen such that the conditions are valid for large particles. The maximal allowable capillary radius decreases exponentially with increasing voltage. It can clearly be seen that, under typical CE conditions, Joule heating does not affect efficiency, if the buffer concentration is 10 mM and the diffusion constant of the analyte is 10^{-7} $cm \cdot s^{-2}$ (Fig. 3.12.d). The same holds for Fig. 3.12.b and 3.12.c, if a voltage of 25 kV is not exceeded. Only at high buffer concentration in combination with a lower diffusion constant (Fig. 3.12.a) does Joule heating dramatically limit the maximal allowable capillary radius, e.g. at 20 kV, capillaries exceeding 90 µm in i.d. should not be used, at 25 kV even diameters greater than 60 µm should not be used anymore. It can be concluded that Joule heat affects the separation efficiency of large analytes much more than that of smaller ones and that temperature effects are negligible for narrow bore tubes having diameters smaller than 50 µm.

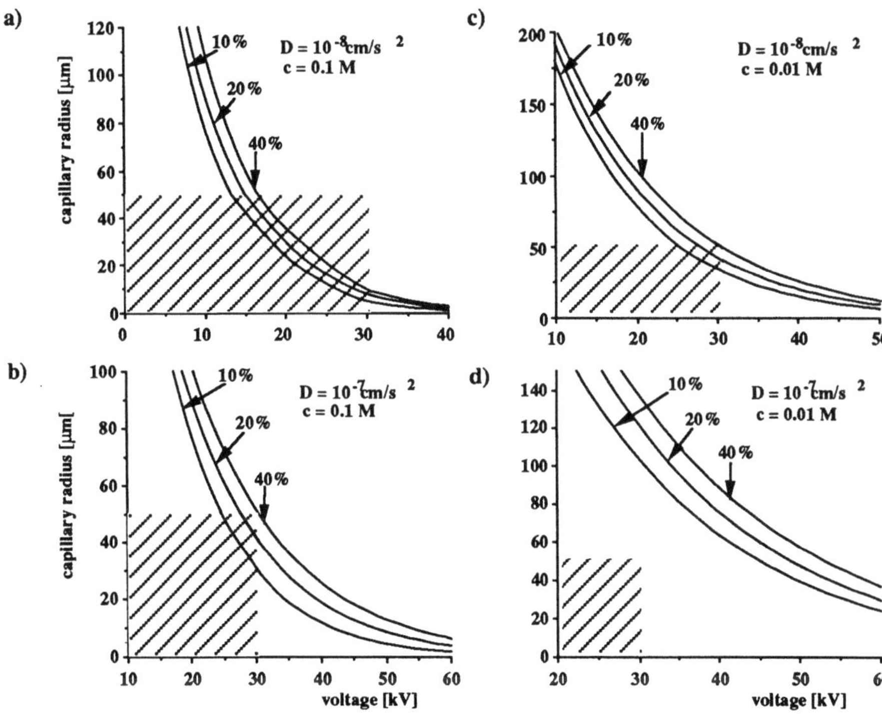

Fig. 3.12. Maximal allowable capillary radii as a function of operational voltage and electrolyte concentration. Supposed conditions: fused silica capillary, 100 cm x 375 µm o.d., electrolyte concentration 0.1 M (**a, b**) and 0.01 M (**c, d**), limiting conductance of the buffer 150 $cm^2 \cdot mol^{-1} \cdot \Omega^{-1}$, diffusion constant of the analyte 10^{-8} $cm^2 \cdot s^{-1}$ (**a, c**) and 10^{-7} $cm^2 \cdot s^{-1}$ (**b, d**). The 3 curves represent the radii at 10%, 20% and 40% plate height increase above the minimum theoretical value. The *dashed region* shows CE conditions typically used (according to Ref. 34 and with permission of the American Chemical Society)

On the other hand absorption detection sensitivity decreases with smaller i.d.'s because of a shorter pathway of the light beam in accordance with Lambert-Beer's law (see Sect. 4.2.3). Another fact which is often neglected is increased adsorption. The ratio of sample volume to surface area decreases with smaller tube diameters leading to more active adsorption sites per sample volume. This fact can be most important if compounds tend to adsorb strongly on the silica. Additionally wider tubes allow easier injection procedures and larger sample volume loads.

The impact of capillary dimensions on the performance of optical on-column detection is discussed in Sect. 4.2.3.

3.2.3 Temperature

Temperature programming and gradient techniques are routine methods to solve elution problems in gas chromatography and to improve both speed and efficiency in HPLC. In electrophoretic separation methods, however, temperature is usually considered in a negative context due to the loss in efficiency by excessive Joule heating. Hence, temperature control is commonly used to provide efficient heat removal. So far, only a few contributions have been made dealing with the impact of column temperature on the selectivity in capillary electrophoretic techniques. Among these, there are several promising aspects of the influence of temperature in CE that need to be considered.

As already mentioned in sects. 2.3 and 3.1.3 both electrophoretic and electroosmotic mobility increase with increasing temperature due to temperature-induced viscosity changes. Thus, the major effect of temperature is to shorten the analysis time in the presence of electroosmotic flow. A temperature increase of 10 K enhances the EOF approximately 20%. Because the electrophoretic mobilities are also enhanced, the analysis time is not reduced by the same value for species migrating in the opposite direction of the EOF. One should be aware that exploiting the reduction of analysis time by working at higher temperatures requires efficient thermostation of the system because of the higher Joule heat that has to be removed. Additionally, shorter analysis times can go at the expense of resolution.

Issaq et al. [53] have shown that the separation of a series of 7 dipeptides at pH 2.5 deteriorates as the temperature is increased from 25 °C to 45 °C. This can be due to the decrease in migration time which leads to a co-elution of closely neighbouring peaks.

Demorest and Dubrov [54] have investigated the temperature influence on the migration times and efficiency in capillary gel electrophoresis of oligonucleotides. In the absence of an EOF, an increase of the column temperature from 30 - 50 °C results in a decrease of the migration time by 1.1% per one degree K. This value is almost half the value estimated above. This can be due to additional temperature effects of the highly cross-linked gel matrix.

Guttman and Cooke [55] have studied the effect of temperature on the separation of DNA restriction fragments in terms of migration times t_i and resolution R in capillary gel electrophoresis. They have found that a maximum migration time and a maximum resolution exist when working in the isorheic mode (I = constant). In contrast, applying a constant voltage (isoelectrostatic mode) leads to a continuous increase of t_i and R with increasing temperature from 20 - 50 °C. This behavior can be explained by the following reflection. Combining Eqs. 2-10 and 2-19 leads to

$$t_i = \frac{L_D}{v_i} = \frac{6\pi \cdot \eta \cdot r \cdot L_D}{z_i \cdot e_0 \cdot E} \tag{3-29}$$

Because of Eq. 2-12, it follows that

$$\ln t_i = \ln\left(\frac{6\pi \cdot r \cdot L_D}{z_i \cdot e_0 \cdot E}\right) + \frac{E_A}{R \cdot T} \tag{3-30}$$

In the isoelectrostatic mode only the second term on the right side of Eq. 3-30 is temperature dependent and ln t_i increases linearly with $1/T$. On the other hand, in the isorheic mode both terms of Eq. 3-30 are temperature dependent, because now, E decreases with temperature if I remains constant. If the temperature is raised the effect of field strength and viscosity will first be partially balanced, leading to a maximal t_i. At a further increase of temperature the lower field strength will prevail over the lower viscosity leading to a slower migration as in the isoelectrostatic mode. The authors have realized further that, in both modes, the plate number decreases as T increases in particular for the high molecular weight fragments. This can be due to conformational changes at higher temperatures.

Another interesting aspect of the impact of column temperature are thermally induced changes of protein structures. Recently, the influence of column temperature on the electrophoretic behavior of horse heart myoglobin and bovine α-lactalbumin type III have been studied [56]. Myoglobin is stable in the 20 - 45 °C temperature range, which is typical for small globular proteins. α-Lactalbumin has a conformational transition at 32 °C. Conformational transitions result in effective charge or hydrodynamic shape changes which can result in variations in migration time and peak shape. This leads to asymmetric peaks and sigmoidal mobility plots versus temperature in the transition region. The authors conclude that broadened or multiple peaks do not necessarily mean that a protein sample is impure. They suggest subambient temperature control (4 °C) as is also used in slab gel electrophoresis.

Probably the most impressive effect of the column temperature is its relation to chemical equilibria, such as metal chelation, micelle partitioning, complex formation and dissociation. Both the mass action constant as well as the rate

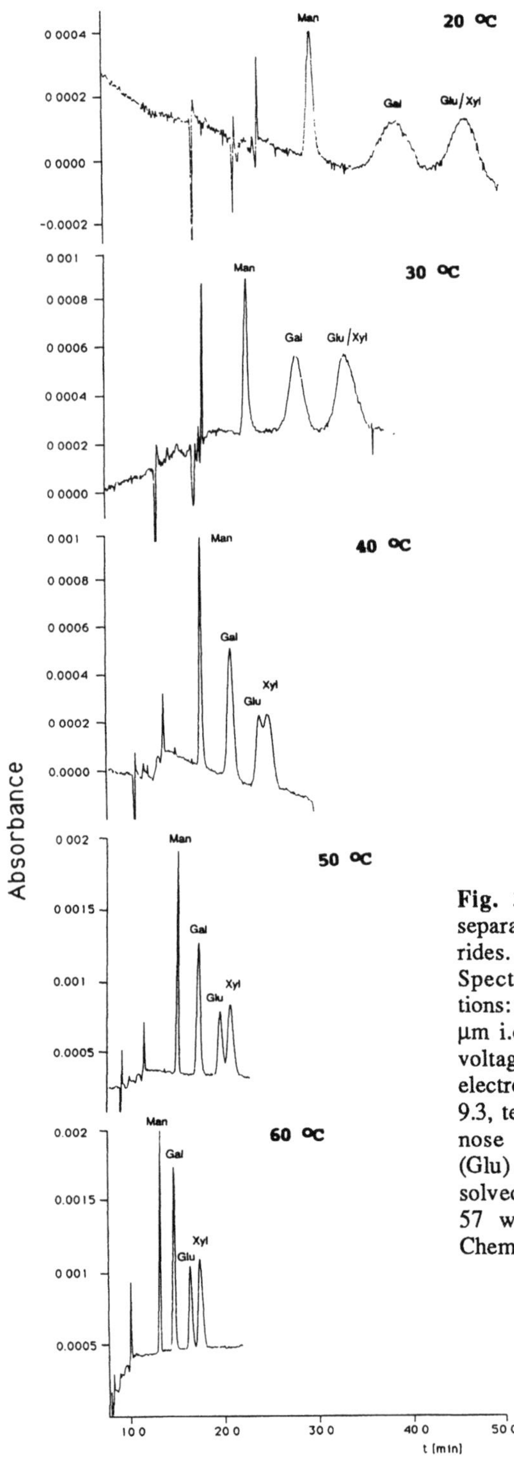

Fig. 3.13. Effect of temperature on the separation of underivatized monosaccharides. Instrument: SpectraPhoresis 1000, SpectraPhysics; experimental conditions: fused silica capillary, 94 cm x 75 μm i.d., hydrodynamic injection for 1 s, voltage 20 kV, UV detection at 195 nm, electrolyte system 50 mM $Na_2B_4O_7$, pH 9.3, temperature 20- 60 °C; sample: mannose (Man), galactose (Gal), glucose (Glu) and xylose (Xyl), each 10 mM, dissolved in water. (Reproduced from Ref. 57 with permission of the American Chemical Society)

constant are temperature dependent. Thus, chemical equilibria can be manipulated by the proper control of temperature.

According to the Arrhenius equation, the rate constant is indirectly proportional to the natural logarithm of the temperature. As a rule of thumb, a temperature increase of 10 °C doubles the reaction rate. This means that the reaction equilibrium is reached faster at elevated temperatures. Hoffstetter-Kuhn et al. [57] have studied the influence of temperature on the electrophoretic behavior of carbohydrates in the presence of borate. They have found out that the temperature dependence of the reaction rates of the different dynamic equilibria of carbohydrates in borate solutions affects the electrophoretic separation dramatically. This is illustrated in Fig. 3.13. with the separation of a mixture of four monosaccharides at elevated temperatures. At 20 °C the peaks are very broad with baseline widths up to 7 minutes. Glucose and xylose are not separated. As shown in the successive electropherograms, by increasing the temperature in the capillary, separation efficiency is enhanced dramatically. The equilibrium is reached faster, resulting in narrower peak shapes. Besides efficiency, sensitivity is significantly improved by raising the temperature (see also Sect. 7.8).

An example for the impact of column temperature on the mass action constant of dissociation is the use of temperature programming to establish a pH step or gradient in situ within the capillary. For this purpose, the temperature is varied as a function of time (dynamic mode) or as a function of position (static mode) during electrophoresis. To get a high temperature gradient a buffer system with a large temperature coefficient dpH/dT has to be used. Among commonly used buffer systems, TRIS buffer has the largest pH shift with temperature [58]. If the pH (1) of a TRIS buffer at a certain temperature T_1 is known, the temperature coefficient dpH/dT can be used to estimate the pH (2) of the buffer at a different temperature T_2 by the following equation:

$$pH_2 = pH_1 - 0.0248\,(T_2 - T_1) \qquad (3\text{-}31)$$

In the range of 20 - 60 °C a plot of pH values versus temperature is linear with a correlation coefficient of 0.999. The electrophoretic behavior of weak acids in the presence of both static and dynamic pH gradients has been investigated [59]. The effect of pH is most significant if the range of pH change matches the negative logarithm of the acidity constant, pK_a, of the analytes. In the above mentioned publication, five fluorescent dyes were isothermally separated at 20 °C and at 70 °C. At 20 °C only three dyes are resolved. Although the analysis time has been shortened by 25%, raising the temperature does not improve the separation. By employing a step change of temperature from 70 °C to 20 °C and thus pH and viscosity, after 2 min of running time the resolution has been significantly improved. Nevertheless, the baseline was not stable, probably due to refractive index changes caused by the sudden temperature change. By creating a temperature gradient from 70 °C on the injection side to 20 °C at the end of the capillary, which in turn induces a pH gradient of 6.5 to 7.3 along the capillary, the separation has been further improved and the baseline is stable. This method can be applied to complex samples, where inadequate separation is obtained at any single buffer pH.

3.3 Electrolyte System

The electrolyte system plays another important role in CE performance. Properties like pH, ionic strength and the composition affect both selectivity and efficiency tremendously. The pH, for instance, is the most important selectivity factor in electrophoretic separation of ampholytes. The ionic strength of the electrolyte system does not only determine the degree of Joule heating at constant voltage, but also has a marked influence on both electroosmotic flow and electrophoretic mobility. In addition, the buffer composition can improve efficiency as well as selectivity in several ways. Last but not least, buffer additives such as complexing agents and organic solvents can improve or even facilitate separation. In the following we will give an overview of all important factors and facts dealing with the electrolyte system.

3.3.1 Basic Requirements

The transportation of current in electrophoresis is accomplished by using ionizable salt solutions as running electrolytes. Owing to the strong influence of the pH on electroosmotic and electrophoretic mobility, buffer substances are commonly employed as electrolytes. In general, these buffer systems should fulfill a number of requirements (see also Sect. 3.1.6):

- The buffer system should have no negative effect on the separation.
- A high buffer capacity over a broad pH range must be guaranteed.
- The pH should show a low variation with temperature.
- In the case of UV-VIS absorbance as detection mode, the buffer should show low UV absorbance at the wavelength of interest.
- The mobility of the buffer ion should be similar to the analytes to minimize electrophoretic dispersion.
- The electrophoretic mobility of the counterion should be as low as possible to minimize heat generation and to allow high voltages to be applied.

3.3.2 pH

The pH value of the electrolyte solution in CE is the most important separation parameter for changing the selectivity of the system. In general, separation in electrophoresis is based on differing mobilities of the analytes, which in turn depend on their size and net charge. The size of an ion is related to the molecular mass and the degree of hydration depending again on factors like ionic strength and polarity of the solution. The net charge of the ion is dependent on the degree of ionization given by the pK value of the acid or basic functional group and the pH of the solution. In the particular case where the substances to be separated are entirely charged over the pH range of interest, variation of the pH does not influence the net charge of the ions. This holds usually for ions of strong acids or bases like chloride, nitrate, sodium or potassium. If, however, the ionizable functional groups of the analytes are weak acids or bases, the pH of the electrolyte shows a strong influence on the net charge. Fig. 3.14. depicts the electrophoretic separation of four nucleosides at different pH values ranging

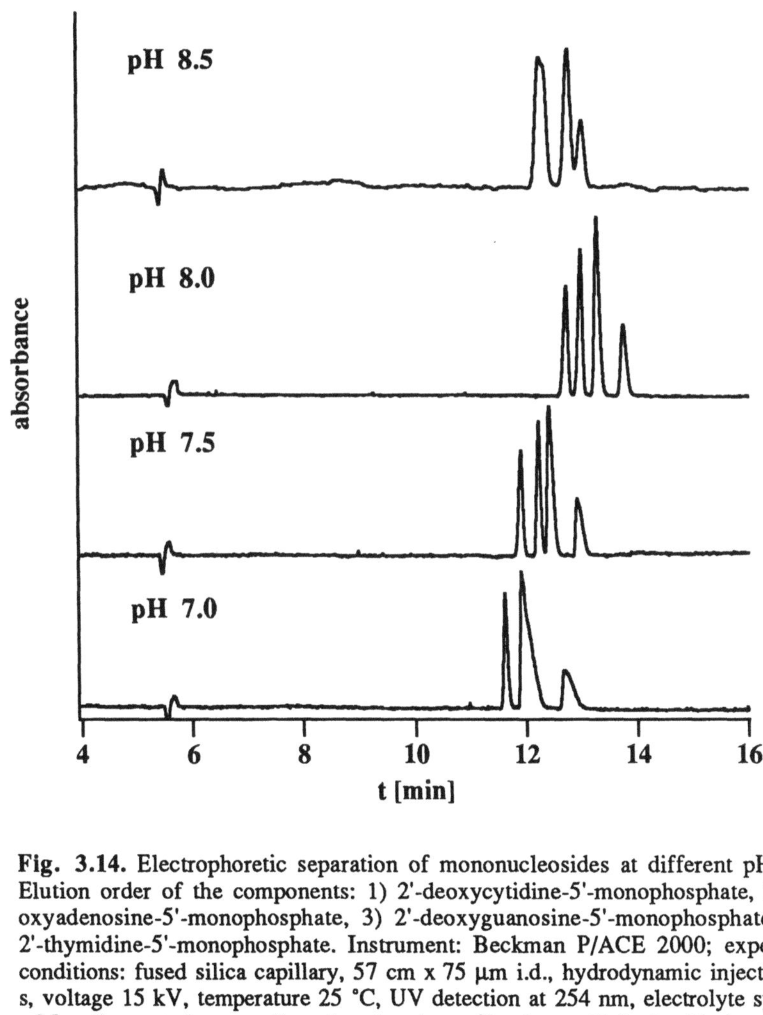

Fig. 3.14. Electrophoretic separation of mononucleosides at different pH values. Elution order of the components: 1) 2'-deoxycytidine-5'-monophosphate, 2) 2'-deoxyadenosine-5'-monophosphate, 3) 2'-deoxyguanosine-5'-monophosphate and 4) 2'-thymidine-5'-monophosphate. Instrument: Beckman P/ACE 2000; experimental conditions: fused silica capillary, 57 cm x 75 µm i.d., hydrodynamic injection for 1 s, voltage 15 kV, temperature 25 °C, UV detection at 254 nm, electrolyte system 25 mM sodium tetraborate, adjusted to the given pH values with hydrochloric acid

from pH 7 to 8.5. The difference in charge depends on the degree of dissociation of the amine functionalities and the phosphate groups. As one can readily see, even small changes of the pH have a dramatic influence on the resolution of the four components. For this reason it is necessary to use properly adjusted buffers as electrolyte solutions.

On the other hand, it becomes obvious from Fig. 3.14. that resolution can be optimized very easily by finding the optimum pH value for a specific separation. Nevertheless, a rough screening of the resolution at a few pH values can suggest poor separation over a wide pH range, because the optimum is not found by the screening conditions. Therefore, a more precise procedure to find the pH optimum is presented in the following. For this purpose, the net charges of the analytes in dependence of the pH value of the buffer solution have to be found first. Because we are only interested in the relative change of the net charge with the pH it is not necessary to determine absolute net charges. The relative net charge can be calculated from the quotient of the concentrations of the charged and the neutral species according to the Hendersson-Hasselbalch equation (Eq. 3-32), which is derived from the chemical reaction of dissociation in its general form for e.g. an acid (proton donor):

$$HA + H_2O \rightleftarrows A^- + H_3O^+$$

where HA represents the undissociated acid and A^- the conjugate base.

$$pH = pK_a + \log\frac{[A^-]}{[HA]} \tag{3-32}$$

[HA] concentration of the undissociated acid [M]
$[A^-]$ concentration of the conjugate base [M]
pK_a negative common logarithm of the dissociation constant K_a of the acid ($-\log K_a$)

An analogous equation can be set up for a base (proton acceptor) as follows:

$$A^- + H_2O \rightleftarrows HA + OH^-$$

$$pOH = pK_b + \log\frac{[HA]}{[A^-]} \tag{3-33}$$

where A^- represents the base and HA the conjugate acid.

$[A^-]$ concentration of the base [M]
[HA] concentration of the conjugate acid [M]
pK_b negative common logarithm of the dissociation constant K_b of the base ($-\log K_b$)

Transforming Eq. 3-33 to the pH equals

$$pH = 14 - pK_b - \log\frac{[HA]}{[A^-]} = pK_a - \log\frac{[HA]}{[A^-]} \tag{3-34}$$

The relative net charge α^+ and α^- of the charged species of a proton acceptor and a proton donor are derived from the dissociation constant α which is defined as

$$\alpha = \frac{\chi}{c} \tag{3-35}$$

χ molar fraction of the charged species [M]
c total concentration of the species [M]

Now, it follows from Eqs. 3-32 to 3-35 for a proton donor that

$$\alpha^- = \frac{10^{(pK_a - pH)}}{10^{(pK_a - pH)} + 1} - 1 \tag{3-36}$$

For the relative net charge α^+ of a proton acceptor it follows from Eq. 3-35:

$$\alpha^+ = \frac{10^{(pK_a - pH)}}{10^{(pK_a - pH)} + 1} \tag{3-37}$$

In the case of zwitterions or ampholytes which possess i weak acidic and j weak basic groups in one molecule the total net charge Z has to be calculated as follows:

$$Z = \sum_{i=1}^{n} \alpha_i^- + \sum_{j=1}^{m} \alpha_j^+ \tag{3-38}$$

Based on Eqs. 3-36 to 3-38 the net charges of two weak acid/base pairs as well as of an amino acid and a dipeptide have been calculated in the pH range 0 - 14. The results are shown in Fig. 3.15.a. Weak monobasic acids like acetic acid exhibit infinitely small negative charges approaching pH 0 and one negative charge approaching pH 14. At the point where the pH is equal to the pK_a value, 50% of the molecules are charged leading to an overall charge of 0.5. An analogous curve is obtained for weak bases like TRIS with one positive charge approaching pH 0 and small positive charges in the upper basic pH range. Ampholytes have zero charge over a given region like amino acids or exhibit charge inversion at a particular pH, the so-called isoelectric point (pI), as most peptides and proteins do. For instance, alanine has an acidic carboxylic group

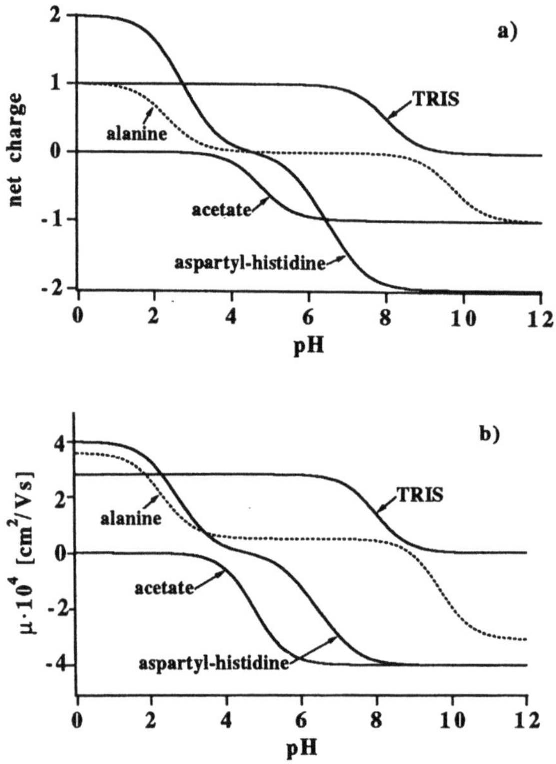

Fig. 3.15. Net charge (**a**) and net mobilities (**b**) of weak acids and bases dependent on the pH. The absolute mobilities in $cm^2 \cdot V^{-1} \cdot s^{-1}$ are $-4 \cdot 10^{-4}$ for acetate, $2.8 \cdot 10^{-4}$ for TRIS, $-3.6 \cdot 10^{-4}$ and $3.1 \cdot 10^{-4}$ for alanine, and $\pm 2.0 \cdot 10^{-4}$ for aspartyl-histidine

with a pK_{a1} value of 2.43 and a basic amine functionality with a pK_{a2} value of 9.69. Depending on the buffer pH of the solution the following reactions occur:

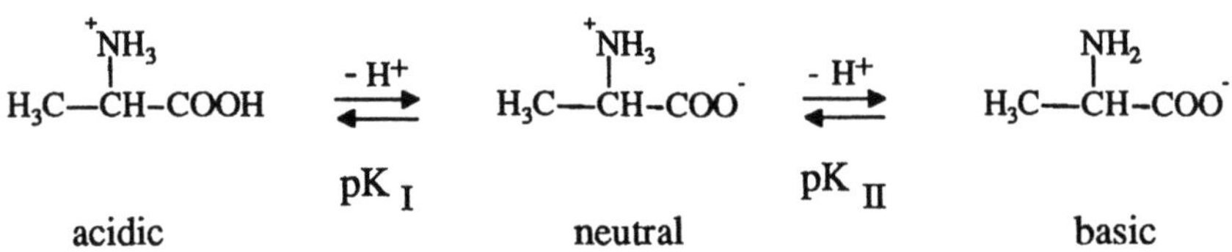

According to the selection of the pH, alanine is either positively, neutral or negatively charged. We see from Fig. 3.15.a that the net charge of alanine is practically zero over a wide pH range. In this pH region alanine behaves like a neutral component and possesses no electrophoretic mobility. On the other hand, the curve shape of aspartyl-histidine is typical for a larger peptide or a protein which has well defined isoelectric points, theoretically, in a single pH value. In general, for those ampholytes where the following relation holds

$$pK_{a_1} - pK_{a_2} < 4 \tag{3-39}$$

the isoelectric point is determined by a particular pH. But if

$$pK_{a_1} - pK_{a_2} \geq 4 \tag{3-40}$$

the isoelectric condition is fulfilled over a broader pH range. However, even in these cases a single pH value is given in the literature. For an ampholyte with various acidic and basic groups the terms on the left side of Eqs. 3-39 and 3-40 can be written in the general form

$$\frac{1}{n}\sum_{i=1}^{n} pK_{a_i}\,(\text{acids}) - \frac{1}{m}\sum_{j=1}^{m} pK_{aj}\,(\text{bases})$$

Whereas an isoelectric point is found again at one particular pH for the term being < 4, an isoelectric range exists if the term becomes ≥ 4.

The isoelectric point of an ampholyte consisting of one acidic and one basic functionality can be calculated from the dissociation of a zwitterion according to Eq. 3-41.

$$pI = \frac{1}{2}\cdot\left(pK_{a_1} + pK_{a_2}\right) \tag{3-41}$$

If no other acidic or basic functional groups are present, Eq. 3-41 can also be used for the calculation of the pI of oligopeptides. For an ampholyte with more than two ionizable groups, only the dissociation constants proximal to its pI are of consequence. For instance glutamic acid has three pK values of 2.19, 4.25 and 9.67, respectively. The pI is to be calculated as 3.22 taking the two acidic pK values into account. Another possibility, especially for larger polypeptides and proteins, is to determine the pI by a mobility curve which can be made by IEF on a polyacrylamide slab gel followed by ZE in the vertical direction of IEF. A large number of pI values of proteins and polypeptides can be found at Righetti [59, 60].

Since the isoelectric point shows a marked dependence upon the ionic strength of the solution [61] in the sense that the pI increases with falling salt concentrations, special care has to be taken in choosing the optimal buffer concentration.

In the same degree as the net charge changes with the pH, the electrophoretic mobility of the species changes as well. The net mobility μ_i^n of a weak electrolyte i is given by the product of the absolute mobility times the dissociation degree of the ion α at the particular pH:

$$\mu_i^n = \mu_i^0 \cdot \alpha \tag{3-42}$$

Thus, the net mobilities can easily be calculated with the aid of Eqs. 3-36 and 3-37. In Fig. 3.15.b the pH dependence of the net mobilities instead of the net charges is shown. As already mentioned in Sect. 2.3 ionic interactions do not play such an important role for weak electrolytes as in the case of strong electrolytes, because the mobility is far more affected by the degree of dissociation. Nevertheless, one has to keep in mind that experimentally determined mobilities by CZE are never absolute mobilities, even if the pH value is chosen in the way that the weak electrolyte is totally dissociated.

From the point of view of separation we are more interested in the difference in net mobility of two components rather than in absolute values because the difference in mobility of the analytes brings about the separation. From a plot as shown in Fig. 3.15.b the optimum pH range can be estimated. If absolute mobilities which are needed for the calculation of the mobility curve are unknown, relative mobilities can either be determined experimentally by referring them to the known mobility of an internal standard or they can be calculated from Offord's empirical equation [62, 63]:

$$\mu_{rel} = \frac{Z}{\sqrt[3]{M^2}} \tag{3-43}$$

μ_{rel} relative mobility
Z total net charge
M molar mass [$g \cdot mol^{-1}$]

Equation 3-43 can also be used to correlate the mobility with charge-to-size parameters [53]. The factor $\sqrt[3]{M^2}$ arises from the assertion that an ion moving through a conducting medium would experience a retarding force that is proportional to its surface area rather than its radius. This implies that the mobility would be inversely proportional to $\sqrt{r}$ and, hence, $\sqrt[3]{M^2}$, rather than to r and $\sqrt[3]{M}$, as it is assumed by Stokes' law.

An exact mathematical treatment to calculate the optimal pH value for a separation has been shown by Consden et al. [64] and is presented in the following. Consider two acid/base pairs HA/A$^-$ and HB/B$^-$ having dissociation constants K_A, K_B and absolute electrophoretic mobilities μ_A^0, μ_B^0 of the fully dissociated ions. The net mobility μ_A^n will be:

$$\mu_A^n = \frac{\mu_A^0 \cdot [A^-]}{[HA] \cdot [A^-]} - \frac{\mu_A^0 \cdot K_A}{[H^+] + K_A} \tag{3-44}$$

An analogous equation can be derived for component HB/B$^-$. The difference of the mobilities μ_A^n and μ_B^n is then

$$\mu_A^n - \mu_B^n = \frac{\mu_A^0 \cdot K_A}{[H^+] + K_A} - \frac{\mu_B^0 \cdot K_B}{[H^+] + K_B} \tag{3-45}$$

By transforming Eq. 3-45 it can readily be shown that the difference in net mobility is at the maximum when

$$pH = \frac{pK_A + pK_B}{2} - \log\left(\frac{\sqrt{\frac{\mu_A^0}{\mu_B^0}} - \sqrt{\frac{K_A}{K_B}}}{1 - \sqrt{\frac{\mu_A^0 \cdot K_A}{\mu_B^0 \cdot K_B}}}\right) \tag{3-46}$$

In the case of an anionic separation the acid with the higher pK_a must be called HB. In analogy, for the cationic separation of two acid/base pairs AH^+/A and BH^+/B the base with the higher pK_a must be called A. In order to determine the optimum pH, the sign of the addition of pK_B must then be reversed.

If two analytes have the same absolute mobilities, e.g. optical isomers, Eq. 3-46 will become:

$$pH = \frac{1}{2} \cdot (pK_A \pm pK_B) \tag{3-47}$$

Wren [65] has investigated the validity of Eq. 3-47 for the separation of me–thylpyridines and has found good agreements of theory and practice. This simple equation can also be used without making a significant mistake if the absolute mobilities differ only slightly from each other.

Equation 3-44 can also be applied to determine the dissociation constants of weak acids and bases by CZE [66]. Rearranging leads to the following relationship for an acid/base pair where the conjugate base is dissociated:

$$\frac{1}{\mu_i^0} = \frac{1}{K_a \cdot \mu_i^n} \cdot [H^+] + \frac{1}{\mu_i^n} \tag{3-48}$$

A plot of $1/\mu_i^0$ against $[H^+]$ should be a straight line with a slope equal to $1/K_a \cdot \mu_i^n$ and an intercept equal to $1/\mu_i^n$. The ratio of the slope to the intercept should be equal to $1/K_a$. Hence, pK_a as well as the absolute mobility can be determined from this plot. In a similar way the following equation can be used to determine pK_a and the absolute mobility of an acid/base pair where the conjugate acid is dissociated:

$$\frac{1}{\mu_i^0} = \frac{K_a}{\mu_i^n} \cdot \frac{1}{[H^+]} + \frac{1}{\mu_i^n} \tag{3-49}$$

3.3.3 Choice of Buffer

The choice of the buffer system used as electrolyte solution has a major influence on the separation and should be made carefully. Most buffer systems have sufficient buffer capacity only in a limited pH range. Due to the logarithmic definition of pH, buffer capacity decreases by a factor of 10 for every pH unit away from the pK. Thus the buffer capacity of an ampholyte near its pI value depends on the quantity pI - pK_1. Svensson rejected as unsuitable all ampholytes with (pI - pK) > 2.5. Particularly useful are those buffers having a high buffer capacity at simultaneously low conductance resulting in a low current and heat generation. A broad variety of buffer systems can be used in CE. Most of them have evolved empirically to be suited for a specific separation problem, while others are taken from conventional free-flow or gel electrophoresis. Phosphate and borate, often in combination with TRIS, are certainly the most frequently used buffer systems. In Appendix 8.1, often used buffers, their pK values and their useful pH ranges are summarized.

Biological or Good buffers, e.g. ACES, HEPES, TRIS etc. are also common, in particular for separations of peptides and proteins. Their properties are also listed in Appendix 8.1.

3.3.4 Ionic Strength

Besides the pH, the ionic strength is an important tool that we can use to improve efficiency, resolution and sensitivity of the separation system. There are many publications dealing with the influence of the buffer's concentration on both electroosmotic and electrophoretic mobility in CE, among others those cited in [14, 67, 68]. Since electrophoresis and electroosmosis are based on the same principles, variation of the salt concentration should have identical effects on μ_i and μ_{eo}. The investigation of the impact of the ionic strength on electromigration, however, is a critical task, because no general rule exists for the different species which can be separated by electrophoresis. The charged compound can be a small ion, a polymer or a whole cell. Furthermore, it can behave like a weak or a strong electrolyte. In the case of proteins, the salt concentration can have additional effects on its ternary structure. Finally, not only the concentration, but also the kind of electrolyte solution can play an important role in some cases. Consequently, a complicated combination of factors has to be considered. Therefore, the reader is requested to check the impact of ionic strength on his individual separation and consider the following reflections as helpful guidelines.

Based on the theory of Debye, Hückel and Onsager, Henry introduced the following formula which takes account of ionic strength effects of the electrophoretic mobility:

$$\mu_i = \frac{Q_{eff}}{4\pi\eta r(1+\kappa r)} \cdot f(\kappa r) \tag{3-50}$$

where κ^{-1} is identical to the radius of the ionic atmosphere or double layer thickness ß of the molecule and Q_{eff} represents its effective charge (compare to Eq. 2-15). The term 1 + κr takes into account the ionic atmosphere surrounding the charged species. The so-called Henry's function f(κr) is a measure for the deformation of the diffuse double layer by the counterions (the relaxation effect) and can be calculated from curves given by Henry for different values of ionic strength [69]. The validity of this formula has been proven by Tiselius and Svensson for the dependence of the effective mobility of egg albumin on the ionic strength of phosphate buffer [62]. For this purpose, they have compared their experimentally observed values for the mobility of the protein with those calculated on the basis of the Debye-Hückel-Henry theory by plotting the mobility versus the square root of ionic strength (Fig. 3.16.). The observed data fit very well with the calculated curve which shows a logarithmic decrease of the mobility with increasing $\sqrt{I}$. In the same publication, the authors have calculated "ideal" mobilities of egg albumin by means of Eq. 2-11. This approach is based on the assumption that no ionic interactions occur and that there is no double layer surrounding the protein. At pH 7.1 the protein has an overall negative net charge. Some of the positive charges are neutralized by phosphate ions. Thus, the charge of the protein will increase with increasing ionic strength and thus, also, the mobility increases. The ideal and the observed mobilities approach each other at low ionic strengths, and the values extrapolated to I = 0 are almost identical, suggesting that the protein migrates as a free ion at zero ionic strength.

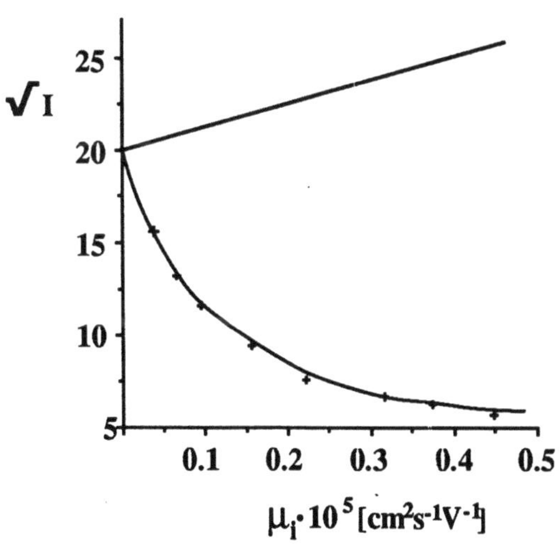

Fig. 3.16. Mobility of egg albumin at varying ionic strengths. *Upper curve*: ideal mobility, calculated on the assumption of free ionic migration. *Lower curve*: calculated mobilities on the basis of the Debye-Hückel-Henry theory. The *crosses* represent the observed values. Electrolyte system phosphate buffer, pH 7.1. (With permission from Ref. 61)

Wieme [70] has derived a formula for μ_i and the complementary μ_{eo} which accounts for the influence of both ionic strength and effective charge on the mobility:

$$\mu_i, \mu_{eo} \approx \frac{3 \cdot 10^{-8} \cdot Q_{eff}}{\eta \cdot \sqrt{I}} \tag{3-51}$$

According to this equation which has its original also in the Debye-Hückel theory both electrophoretic and electroosmotic mobility should be directly proportional to the charge at the surface and to the reciprocal of the square root of I. The mobility should be doubled if the ionic strength is increased 4-fold. Although several authors have confirmed this relationship it cannot be used universally.

A similar relationship already presented in Sect. 2.5 has been developed by Salomon et al. [14] for the impact of ionic strength and buffer composition on the electroosmotic mobility. They have related the charge density Q at the silica surface with the equilibrium constant K_W which describes the reversible adsorption of cations on the silica surface (see Eq. 2-27). A plot of $1/\mu_{eo}$ versus buffer concentration provides a curve which fits well with the theoretical curve calculated by means of Eq. 2-27. They conclude that μ_{eo} does not only depend on the double layer thickness, but also on the charge density at the silica surface. By using Eq. 2-27 they have also calculated the thickness of the rigid layer d, the adsorption constant K_w and Q_0 for a number of different electrolyte solutions. Their data are summarized in Table 3.8. The values for d suggest that the rigid layer is more than a single layer of cations. Hydrated Na^+ ions, for instance, have a diameter of 3 Å, whereas the thickness of d has been calculated to 39 Å. Hence, a better description involves several layers of ordered hydrated cations interspersed among buffer molecules and buffer anions. Furthermore, it can be calculated from the total number of ionized groups Q_0 that only 0.3% of the surface SiOH groups are ionized, assuming that there are between $4 \cdot 10^{18}$ and $5 \cdot 10^{18}$ SiOH groups per square meter. For an additional discussion of Table 3.8. see Sect. 3.3.5.

Table 3.8. Thickness of the rigid layer d, surface charge Q_0 and adsorption constant K_W for different buffers. Data are calculated by means of Eq. 2-27 and are taken from Ref. 14.

Buffer Composition	d [Å]	$Q_0 \cdot 10^{-16}$ [sites·cm^{-2}]	K_W [mM^{-1}]	$\mu_{eo} \cdot 10^4$ [cm$^2 \cdot$V$^{-1} \cdot$s^{-1}]
MES/LiOH, pH 6.05	16	1.4	0.017	5.03
MES/NaOH, pH 6.05	39	1.1	0.02	8.0
MES/KOH, pH 6.05	32	0.9	0.034	4.98
MES/RbOH, pH 6.05	28	1.4	0.023	7.43
MES/His, pH 6.05	9.5	2.2	0.034	5.55

We have also investigated the influence of the ionic strength on the mobility of the EOF and some benzoyl derivatives. If the separation is performed under conditions where Joule heating can be disregarded, we have also found, that both electroosmotic as well as electrophoretic mobilities decrease linearly with $1/\sqrt{I}$. The graphs are shown in Fig. 3.17.

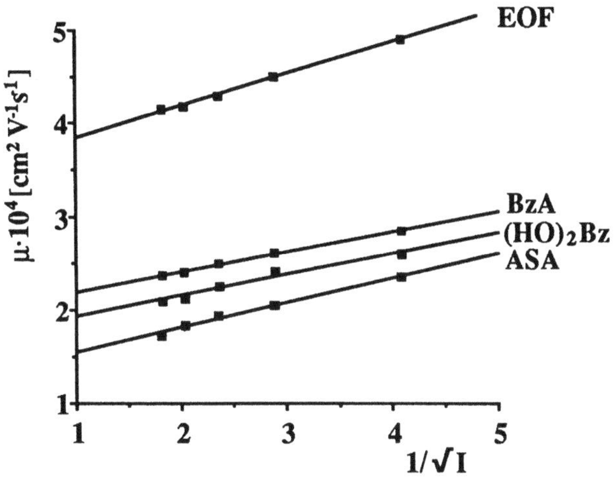

Fig. 3.17. Plot of electroosmotic and electrophoretic mobility versus the ionic strength for the separation of benzyl alcohol (EOF), acecylsalicylate (ASA), dihydroxybenzoate $(HO)_2Bz$ and benzoate (BzA) at different sodium phosphate concentrations. Instrument: Beckman P/ACE 2000; experimental conditions: fused silica capillary, 57 cm x 75 μm i.d., hydrodynamic injection for 1 s, voltage 5 kV, temperature 25 °C, UV detection at 200 nm, electrolyte system 20 - 100 mM sodium phosphate, pH 9.0

In practice, however, when working at higher field strengths, variation of the ionic strength induces several effects, i.e. temperature increase and viscosity changes, which in turn influence the mobility. Fig. 3.18.a shows the anionic separation of three benzoic acid derivatives at several buffer concentrations. Benzyl alcohol has served as electroosmotic flow marker and is represented by peak 1. While EOF is only slightly decreased with rising buffer concentration, the migration time of the analytes declines noticeably with increasing buffer concentration. In addition, higher resolutions are obtained with increasing buffer concentration which can be caused by a better suppression of electrophoretic dispersion. The plate number of peak 4, however, enhances from 20 000 to 50 500 at 75 mM, but declines again to 38 500 at 100 mM phosphate. The decrease might be attributed to band broadening caused by excessive Joule heat. If the mobilities calculated from Fig. 3.18.a are plotted versus the ionic strength, straight lines are found with correlation coefficients better than 0.998 (Fig. 3.18.b). Interestingly the mobilities of the solutes increase continuously with

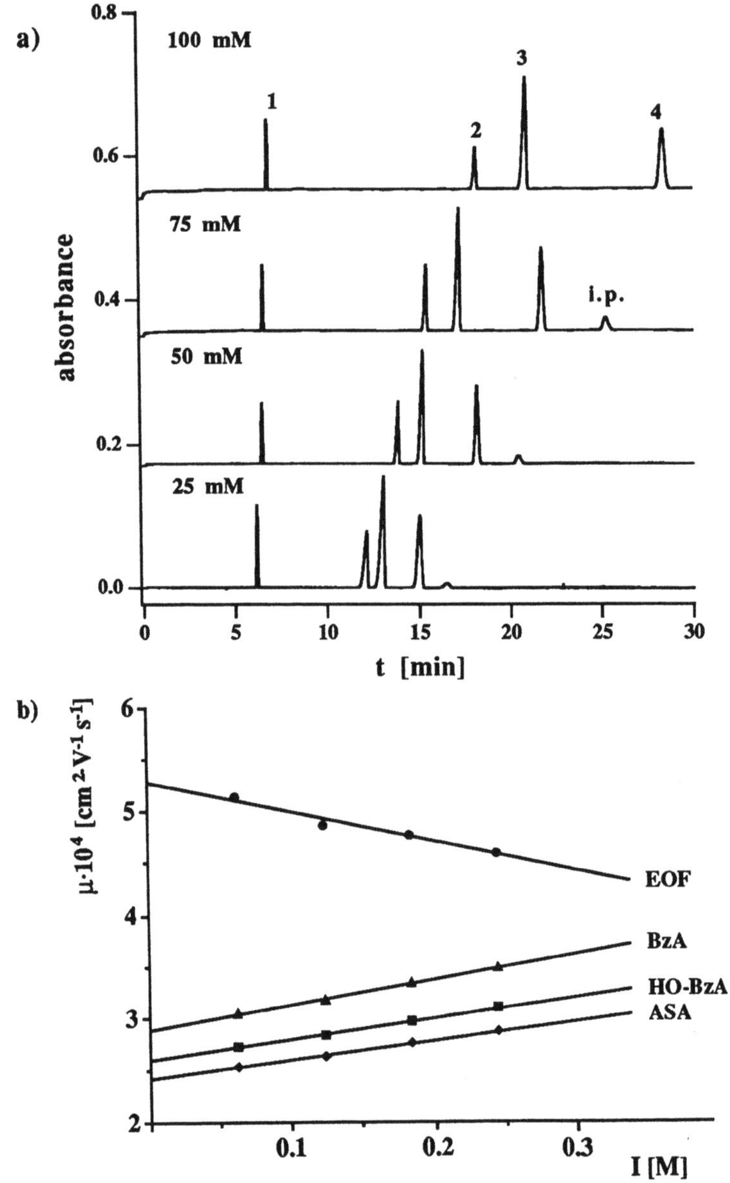

Fig. 3.18. (**a**) Electrophoretic separations of benzyl alcohol (**1**), acetylsalycilate (**2**), benzoate (**3**) and *p*-hydroxybenzoate (**4**) at different sodium phosphate concentrations. The impurity (i.p.) is probably salycilate produced by hydrolysis of acetylsalicylic acid. Experimental conditions as in Fig. 3.17. except for the voltage which is 15 kV. (**b**) Plot of electroosmotic and electrophoretic mobility versus the ionic strength of the separation shown in (**a**): benzyl alcohol as EOF marker (eo), acetylsalicylate (ASA), p-hydroxybenzoate (HO-BzA) and benzoate (BzA)

higher ionic strengths while the electroosmotic mobility decreases. The slope of the straight lines can be regarded as an increment of the buffer on the selectivity of the system.

3.3.5 Impact of Buffer Composition

The mobility of the buffer ions not only has an effect on electrophoretic dispersion as discussed in Sect. 3.1.4, but also on the resulting current at a given field strength. This is demonstrated in Fig. 3.19. by an Ohm's law plot of different salt solutions. Because the equivalent concentration of all salts is 50 mM with sodium as the common cation, differences in the slopes are effected solely by the different ionic equivalent conductance of the anions. For instance, chloride has a higher equivalent conductance (76.4 $cm^2 \cdot \Omega^{-1} \cdot M^{-1}$) than fluoride (55.4 $cm^2 \cdot \Omega^{-1} \cdot M^{-1}$). The influence of heat generation on the curve shape is more pronounced for solutions like sodium chloride, which has a high conductance, than for sodium dihydrogenphosphate. Thus, if high field strengths are desired, it can sometimes be advantageous to choose the ion with the lower mobility.

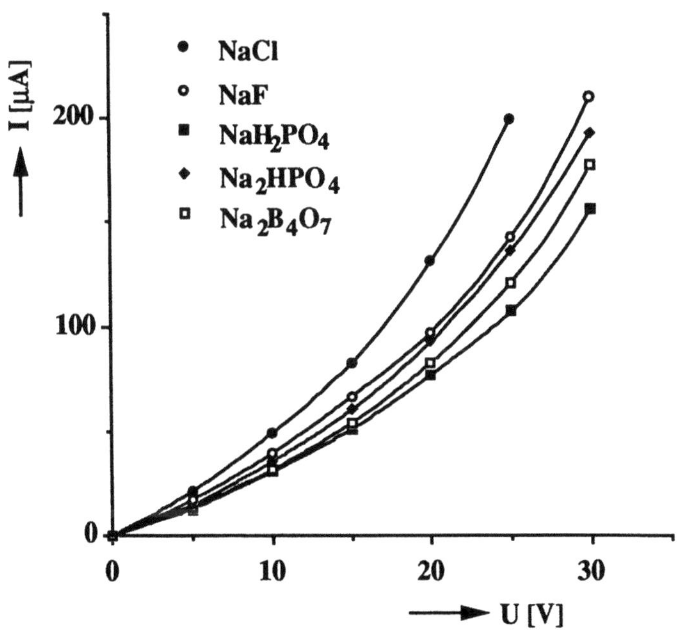

Fig. 3.19. Plots of applied voltage versus resulting current for different salt solutions. The concentration of each solution is 50 mequiv. Instrument: Beckman P/ACE 2000; experimental conditions: fused silica capillary, 57 cm x 75 μm i.d., temperature 25 °C

Hint: *The maximum buffer concentration and voltage, beyond which the generated Joule heat is excessive for the instrument and the column used, can be determined as follows:*
For each buffer concentration an Ohm's law plot as described in Sect. 3.2.1 is performed. From each plot the maximum voltage at which the curve begins to deviate from linearity is recorded. The buffer concentration is then plotted versus the maximal voltage. The resulting curves which can be made for different kinds of buffer solutions represent a practical guide for the selection of buffer concentration and applied voltage.

Note, that the ionic strength of the different buffers can be identical, but the current produced at different field strengths can vary significantly. Thus, not only the concentration or ionic strength of the buffer is the critical parameter deteriorating separation performance at high voltages, as often stated, but the mobilities of the buffer constituents, therefore its specific conductance. Good buffers, for instance, have very low conductivities even at high ionic strengths.

Hint: *Sometimes it is desirable to work at higher conductivities to reduce peak distortion due to electrophoretic dispersion and adsorption processes at the silica wall. If employing Good buffers or other buffer systems with excellent buffer capacity but low specific conductance, the ionic strength can be very high without increasing the conductivity of the solution significantly. In these cases the conductivity of the buffer system can be increased by adding a neutral salt such as NaCl.*

The effects of buffer anions on electrophoretic mobility and separation of dansylamino acids have been studied by Atamna et al. [71]. While the influence of the buffer on the separation factors α of the analytes are more or less similar for phosphate, acetate, citrate, bicarbonate and tetraborate, migration times differ over 50%. In this case, selectivity remains constant because buffer-analyte interactions are negligible. The changes in migration time originate from the interaction of the salt with the silica surface which alters the zeta potential and the electroosmotic flow. Tetraborate exhibits by far the lowest electroosmotic mobility indicating that strong interactions between boron and silica occur.

As shown in Table 3.8., the choice of the buffer's cation influences the value of μ_{eo} under otherwise constant conditions. The values do not show a continuous change from Li^+ to Rb^+ and the highest EOF is found for Na^+. This indicates that a combination of factors must be responsible for the change in EOF.

The influence of the kind of cation on the EOF has also been investigated by Atamna et al. [72]. Their results contradict to those made by Salomon. In buffers of lithium, potassium, sodium, rubidium and cesium acetate the migration times of dansyl-alanine increases linearly with the reciprocal hydrated radii of the cations due to the increase of EOF with increasing radius of the hydrated cation. The highest electroosmotic mobility is obtained in lithium acetate which exhibits the largest hydrated ionic radius. It has also been shown by the same authors that sodium phosphate gives much better resolution and higher separation factors than potassium phosphate under otherwise the same experimental conditions [73].

Besides their impact on the EOF, on the current produced by the system and on the pH, buffers may exert an influence on the conformational stability of macromolecules. More than 100 years ago, Hofmeister showed that simple salts can influence the solubility and interactions of biopolymers. According to their ability to improve the solubility he arranged numerous anions and cations in an order as shown below:

$$PO_4^{3-} < SO_4^{2-} < CH_3COO^- < F^- < Cl^- < Br^- < NO_3^- < I^- < SCN^- < CCl_3COO^-$$
$$(CH_3)_4N^+ < NH_4^+ < K^+, Na^+, < Li^+ < Mg^{2+} < Ca^{2+} < Ba^{2+}$$
$$(CH_3)_4N^+ < (C_2H_5)_4N^+ < (C_3H_7)_4N^+ < (C_4H_9)_4N^+$$

chaotropic effect ⇒
⇐ taxigenic effect
favors hydrophobic interactions ⇒

Ions on the right side are strongly chaotropic, e.g. iodide, trichloroacetate and alkaline earth ions. These ions have a large ionic volume (low charge density) and are easily to be polarized. This property favors interactions with charged and uncharged compounds. In general, cations are less effective than anions. Combinations of anions and cations of the right side result in salts with strong chaotropic effects which break water structures and, thus,

> reduce the self association of water
> decrease molecule clusters
> decrease the viscosity of the solution.

As a consequence the penetration of extraneous molecules into the water association is facilitated (salt-in effect). In contrary to that, taxigenic ions, e.g. fluoride, ammonium, favor water structures. They reduce the amount of non-associated water molecules that are no longer available for hydratization (salt-out effect). Unfortunately, these effects are significant only at high salt concentrations which are commonly not feasible in CE.

Instead of strong electrolytes, several Good buffers are also chaotropic or taxigenic, respectively. The following series gives some examples:

$(NH_4)_2SO_4$ >> TRICINE > BICINE < HEPES < CAPS << LiBr

⇐ chaotropic | taxigenic ⇒

Instead of chaotropic ions special non-electrolytes are frequently used in electrophoresis to improve the solubility or reduce interactions. In analogy to the Hofmeister series a similar order can be set up as follows:

nicotinamide > methylacetamide > urea | sorbitol < fructose < saccharose

⇐ chaotropic | taxigenic ⇒

Because these compounds do not contribute to the electric current in CE, they can easily be used even at extremely high concentrations. For instance, urea is often used in concentrations as high as 8 M to break hydrogen bonds of proteins and nucleic acids and to improve their solubility.

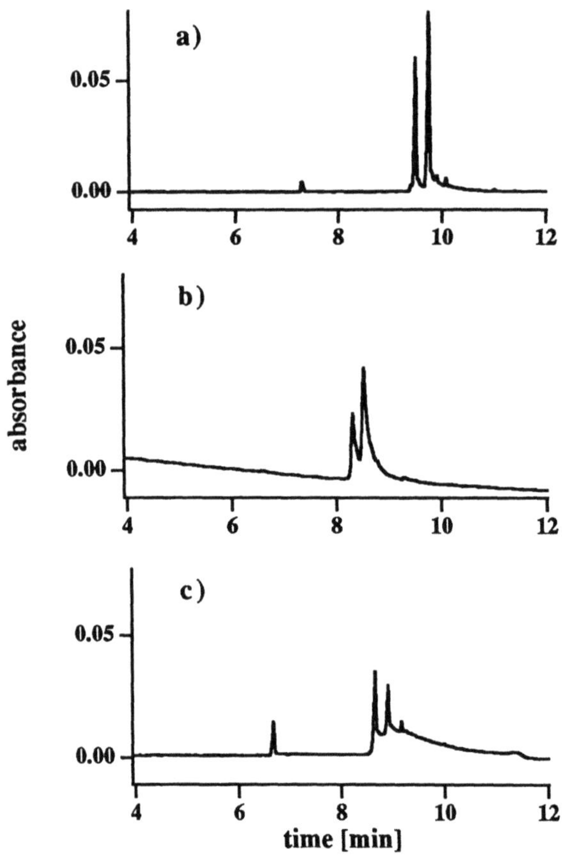

Fig. 3.20. Influence of the age of the buffer in the electrode vessels on the separation of two peptides. Instrument: Beckman P/ACE 2000; experimental conditions: fused silica capillary, 57 cm x 75 µm i.d., hydrodynamic injection for 1 s, voltage 20 kV, temperature 20 °C, UV detection at 200 nm, volume of the electrode vessels ca. 2 mL, buffer system: 50 mM $Na_2B_4O_7$, pH 9.3. (**a**) new buffer, (**b**) buffer after 20 subsequent runs and (**c**) buffer after 22 subsequent runs

Last, but not least, it has to be mentioned that the change of buffer composition in the electrode vessels in the course of subsequent runs can lead to deteriorated electropherograms even if the capillary is reconditioned after each run. This is illustrated in Fig. 3.20. where 3 electropherograms of the same separation are shown without changing the buffer solutions in the electrode reservoirs. Therefore it is suggested to change the buffer in the electrode reservoir at least once a day.

3.3.6 Complex Formation

One major disadvantage of capillary electrophoresis is its inherent lack of selectivity for neutral analytes. This lack can be overcome by adding buffer constituents which are able to interact with the analytes. Besides the use of micelles in micellar electrokinetic capillary chromatography, complexing agents can be added to the background electrolyte introducing a charge to the neutral analyte. Complexing agents also represent an elegant way to improve selectivity between already charged, but closely related compounds. Separation takes place based on different electrophoretic mobilities of the associated or complexed species arising from different complex formation constants with the analytes. Depending on the species of the buffer additive a distinction can be made between four fundamental approaches:

> Borate complexation
> Ion pairing
> Agents forming inclusion complexes
> Metal complexation

For the sake of simplicity solvation processes will be omitted in the following discussion, although one has to be aware that they may play an important part in complexation reactions. In the simplest case complex formation can be described by the following, general reaction:

$$A + L \underset{k_{-1}}{\overset{k_1}{\rightleftarrows}} [A \cdot L]$$

where A is the analyte, L is the ligand, [A·L] is the binary complex and k_1 and k_{-1} are the rate constants. The association constant K_A is defined according to the law of mass action as

$$K_A = \frac{k_1}{k_{-1}} = \frac{[[A \cdot L]]}{[A] \cdot [L]} \tag{3-52}$$

If more than one ligand is bound to the central ion, complex formation will take place stepwise such that there are as many formation constants as complexes. For simplicity reasons it is assumed that only a one-to-one complex is formed in the following.

The degree of complexation is given by the association degree α which represents the ratio of the concentration of the complexed species [A·L] to the total concentration of the analyte $[A]_{total}$:

$$\alpha = \frac{[[A \cdot L]]}{[A]_{total}} \tag{3-53}$$

For the resulting net mobility μ_A^n of analyte A due to complexation the following relationship can be described:

$$\mu_A^n = \alpha \cdot \mu_{[A \cdot L]}^0 + (1-\alpha) \cdot \mu_A^0, \tag{3-54}$$

where μ_A^0 and $\mu_{[A \cdot L]}^0$ represent the absolute mobility of A and the mobility of the fully complexated form, respectively. For a constant concentration [A] and varying ligand concentrations [L], the rate of complexation will be proportional to the fraction (1-α) of A which is not complexed. In turn the rate of dissociation of the complex is directly related to the complexed fraction. At equilibrium the following equation holds:

$$k_1 \cdot (1-\alpha) \cdot [L] = k_{-1} \cdot \alpha \tag{3-55}$$

Solving this equation to α gives

$$\alpha = \frac{k_1 \cdot [L]}{k_{-1} + k_1 \cdot [L]} = \frac{K_A \cdot [L]}{1 + K_A \cdot [L]} \tag{3-56}$$

According to Eq. 3-54 the net mobility is proportional to α. By inserting Eq. 3-54 into 3-56 we obtain

$$\mu_A^n - \mu_A^0 = \frac{K_A \cdot [L] \cdot \left(\mu_{[A \cdot L]}^0 - \mu_A^0\right)}{1 + K_A \cdot [L]} \tag{3-57}$$

Rearranging Eq. 3-57 gives

$$\frac{1}{\mu_A^n - \mu_A^0} = \frac{1}{K_A \cdot \left(\mu_{[A \cdot L]}^0 - \mu_A^0\right)} \cdot \frac{1}{[L]} + \frac{1}{\left(\mu_{[A \cdot L]}^0 - \mu_A^0\right)} \tag{3-58}$$

A plot of $1/\mu_A^n - \mu_A^0$ versus 1/[L] should be linear with a slope of $1/K_A \cdot \left(\mu_{[A \cdot L]}^0 - \mu_A^0\right)$ and an y-intercept of $1/\left(\mu_{[A \cdot L]}^0 - \mu_A^0\right)$, from which $\mu_{[A \cdot L]}^0 - \mu_A^0$ and K_A can be calculated. μ_A^0 can be calculated from an electrophoretic run of A without any complexing agent in the buffer system. Although K_A is not a proper equilibrium constant, it is a measure for the interaction of the ligand with the analyte. $\mu^0_{[A\cdot L]} - \mu^0_A$ mirrors the difference between the mobility of the complexed and the free analyte. The double reciprocal plot given in Eq. 3-58 corresponds to the Lineweaver-Burk plot which is derived from the Michaelis-Menton equation of enzyme kinetics.

3.3.6.1 Borate Complexes

The complex formation of borate with polyols in alkaline solutions represents a well-known example which is used extensively in the separation of sugars and catechols. The complex formation can be described by the following equations:

$$(1) \quad B^- + L \rightleftharpoons [BL]^- + H_2O$$

$$(2) \quad [BL]^- + L \rightleftharpoons [BL_2]^- + H_2O$$

where L is the polyol ligand and B^- represents tetrahydroxyborate, $[B(OH)_4]^-$. In a pH range of 8 - 12, aqueous borate solutions contain not only tetrahydroxyborate but also more highly condensed polyanions such as triborate $[B_3O_3(OH)_5]^{2-}$ and tetraborate $[B_4O_5(OH)_4]^{2-}$. It should not matter what kind of borate buffer is dissolved, because the equilibrium between the different species depends only on the pH and the total boron concentration.

The structure of the polyol-borate complex is closely related to its stability in aqueous solution. Since equilibria (1) and (2) are dynamic, all polyols will be associated with a negative charge. The magnitude of this charge is determined by the position of the equilibrium and therefore by the stability of the complex. Given a constant amount of polyol ligand, the complex concentration rises with increasing borate concentration according to the law of mass action as well as with increasing pH due to the higher concentration of tetrahydroxyborate in alkaline solution.

For any particular polyol ligand, the complex stability depends strongly on the concentration, the number of hydroxyl groups and the presence of substituents [57]. From electrochemical measurements and ^{11}B-NMR spectroscopy it can be concluded that

- Polyols form 1:1 and 1:2 complexes
- Not only hydroxyl groups on adjacent carbon atoms, but also those on alternate carbon atoms are involved in complexation, leading to five and six-membered rings, respectively
- *cis*-1,2-Diols are preferred in complexation rather than *trans*-1,2-diols
- An increase in the number of hydroxyl groups increases the stabilization
- Coulombic repulsion destabilizes the complex
- Cyclic carbohydrates can form stable complexes only with vicinal hydroxyl groups with cis configuration

The principle of the complexation of polyols by borate is not limited to organic diols and carbohydrates but can also be successfully applied to glycopeptides. Fig. 3.21. shows the separation of a hexapeptide from its N-acetylglucosamine derivative at pH 11. Without borate the two peptides cannot be separated from each other. Complexation of the glycopeptide with borate is required to resolve both peptides as shown in Fig. 3.21.b. The complex formation is associated with an additional negative charge, whose magnitude is determined by the stability of the complex. The electrophoretic mobility of the glycopeptide is modified and the component elutes as the later peak. Since complexation of polyols with borate is strongly dependent on the buffer pH and the borate concentration, both parameters can be adjusted for optimization of the separation such that higher pH and higher borate concentration result in a higher proportion of the complexed species and in a more negative net charge.

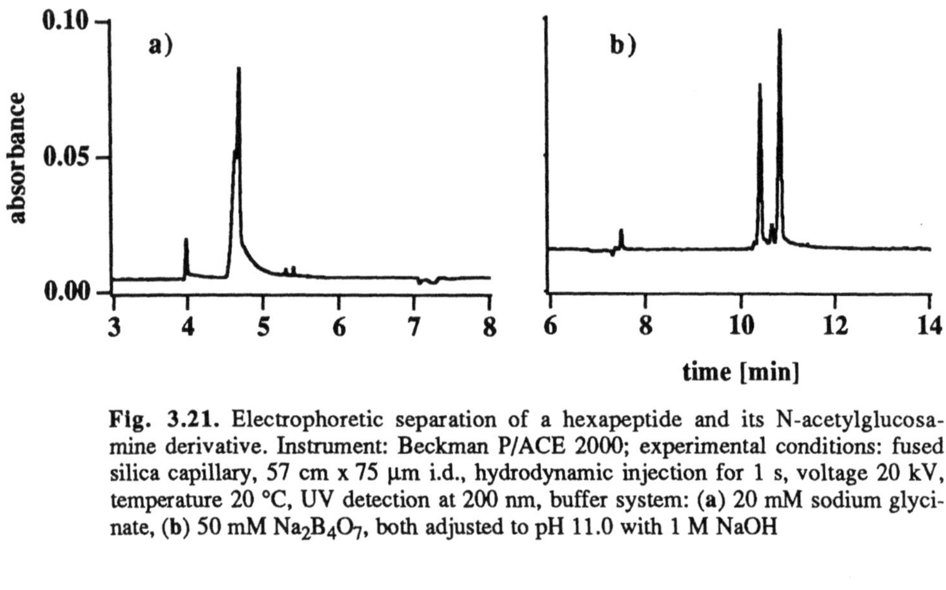

Fig. 3.21. Electrophoretic separation of a hexapeptide and its N-acetylglucosamine derivative. Instrument: Beckman P/ACE 2000; experimental conditions: fused silica capillary, 57 cm x 75 µm i.d., hydrodynamic injection for 1 s, voltage 20 kV, temperature 20 °C, UV detection at 200 nm, buffer system: (**a**) 20 mM sodium glycinate, (**b**) 50 mM $Na_2B_4O_7$, both adjusted to pH 11.0 with 1 M NaOH

3.3.6.2 Ion Pairs

Besides borate there exist a variety of complexing agents based on ion pairing or better solvophobic association for the separation of neutral components. Common ion pairing agents are tetraalkylammonium salts, e.g. tetrabutylammonium (TBA) sulfate and tetrahexylammonium (THA) perchlorate. The general reaction scheme of analyte A with THA is given in the following:

(1) $A + THA^+ \rightleftharpoons [A \cdot THA]^+$

(2) $[A \cdot THA]^+ + THA^+ \rightleftharpoons [A \cdot (THA)_2]^{2+}$

Owing to Coulombic repulsion forces the formation of a ternary complex as shown in the second reaction step is rather unlikely. Walbroehl and Jorgenson [74] have reported the separation of five polyaromatic hydrocarbons in 25 mM THA perchlorate dissolved in 50% aqueous acetonitrile.

McLaughlin and coworkers [75] have used hexanesulfonic acid to alter the retention time of cationic peptides by hydrophobic pairing. Using a peptide mapping of a tryptic digest, they have realized some selectivity changes and significant increases in resolution. The hydrophobic alkyl chain of the hexanesulfonic acid interacts with hydrophobic moieties of the peptides, thereby adding negative charges. It should be noted that the longer retention times observed may also be a result of the change of zeta potential.

3.3.6.3 Inclusion Complexes

To enhance selectivity, host-guest complexation or inclusion complexation has attracted increased attention. For this purpose cyclodextrins and crown ether have shown to be particularly useful as complexing agents. They are able to form highly selective complexes, which can be utilized even for enantio-separation.

a)

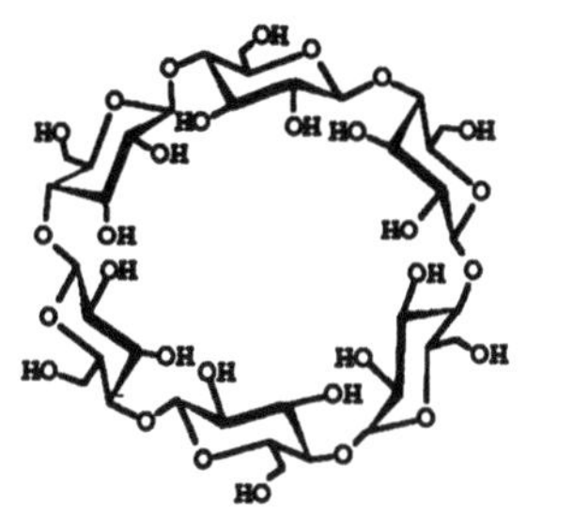

c)

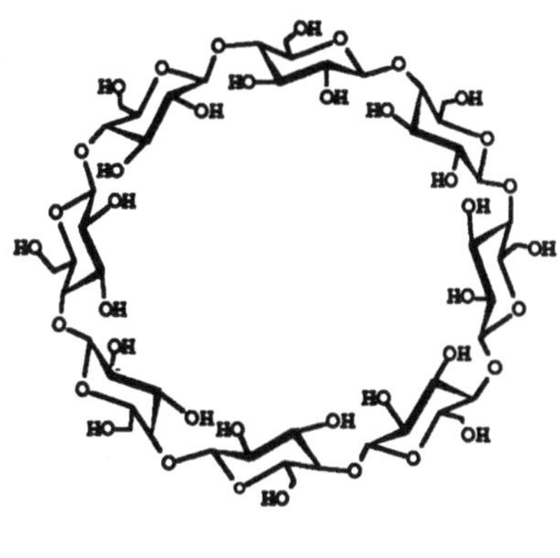

b)

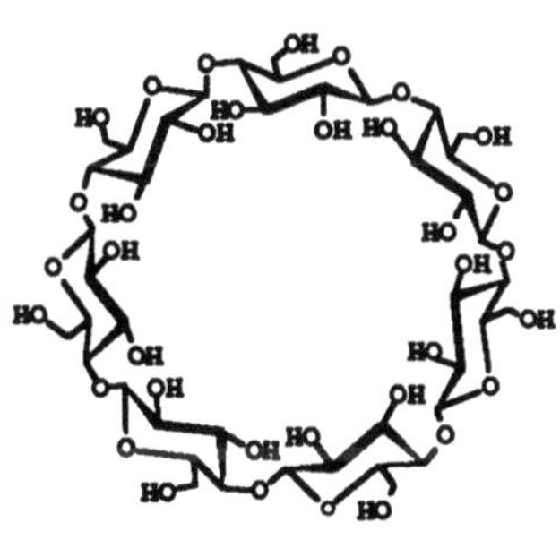

Fig. 3.22. Chemical structures of α- (**a**), β - (**b**) and γ-cyclodextrin (**c**)

Cyclodextrins (CD) are cyclic oligosaccharides consisting of six, seven or eight glucose units corresponding to the names α-, β- or γ-cyclodextrin (Fig. 3.22.). CD's form a truncated cone with a rim of secondary hydroxy groups at the opening with the larger diameter. The internal cavity contains no hydroxy functions and exhibits a hydrophobic character. Owing to this hydrophobic nature cyclodextrins are able to form inclusion complexes with aromatic or alkyl groups. Although cyclodextrins are basically neutral, they can alter the electrophoretic mobility of an analyte based on differing molecular masses of the complexed and the free species. Thus, the electrophoretic mobility is influenced by the complex formation constant. The complex stability between CD and its guest molecule is governed by factors like van-der-Waals interactions (hydrophobic), solvation effects and hydrogen bonds. Substantial differences of the complex formation constant are found even for structurally similar compounds like enantiomers. For this reason, cyclodextrins are successfully employed for the separation of optical isomers (see Sect. 7.10). Some structure-relevant data of α-, β- and γ-cyclodextrin are summarized in Table 3.9.

Table 3.9. Physical data of α-, β- and γ-cyclodextrin [76]

Cyclodextrin type	α	β	γ
Number of glucose units	6	7	8
Molecular mass	973	1135	1297
Inner diameter of cavity [nm]	0.47 - 0.52	0.60 - 0.64	0.75 - 0.83
Outer diameter of cavity [nm]	1.46	1.54	1.75
Equivalents of H_2O bound in cavity	6	11	17
Specific optical rotation $[\alpha]_D^{25}$	150.5 ± 0.5	162.5 ± 0.5	177.4 ± 0.5
Melting point [°C]	278	299	267

While aqueous solubilities of α- and γ-CD are high, the low solubility of β-CD may create problems if one wants to work at higher concentrations. This problem can be overcome in several ways. Derivatives of β-cyclodextrin (i.e. methylated or hydroxyalkylated β-cyclodextrins) are much more soluble in water than the parent compound. Alternatively urea up to 8 M or organic solvents can be used to enhance water solubility. Tables 3.10. and 3.11. outline some solubility data of α-, β- and γ-CD.

Terabe et al. [81] have separated positional isomers of substituted benzenes like nitroaniline, dinitrobenzene, etc., by using 2-O-carboxymethyl-β-CD. Derived from micellar electrokinetic capillary chromatography (MECC), the authors have developed a mathematical model which allows the calculation of the

Table 3.10. Molar solubility of cyclodextrins in aqueous media at 25 °C [77, 78]

Cyclodextrin type	Water	4 M Urea	8 M Urea	0.2 M NaOH
α-CD	0.12	0.06	0.09	≥ 0.20
β-CD	0.017	0.09	0.23	≈ 0.15
γ-CD	0.18	0.37		≈ 0.20
dimethyl-β-CD	0.43			
trimethyl-β-CD	0.22			
hydroxypropyl-β-CD	0.42			

Table 3.11. Solubility [g·L^{-1}] of β-cyclodextrin in water-organic solvent mixtures at 20 °C according to Refs. 79 and 80

Organic Solvent [%]	MeOH	EtOH	CH_3CN
5	11.4		19.2
10	9.2	18.2	22.8
20	6.0	20.1	31.0
30	6.4	22.0	26.7

partition coefficient between the cyclodextrin pseudophase and the aqueous phase. Their derivation, however, holds only for a charged species of CD.

Figure 3.23. demonstrates the influence of β-cyclodextrin in CZE to enhance the resolution of positional isomers of dihydroxybenzoic acids. The complexation by β-cyclodextrin decreases the charge density and the electrophoretic mobility of the analytes. Consequently, the analytes elute faster. The selectivity of the separation systems changes owing to the different complex stabilities.

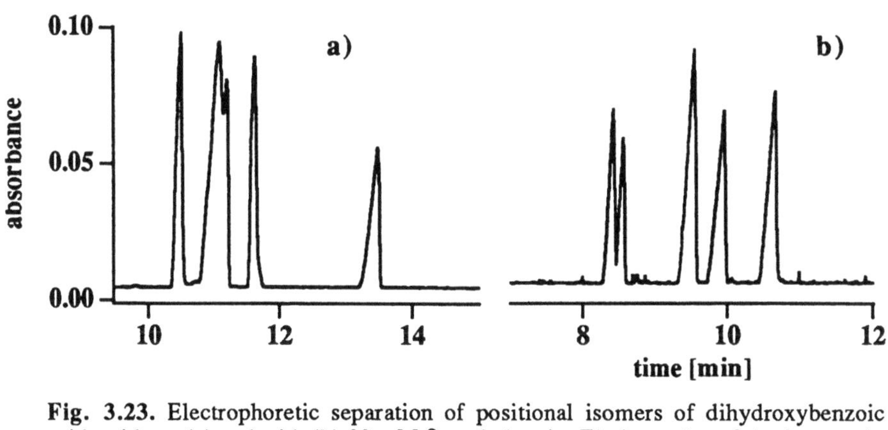

Fig. 3.23. Electrophoretic separation of positional isomers of dihydroxybenzoic acids without (**a**) and with (**b**) 20 mM β-cyclodextrin. Elution order of the isomers in (**a**) 2,4-, 2,3-, 2,6-, 3,4- and 2,5-dihydroxybenzoic acid; (**b**): 2,4-, 2,6-, 2,3-, 2,5- and 3,4-dihydroxybenzoic acid. Instrument: Beckman P/ACE 2000; experimental conditions: fused silica capillary 57 cm x 75 μm (i.d.), hydrodynamic injection for 1 s, field strength 300 V·cm^{-1}, temperature 25 °C, electrolyte system 25 mM sodium phosphate, pH 8.0, UV detection at 214 nm

A reversed approach has been described by Nardi and coworkers [82]. They have utilized the complex formation ability of CDs with benzoic acid to separate α-, β- and γ-CD. Benzoate added to the running buffer provides the separation by complex formation and serves as a tracer for the indirect detection of the tion by complex formation and serves as a tracer for the indirect detection of the analytes. From the elution order of the cyclodextrins one can clearly see that γ-CD forms the weakest complex (elutes first) followed by α-CD and β-CD which is most retained by the strongest complex.

The second class of compounds which are able to form stable inclusion complexes in aqueous solutions are the macrocyclic polyethers. Crown ethers are the best known representatives of this class. The polyether ring contains a number of electron-donor heteroatoms, i.e. O, N, S, in the structure. Depending on the geometry and the heteroatoms, the cavity shaped by the ring system is able to complex not only metal cations but also organic amines, diazonium cations, etc. The complex formation is based on dipole-ion/dipole interactions between host and guest. The magnitude of the complexation depends on the size of the cavity and of the guest ion. Maximal stability is obtained if the ion fits exactly into the cavity.

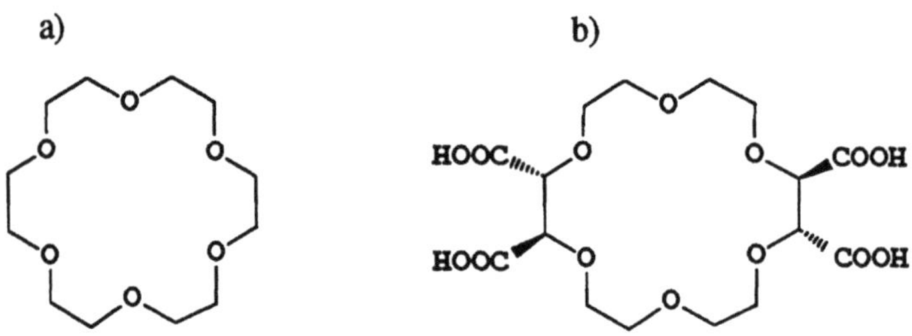

Fig. 3.24. Chemical structures of (**a**) 18-crown-6 and (**b**) 18-crown-6-tetracarboxylic acid

According to the nomenclature for crown ethers, the compound shown in Fig. 3.24.a is commonly called 18-crown-6 (18C6), where 18 represents the total number of atoms in the ring system and 6 indicates the number of oxygens. 18C6 forms stable complexes particularly with potassium, ammonium and protonated primary amines. The latter two, held inside the cavity, are bound via three $^{+}$NH···O hydrogen bonds in a tripod arrangement. Derivatives of 18C6 such as 18-crown-6-tetracarboxylic acid (18C6H$_4$) as shown in Fig. 3.24.b are successfully employed as chiral selector for enantiomeric separation (see Sect. 7.10).

3.3.6.4 Metal Complexes

The potential of metal complexation for enhancing the selectivity in capillary electrophoresis has not been widely explored. Changes in electrophoretic mobility of an analyte is based on the interaction with multicomponent chelate complexes consisting of a central ion, i.e. Cu(II), Zn(II), and at least two bi-

$$[L]_n \cdot [M] + [A] \rightleftarrows [L]_{n-1} \cdot [M] \cdot [A] + [L]$$

where [L] is the ligand, [A] the analyte and [M] is the metal ion. Zn(II) and Cu(II) ions can interact with certain N, O or S atoms on amino acids, peptides or proteins. A schematic representation of a complex consisting of a copper cation and two amino acids is shown in Fig. 3.25.

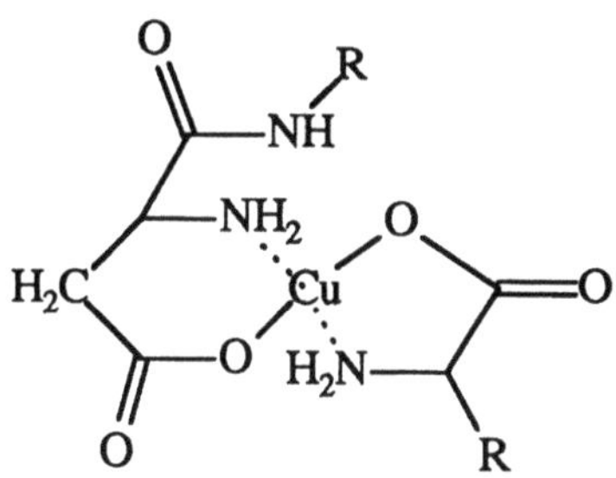

Fig. 3.25. Schematic model of a ternary metal complex consisting of Cu(II) as central ion and two amino acids as ligands

Mosher [83] has separated histidine-containing peptides by adding Cu(II) and Zn(II) to the running buffer. Samples which comigrate in the absence of metal ions are well resolved under these conditions. An interesting application of metal complexation has been presented by Bocek and coworkers [84]. Enantiomeric and diastereomeric Co(III) complexes with ethylenediamine, phenantroline and several amino acids are separated by using sodium tartrate as the mobile phase.

The addition of a weakly complexing agent to the electrolyte solution can markedly improve the separation of closely related metal cations sharing almost identical effective mobilities. Examples are the use of α-hydroxyisobutyric acid as a weakly chelating agent for the separation of mixtures of rare earth metals and other heavy metals and citric acid as a complexing agent for alkaline earth metals (see Sect. 7.1).

In order to complex undesirable trace metals present in the electrolyte system or in the sample, small amounts of ethylendiamine tetraacetate (EDTA) are sometimes added to the buffer solution.

3.3.7 Organic Modifiers

A number of modifiers are used in capillary electrophoresis to change the physico-chemical nature of the separation system. Organic solvents added to the buffer electrolyte alter the polarity and the viscosity of the mobile phase. As a consequence both the electroosmotic flow and the electrophoretic mobility of buffer electrolyte alter the polarity and the viscosity of the mobile phase. As a consequence both the electroosmotic flow and the electrophoretic mobility of the analytes are effected. Additionally, organic solvents are widely used as solubilizing agents for analytes which are poorly soluble in aqueous solutions. Fujiwara and Honda [85] have investigated the influence of methanol and acetonitrile on electroosmosis. As they have shown, the electroosmotic mobility declines almost linearly to approximately 70% of the original value by increasing the methanol concentration from 0 to 60% (v/v). The addition of acetonitrile, however, has caused a slight increase of the EOF. The effect of organic solvents on the electrophoretic mobility is difficult to predict. In general, both methanol and acetonitrile seem to raise electrophoretic mobility and to enhance differences in mobility. As a result resolution should be improved. On the other side, Zhu et al. [86] have investigated the influence of acetonitrile on the separation pattern of a set of nine peptides. They have found that the separation patterns are not significantly different.

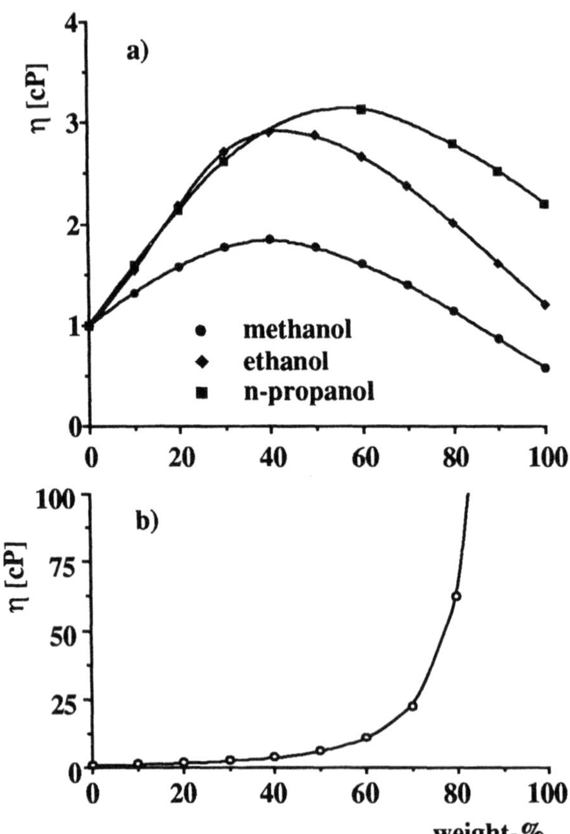

Fig. 3.26. Change of viscosity of alcohol-water mixtures at 20 °C. Aqueous solutions of (**a**) methanol, ethanol and *n*-propanol, (**b**) glycerin

Besides acetonitrile and methanol, higher alcohols like ethanol, *n*-propanol and even diols and polyols are also suitable for adjusting the electroosmotic flow by modifying the viscosity of the mobile phase. While the viscosity of glycerol-water mixtures increases exponentially with an increasing proportion of glycerol (Fig. 3.26.b), aqueous solutions of the other alcohols show a maximum viscosity between 40 to 60% as shown in Fig. 3.26.a. The dependence of EOF on the chain length of alcohols follows the order: methanol < ethanol < propanol < butanol [87]. Diols or polyols are more efficient in changing the viscosity of the mobile phase: ethylene glycol < glycerol < PEG 400.

Other organic modifiers which are frequently used to reduce the zeta potential and, hence, electroosmosis are methyl cellulose, hydroxypropylmethyl cellulose (HPMC) and other hydrophilic linear polymers derived from cellulose. Low concentrations (0.01 - 0.03%) of these additives decrease not only the zeta potential by shielding the silanol groups of the wall, but also prevent undesirable adsorption. As one can see from Eq. 2-26, increasing the viscosity of the buffer electrolyte results in a decrease of the electroosmotic mobility. Hence, higher concentrations of these linear polymers (> 0.1%) reduce the EOF by increasing the viscosity of the buffer and cause a molecular sieving effect through the formation of an entangled polymer network (see Sect. 5.2.3.3).

Urea is particularly useful for the electrophoretic separation of nucleotides [88] or proteins that are insoluble at the low ionic strengths required in electrophoresis [89]. The use of urea between 5 and 8 M involves the denatured state of the biopolymers by breaking hydrogen bonds in the tertiary structure (see also Sect. 3.3.5). Sometimes a small amount of mercaptoethanol or dithiothreitol is added to cleave disulfide bonds between polypeptide chains.

4 Instrumentation

This chapter is devoted to the equipment for capillary electrophoresis. Basic components of the instrumental set-up will be described insofar as they are already commercially available or if they have potential for use in the near future.

Capillary electrophoresis is a simple technique with regard to the instrumentation needed. One can perform CE with a device consisting of a fused silica capillary, a power supply, a detector, e.g. an HPLC detector modified with fiber optics for on-column detection, and a recorder. Modern CE instruments are additionally equipped with an autosampler for sample injection allowing series analysis, column thermostating and a computer for instrumental control and data acquisition. In the following discussion, particular emphasis is laid on detection systems, which are the most important and most critical component of the instrumentation.

A source list of those manufacturers who are mentioned in the following discourse is given in the Manufacturers' Directory (Appendix 8.4).

4.1 Injection

Sample injection is a crucial procedure in capillary electrophoresis. Even an optimized electrophoretic system produces unsatisfactory separation if injection has been performed carelessly. From the theoretical point of view the sample should be injected as an infinite small volume into the capillary to minimize the initial zone broadness of the analytes. There are two fundamental ways to introduce the sample into the capillary:

> hydrodynamic injection
> electrokinetic injection.

4.1.1 Hydrodynamic Injection

For hydrodynamic injection a pressure drop has to be applied along the capillary either by

> high pressure at the injection side
> vacuum at the detector side
> hydrostatic pressure by utilizing gravity.

In practice, the buffer vial is replaced by the sample vial so that the capillary is immersed in the sample solution. To allow sample introduction the sample vial is then pressurized (pressure injection) or lifted to a defined height for a time period (gravity injection) to create a hydrostatic pressure. Alternatively a vacuum is applied at the end of the capillary to suck up the solution into the tube. After injection is completed the sample vial is replaced once more by the buffer vial and the separation process can start.

The hydrodynamic injection volume introduced into a capillary is a linear function of the applied pressure difference along the capillary and its duration. The injection volume can be calculated by Poiseuille's law for liquid flow through a circular tube:

$$V_i = \frac{\Delta p \cdot \pi \cdot r^4 \cdot t}{8\eta \cdot L_T} \tag{4-1}$$

V_i injection volume [$m^3 = 10^{12}$ nL]
Δp pressure difference [Pa]
r inner radius of the capillary [m]
t injection time [s]
η viscosity [Pa·s = 10^3 cP]
L_T total capillary length [m]

The sample plug length in the capillary can be calculated from the injection volume by dividing by πr^2.

If sample injection is accomplished by gravity the pressure difference is given by the hydrostatic pressure which is defined as

$$\Delta p = \rho \cdot g \cdot \Delta h \tag{4-2}$$

ρ density of the sample solution [$kg \cdot m^{-3}$]
g gravitational acceleration (9.80665 $N \cdot kg^{-1}$)
Δh height difference between liquid levels of sample and buffer vials [m]

Eq. 4-1 shows that the sample volume introduced by hydrodynamic flow can be manipulated by varying the injection time and/or the pressure difference. In practice, an instantaneous pressure application and interruption to the vial would lead to undefined up-ramping and down-ramping of the actual pressure caused by limited pressure source capacity and systemic restrictions. Consequently, irreproducible volumes are injected. The injection volume, given by the product ($\Delta p \cdot t \cdot$constant), is reflected by the area under the curve of a Δp-t-diagram as shown in Fig. 4.1. Reproducible injection has to be accomplished by careful controlling of the compression and decompression process. Fig. 4.1. illustrates both possibilities.

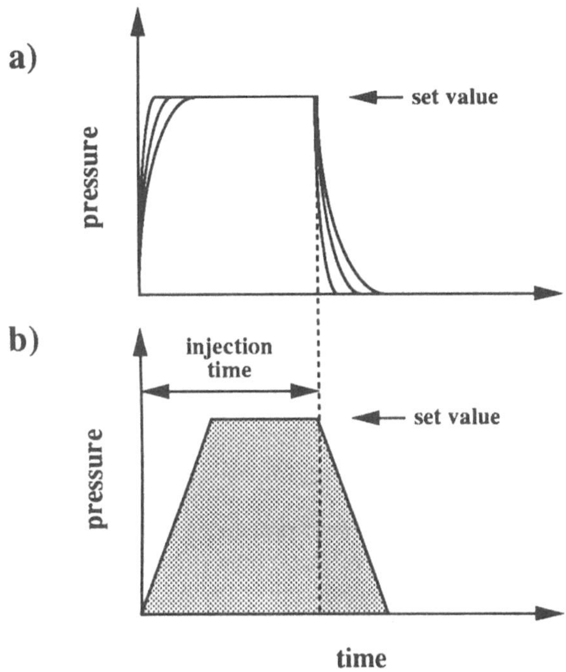

Fig. 4.1. Instantaneous compression and decompression with irreproducible areas (**a**) and controlled up-ramping and down-ramping of the pressure (**b**); according to Ref. 90

From Eq. 4-1 it becomes obvious that the injection volume is dependent to the 4th potency on the inner diameter of the capillary and reciprocal to the tube length. For quantitation it is of great importance that the injection volume depends on the viscosity of the solution. Hence, the injection volume is temperature dependent. If quantitation is to be done by using the external standard method, both the reference and the sample solution must have the same temperature, ideally that of the buffer in the capillary. In Table 4.1. the viscosity of pure water and calculated injection volumes are given as a function of temperature.

4.1.2 Electrokinetic Injection

Electrokinetic injection represents the alternative mode for sample introduction in capillary electrophoresis. This injection mode is based on the fact that voltage causes electrophoretic and electroosmotic movement. If voltage is applied

T [°C]	$\eta \cdot 10^3$ [Pa·s]	$V_{inj.}$ [nL]
20	1.002	2.4
25	0.890	2.6
30	0.798	3.0
35	0.719	3.3
40	0.653	3.6
45	0.596	4.0
50	0.547	4.3

Table 4.1. Viscosity η of water and injection volume $V_{inj.}$ at different temperatures calculated according to Eq. 4-1; injection time 1s, $\Delta p = 3448$ Pa [91]; capillary dimensions: 50 cm x 50 μm i.d. (with permission of Beckman Instruments, Munich)

for a short interval of time, sample is introduced into the capillary due to electrophoretic migration. If, additionally, electroosmotic flow occurs, a sample volume will be introduced into the column. The injected sample volume is then given by

$$V_i = v_{eo} \cdot \pi \cdot r^2 \cdot t \tag{4-3}$$

It should be mentioned that some authors use the sum of v_{eo} and v_i instead of v_{eo} alone to calculate the injected sample volume. Nevertheless, v_i is negligible in comparison to v_{eo}. The quantity of a species i introduced into the capillary by electromigration is

$$Q_i = \frac{(\mu_i + \mu_{eo}) \cdot \pi \cdot r^2 \cdot V \cdot c_i \cdot t}{L_T} \tag{4-4}$$

Q_i amount of species i introduced into the capillary [M]
V voltage [V]
c_i concentration of species i
L_T total capillary length [m]

Thus, the quantity introduced into the capillary can be controlled by varying the voltage and/or the introduction time. With electrokinetic injection the quantity depends on the specific electrophoretic mobility of the component and the electroosmotic mobility. This means that electrokinetic injection discriminates among the ions. Species with high mobilities are drawn in a higher amount into the capillary than those with low mobilities. Reversed charged ions are even repelled from migrating in the capillary. They can only migrate into the capillary if the absolute value of their mobility is smaller than the electroosmotic mobility. Another reason which causes a bias of the sample is related to the electrical conductivity of the sample solvent. Both the electrophoretic and electroosmotic velocity will alter with different sample solutions. Although there have been several attempts at a mathematical description of this phenomenon [92, 93], electrokinetic injection causes extreme problems from the point of view of quantitation. This will be discussed in chapter 6.

In view of the above, hydrodynamic injection is preferable over electrokinetic injection. However, there are occasions where the latter mode is to be preferred, i.e.

> if discrimination of the component of interest from contaminants is desired,
> if polyacrylamide gel-filled capillaries are used where hydrodynamic injection is not feasible,
> if a concentrating of a component from a diluted sample solution is desired (only effective in the absence of EOF).

4.1.3 General Aspects of Injection

One of the major advantages of capillary electrophoresis is the fact, that only minute sample volumes (e.g. 20 μL) are required for analysis but only a few nanoliters are actually injected. In practice, a simple device is needed allowing sample introduction in the microliter scale. However, since small sample volumes may evaporate during analysis times, a vial which can be closed by a cap is strongly recommended. Such a device is shown in Fig. 4.2. [94]. A cap, e.g. of silicone rubber with a die-cut cross opening, seals the glass vial to protect

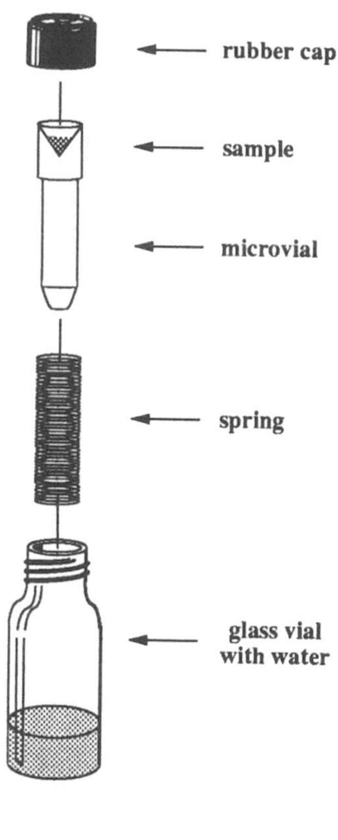

Fig. 4.2. Assembly for sample injection in the microliter scale (with permission of Beckman Instruments, Munich)

the sample and to prevent evaporation of the solvent. A microvial containing approx. 20 µL of sample is assembled in the glass vial with a spring. During injection the microvial is depressed by the capillary and the electrode to make sure that the capillary dips down to the bottom of the sample cone.

Hint: *Evaporation of sample solvent during series analysis is minimized by the addition of some water into the glass vial, which humidifies the head space in the vial.*

Take care that there are no air bubbles in the sample solution.

Failures during operation may arise from a number of causes. Buffers and sample solution can contain dissolved or solid impurities. For example, just traces of heavy metals in the buffer may precipitate proteins or peptides. To minimize this problem buffers and sample should be prepared by using double distilled water and other components of highest purity. Filtering the final buffer solution through a filter with 0.2 - 0.4-µm pores prevents clogging of the tube. Cloudy samples have to be cleared by filtration or centrifugation before analysis. Although we seldom found indications of air bubble formation, degassing of the solutions is also suggested. After injection, precipitation of sample components can occur if sample solvents other than aqueous media are used where the components are least soluble.

Several devices for sample introduction are outlined in the literature. A split-flow sample injection system is described by Teherani et al. [95]. Sample introduction is accomplished with an ordinary HPLC-type microliter syringe. The injected sample is divided proportionally between the capillary and an adjustable split-vent. By varying the length or the internal diameter of the split-vent tubing the injection volume can be manipulated. Olefirowicz and Ewing [96] reported an injection system which is designed to acquire and determine components from cytoplasm of a single nerve cell. The microinjector is prepared from a 5-µm i.d. fused silica capillary. The capillary tip has been etched in HF obtaining a diameter as small as 6 µm. This microinjector is directly positioned into the single nerve cell to sample the cytoplasmatic fluids by electromigration. Rose and Jorgenson [97] have described an autosampler allowing computer controlled automation of the sample injection process in CE. Details of the set-up and performance are presented for both electrokinetic and hydrodynamic injection.

Hint: *Sometimes it is desired that two different sample solutions be analyzed in the same run. This can be done conventionally by mixing both samples and injecting the mixture. However, in the mixing step both sample components are diluted by the additional volume of the other sample. Moreover, and often more important, the samples are contaminated by mixing with the other component. As an alternative to this procedure both samples can be injected individually one after the other and analyzed as usual in one single run. No contamination of the sample solutions occurs and the amount injected can be controlled by the individual injection times. We routinely use this procedure if we want to inject an electroosmotic flow marker in addition to a valuable sample. For small injection volumes (3 - 5 nL) we have been able to show, that the systemic error caused by the method is within the experimental error of the separation.*

4.2 Detection

4.2.1 General Aspects

Detectors for CE and other miniaturized techniques have gained a great deal of attention in recent years as a direct consequence of the low sample capacity to be detected. The maximum allowable detector dead volume V_d is given by [98]:

$$V_d = \frac{\Theta^2 \cdot V_c}{\sqrt{N}} \tag{4-5}$$

Θ^2 fraction of allowable peak broadening
V_c volume of the column [mL] = $\pi r^2 L_T$

If, for instance, a 1-m long, 75-µm i.d. capillary providing 250 000 theoretical plates is used, the detector volume which introduces 10% peak broadening (resulting in $\Theta = \sqrt{0.05} = 0.22$) is about 1 nL. Even smaller detection volumes are required for smaller capillaries. For a 50-cm x 10-µm i.d. capillary, the "allowed" detector volume should be much less than 1 pL. The most important requirements for the design of detectors suitable for capillary separation systems are:

- small volume detection cell
- small contribution to the peak width
- high sensitivity
- large dynamic range
- fast detector response
- good resistance against temperature changes
- reliable and convenient ease of use
- selective detection
- non-selective detection

Obviously, there is no single detector available providing all these properties. For any particular application the appropriate detection mode has to be chosen. In terms of the kind of signal which is obtained, two main types of detectors can be distinguished: non-selective and selective detectors. Whereas non-selective detectors measure differences in physical properties of the analyte relative to that of the whole solution, selective detectors measure a specific property of the analyte. Unselective detectors, including refractive index (RI), conductometric and indirect detection methods, commonly exhibit lower sensitivities and dynamic ranges, but they are more universal than selective detectors, including UV/VIS absorbance, fluorescence, mass spectrometry (MS), Raman, electrochemical and radiometric detectors.

A special kind of non-selective detection gaining more and more acceptance as a universal detection mode in CE is indirect detection. In this mode the signal is derived from the background electrolyte rather than from the analyte itself. During migration through the detection cell the analyte displaces the background electrolyte. For analytes having the same charge as the detectable buffer component, displacement results in a decrease of the signal, because the concentration of the detectable buffer component is lower in the analyte zone compared to its steady-state concentration in the capillary. Thus, the analyte is detected as a negative peak in the electropherogram.

In general, detection of the separated zones is performed on-column or off-column. For on-column detection, the detection cell is part of the separation capillary. Band spreading is minimized because no joints, fittings and connectors are needed. On-column detection, however, is limited to optical detection modes, such as UV, fluorescence and RI detection, and conductometric detection. Off-column or post-column detection of the zones is more difficult to employ. A connection must be made with the end of the capillary permitting current to flow during electrophoresis while not significantly perturbing and broadening the zones. Examples of off-column detection include interfacing to a mass spectrometer and amperometric detection. Since mass spectrometry provides additional information about the separated compounds, e.g. molecular weight and structural characteristics, CE-MS is preferably regarded as a two-dimensional separation technique and will therefore presented in Sect. 5.7.1.

4.2.2 Evaluation of Detector Performance

In order to compare detection systems with respect to their performance or to evaluate the best detection system for a specific separation, the following criteria are useful:

- sensitivity (response factor)
- detection limit
- noise
- linear (dynamic) range
- response index
- time constant (response time)

The sensitivity or response factor refers to input/output and is given by the ratio of a measured signal (current, voltage, absorbance etc.) to an amount (weight [g] or moles of substance [mol]). Sensitivity by itself does not say anything about the minimum detectable quantity. Therefore, it is also necessary to evaluate the baseline noise which can preferably be done by recording the detector response over a time period of about 10 times the peak width. Figure 4.3. shows the noise of an electropherogram for different frequencies. High frequencies arise from incomplete grounding or from the signal amplification system and can usually minimized by filtering. Low frequencies occur due to temperature variations, impurities of the background electrolyte ("chemical noise") or

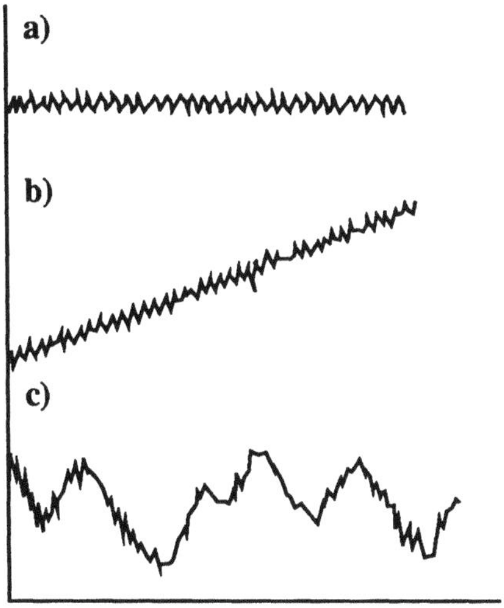

Fig. 4.3. Different forms of the baseline noise: high frequency noise (**a**), baseline drift (**b**) and low frequency noise (**c**)

air bubbles. Very low frequencies are called baseline drift. An amount that would give a peak height of 3 times the baseline noise can still be detected and is therefore a useful reference for the detection limit. (Nevertheless, most authors use a S/N value of 2 to characterize their detection systems). If the sensitivity and the absolute value of the baseline noise are known for a given system, a reasonable detection limit for each substance being determined can be predicted. Instead of the sensitivity and noise, the signal-to-noise ratio (S/N) together with the detection limit of a special substance is often used to characterize the detector sensitivity. Detection limits can be given either as limit of detection (LOD) or concentration detection limit or as mass limit of detection (mass LOD) or mass detection limit. LOD is given in terms of a concentration in g/L or M, mass LOD in terms of grams or moles of solute. The latter is obtained from LOD by multiplication with the injected sample volume. Because of the extreme low sample volumes used in CE, the mass LOD is much lower than the LOD. In many cases, LOD is determined by the so-called static mode where the capillary is filled with the sample solution with no flow and applied voltage. These LOD values are often significantly higher than those which refer to the injected volume of a CE run, because background noise is often increased when voltage is applied and the sample zone migrates through the detection cell. Additionally, the sample is often diluted while separation takes place.

The ideal detector produces a signal which is proportional to the sample concentration over a wide concentration range. Obviously, for real detection systems this linearity is not infinitely high. To evaluate the linearity of a detection system the following equation can be used [99]:

$$\log y = \log A + r \cdot \log c \qquad (4\text{-}6)$$

y detector response
A response factor for the detected substance (sensitivity)
r response index
c sample concentration

r should lie between 0.98 and 1.02. The closer r is to unity the more linear is the response. For r values ≈1 Eq. 4-6 can be written in the form $y = A \cdot c$. The graph of this function is shown in Fig. 4.4. The linear or dynamic range of concentration is now given by the lowest and the highest concentration lying on the straight line.

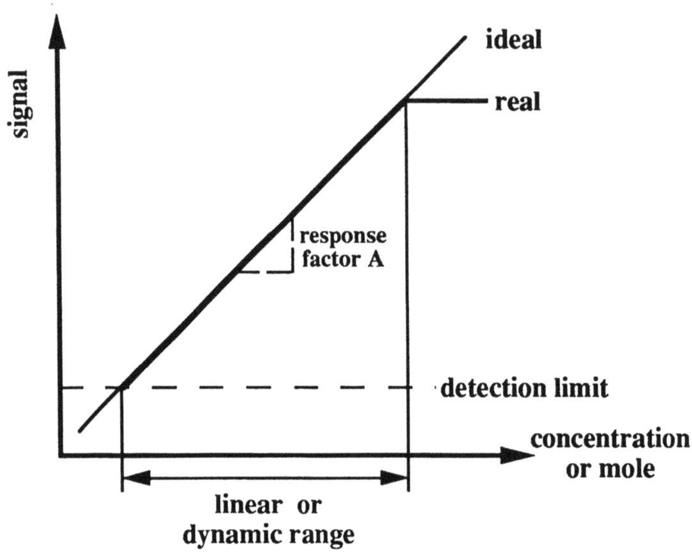

Fig. 4.4. Graph of the function $y = A \cdot c$

Because of dramatic improvements in performance, apparative contributions to band broadening can have significant effects on efficiency and resolution. The more efficient the separation column is, the more concern must be given to the reduction of all other factors which can decrease system efficiency. The most common apparative contribution to distortion is caused by a delay in response to change with a characteristic time constant. No electronic device is capable of responding instantaneously to a changing input or, in other words, can process data from signals of high frequency. Components above a certain frequency contribute no information and are observed as "noise". Vanishing response times imply infinite bandwidth, suggesting a reciprocal relationship:

$$B = \frac{1}{\tau} \tag{4-7}$$

B bandwidth [Hz]
τ response time or time constant of the system [s]

Limited bandwidth is an essential feature of amplifier design and may be nothing more than an intentional increase in the time constant to discriminate against high frequencies. A low-pass circuit like an RC network or an amplifier stage makes an effective noise filter as long as it does not reduce the bandwidth to the extent that frequencies carrying information are attenuated. If this occurs, distortion of the peak shape will become evident. Well-designed detector amplifiers should include a provision for time constant adjustment so that the minimum distortion-free setting can be chosen. To avoid half-width broadening in excess of 5%, the time constant should be less than one third of the temporal variance.

In the on-column detection mode the width of the detection window of the capillary ranges from several hundred microns for absorbance detectors to less than 50 µm for conductometric detectors. Thus, especially for optical on-column detectors, band-broadening due to the width of the detector region can become significant. In this case, the conversion of the temporal width of the analyte zone to the spatial width is given by:

$$w_s = \left(\frac{L_D}{t}\right) \cdot w_t - w_d \tag{4-8}$$

w_s spatial width of the analyte zone [cm]
w_t temporal width of the analyte zone [s]
t migration time of the zone [s]
w_d spatial width of the detector region [cm]

In off-column detection systems, special care has to be taken to minimize both the volume of the detection cell and of the connection between the capillary and the detector. Otherwise the arising dead volumes will lead to a dramatical loss in efficiency.

4.2.3 UV-VIS Absorbance Detection

Owing to its sensitivity to a wide range of compounds and functional groups and its ease of use UV-VIS absorbance has remained the most popular detection principle, although it suffers from low sensitivity compared to other detection modes developed for capillary separation systems, such as electrochemical, mass spectrometric and fluorimetric detection. By UV-VIS absorbance, substances can be registered if they provide at least one of the following functional groups:

- bromine, iodine or sulphur
- two conjugated double bonds
- a double bond vicinal to an atom having a single electron pair
- a carbonyl group
- an aromatic ring

These functional groups do not absorb either with identical intensities or at the same wavelength. The absorbance intensity (extinction) and the position of the absorbance maximum is also influenced by other atoms in the molecule. The Lambert-Beer law describes the intensity of absorbed light in dependence of the concentration c of the analyte and of the optical path length of light through the detection cell l_c [cm]:

$$E = \frac{I_0}{I} = \varepsilon \cdot c \cdot I_c \tag{4-9}$$

I_0 initial light intensity
I light intensity after absorbance
ε molar extinction coefficient or absorptivity $[M^{-1} \cdot cm^{-1}]$

One has to be aware that Eq. 4-9 is deduced for a cell with plane parallel windows. Originally, the extinction E is dimensionless, but it is common to describe an extinction of 1 as 1 absorbance unit (1 AU). 1 AUFS (absorbance unit full scale) means that a full scale of the chart paper corresponds to 1 AU. The wavelengths which can be used to detect absorbing analytes are dependent on the light source and the kind of photometer. In practice, the following wavelengths are commonly used:

- 190 and 200 nm for all kinds of analytes
- 210 and 214 nm for peptides and proteins
- 254 nm for aromatic compounds
- 280 nm for proteins
- 260 nm for nucleic acids

In CE, UV detection, mostly applied in the on-column mode, involves the use of modified commercial HPLC detectors or detectors designed for isotachophoresis. In fact, almost all commercial CE instruments available today employ

UV-VIS absorbance detectors which have been developed for HPLC systems with a modified optical layout. In on-column UV detection, the capillary itself serves as the cylindrical detection cell. Fused silica tubing has a UV cut-off at approx. 170 nm. The polyimide coating at the outside of the capillary can be removed to build the detection window as described in the next section. As a result of the short available light path, which is the inside diameter of the capillary, detection limits lower than 10^{-6} M of the injected material can hardly be realized, because it is very difficult to conduct a sufficient amount of light through the capillary. For the same reason, capillaries having diameters smaller than 50 µm can hardly be used with UV-detection. In addition to the short optical light path, one has to combat shot noise, arising from photon statistics at low light intensity, and nonlinearity due to ill-defined light paths. In other words, detector performance is shot-noise limited. This means that the LOD will deteriorate as the background absorbance increases because less light will reach the photodiode resulting in a higher background or shot noise. Several attempts have been made to maximize the intensity of light passing through the sample.

4.2.3.1 Light Sources for UV-VIS Detection

Firstly, high intensity lamps have to be used. Several types of UV lamps are employed in commercially available equipment. The simplest UV detector is the fixed-wavelength photometer with lamps producing sharp and strong lines at one single wavelength. The most common lamp is the low-pressure mercury lamp which produces a sharp line at 254 nm. Another wavelength used alternately or in tandem with 254 nm is 280 nm, which is obtained from phosphorous excited at 254 nm. Other common lamps used in fixed-wavelength detectors together with the appropriate filters are Zn (214 nm), Cd (229 nm) and As (200 nm) which have been compared by Green and Jorgenson [100] for their usefulness in CE-on-column detection. They have found that the zinc lamp is the best choice for fixed-wavelength operation. Variable wavelength filter photometers employ most often a medium-pressure mercury lamp, which has lines of good sensitivity at 254, 280, 313, 334 and 335 nm. The line of interest is isolated with a narrow bandpass interference filter. Spectrophotometers employ continuous UV sources such as deuterium or xenon-arc lamps, most often in addition to a tungsten lamp providing a continuous spectrum in the visible range. The deuterium lamp produces a continuous UV spectrum in the range of 180 nm to 380 nm with a maximum intensity at 220 nm. A single wavelength of interest is selected by a grating monochromator.

In modern devices such as the SpectraFOCUS (former Linear) multiwavelength detection system from SpectraPhysics, more than one wavelength can be chosen simultaneously. With this feature, analytes absorbing at different wavelengths can be detected in the same run. In addition, absorbance spectra of unknown species can be obtained. Spectrophotometers with continuously variable wavelength design are much more flexible and selective, but also less stable, less sensitive and more expensive than filter photometers. Kobayashi et al. [101] have developed a full 512-element photodiode array detection system for CE which can be used in the range of 200 - 380 nm. With this system, high resolution absorbance spectra of unknown separants can be obtained.

4.2.3.2 Optical Layout of a UV-VIS Detector for CE

Only in a few cases is the light source placed close to the capillary and shielded by appropriate slits. The lamp housing is mostly placed a certain distance away from the capillary placement. Hence, the light beam has to be conducted efficiently from the source to the detection cell. This can be done by a selection of mirrors or by using optical fibers. Figure 4.5. depicts a schematic diagram of two different UV detection devices which can be used for capillary separations.

Figure 4.5.a shows the optical layout of a variable wavelength filter photometer where two mirrors focus and direct the light beam to the wavelength filter. The filter is selected by rotating a bench with a set of different filters in the appropriate position. The beam is then directed through the detection cell of the capillary to a photomultiplier tube (PMT) or a photodiode which sends an electrical signal proportional to the amount of light transmitted to the computer. This arrangement is similar to that used in the P/ACE system of Beckman Instruments.

Figure 4.5.b illustrates the optical path of a UV-VIS spectrophotometer with a double-beam system as used in the SpectraPhoresis device from SpectraPhysics. The light first strikes the grating monochromator, before it is conducted through an optical fiber which bifurcates providing both sample and reference signals (beam splitter). Light from the reference fiber optic is directly focused onto the reference photodiode. Light from the sample fiber optic reaches the detection cell where it is focused by a lens before passing through the capillary. Light emerging from the capillary is focused onto the sample photodiode. In general, double-beam detectors are more sensitive than single-beam detectors due to the apparent lower noise levels.

4.2.3.3 Design of the Detection Cell

Besides the light source and the effective conductance of light to the capillary, the design of the detection cell is of great importance in order to get high sensitivity and low background noise. However, efficient focusing of spatially incoherent light from the light source into capillaries with internal diameters of 100 µm and below is a critical task. Slits or pinholes serving as apertures and focusing lenses made from fused silica or sapphire are commonly used to focus the light beam onto the detection window of the capillary. Figure 4.6. gives a schematic view of a detection cell using a rectangular aperture. The path length is defined by the inner diameter of the capillary. Because of the cylindrical design

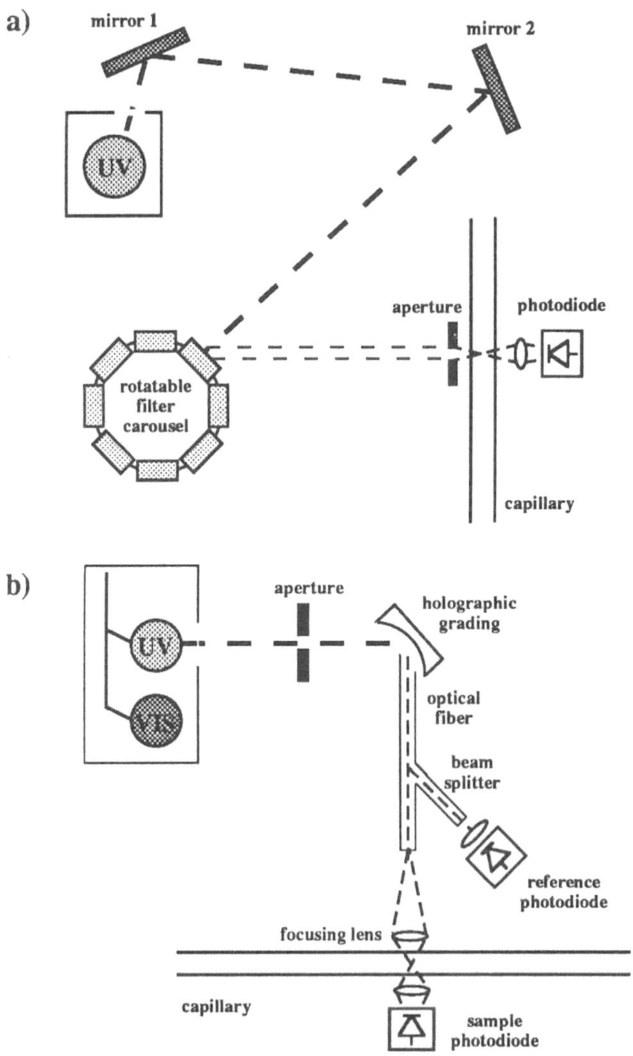

Fig. 4.5. Schematic diagram of two different UV detection devices for CE. (**a**) optical layout of a variable wavelength filter photometer with a single-beam system; (**b**) optical layout of a spectrophotometer with a double-beam system

of the detection cell, the path length decreases from the center of the capillary to the wall. If the aperture is placed completely symmetrically around the center of the capillary, most light rays will pass through the capillary tube near the axis.

Those rays will have a path length given by the capillary diameter. Rays that pass far away from the center of the capillary will have much shorter optical path lengths. The effective path length lies somewhere between the maximum and minimum path length and can be calculated according to Ref. 102. In a first approximation the effective path length is about 0.6 l_c. The effective light path and, hence, the sensitivity is increased with decreasing aperture width. Slight misalignment of the aperture can lead to significant deviations of the calculated effective light path.

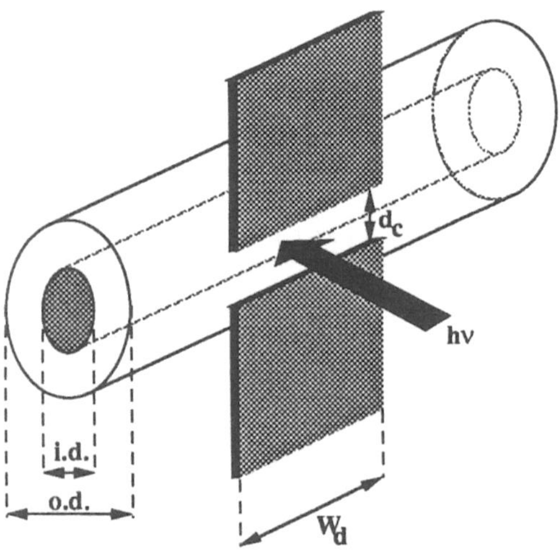

Fig. 4.6. Schematic view of a detection cell using a rectangular aperture. d_c aperture width in perpendicular direction of the capillary axis, w_d aperture length along the axis of the capillary (width of the detection window)

A UV cell with an adjustable rectangular aperture is described by Wang et al. [103]. The aperture body is constructed by sandwiching a shim (25 - 75 μm) between two pieces of metal. A rotatable washer with different slits ranging from 0.2 to 2 mm is placed between the aperture body and the light source. The aperture depends on both the thickness of the shim and the dimension of the slit on the washer. Whereas the length of the aperture along the axis of the capillary can be changed by rotation of the washer, the aperture width perpendicular to the axis is determined by the thickness of the shim.

One of the first reports dealing with the evaluation of UV detectors was from Wahlbroehl and Jorgenson [99]. In their experimental set-up, the reference and light beams pass through 100-μm pinholes. Thus, a spherical aperture with a fixed diameter of 100 μm is given. The positions of the pinholes are adjusted through the use of pinhole positioning mounts. The UV source which is placed close to the capillary is a 7-W cadmium "pen-ray" lamp emitting radiation from both sides. Emission from one side of the lamp can be used as reference beam and the other is used as sample beam. The position of the optical path with respect to the capillary is optimized via the moveable pinhole.

One drawback to these arrangements is the alignment procedure, which must be carried out very carefully to guarantee correct placing of the capillary in the light beam and, hence, good reproducibility and a high S/N ratio. As an alternative, optical fibers can be used to couple the light source with the capillary. A detector of this type which has been developed by Foret et al. [104] consists of a pair of 40-cm long optical fibers with a 200-μm light conducting core that bypass the optical path of a UV-VIS spectrophotometer. One fiber, the source fiber, transfers light from the source directly to the capillary while another fiber, the collecting fiber, carries the transmitted light to the photodiode. The detection cell is constructed from a brass holder, for fixing both the capillary and optical fibers, equipped with two perpendicular grooves into which the capillary and the fibers are inserted. Theoretical optimization of the cell designs utilizing optical fibers has been done by Bruno et al. [105]. They have used a three-dimensional ray tracing algorithm, which simulates the optical phenomena at the air-glass and glass-liquid interfaces. This enables them to choose optical fiber dimensions for a given choice of capillary inside and outside diameters.

The proper choice of the aperture width and the capillary dimensions with respect to inner and outer diameter is very important to reduce or eliminate stray light effects as well as reflection and refraction processes which can occur at the air-glass and glass-liquid interfaces. In order to find the optimal signal-to-noise ratio the optimal balance of aperture dimension and capillary dimension has to be chosen. Bruin et al. [102] have studied the influence of reflection and refraction processes for different detection cells by computer simulations. Light intensity losses due to reflection are very small. Only at the outer edge of the capillary, where the angle of incidence is high, can light reflection occur. Refraction, however, can have a more significant impact. Light rays striking the outer capillary wall will be refracted and would not reach the photocell. Light rays entering the detection window between the center and the inside diameter of the capillary will not undergo much refraction. For light entering the capillary between the inside and the outside diameters, the capillary glass wall works as a converging lens. Figure 4.7. shows some of their results for a cell with an adjustable aperture width (Fig. 4.7.a-c) and a cell with a focusing lens (Fig. 4.7.d,e). It can be observed easily that refraction and reflection can be neglected only at small aperture widths. The fraction of light that enters the capillary between the inside and the outside diameter at the glass wall converges to the center of the capillary. These refracted light rays partly fail to reach the photocell, resulting in a higher background noise. When a sapphire lens is used to focus the light in the liquid flow, the performance depends strongly on the ratio between inner and outer diameter of the capillary and the distance between the lens and the capillary. If the outer diameter becomes too high, a significant fraction of light which does not pass through the detection cell reaches the photocell resulting in a higher noise level.

Stray light around the capillary arising from scattering, which does not take part in the absorbance process, can also reach the photocell, resulting in excessive shot noise. It can be reduced either by employing apertures whose widths are sufficiently small such that only the capillary is illuminated or by placing a focusing lens in front of the detection window. Due to the higher effective path

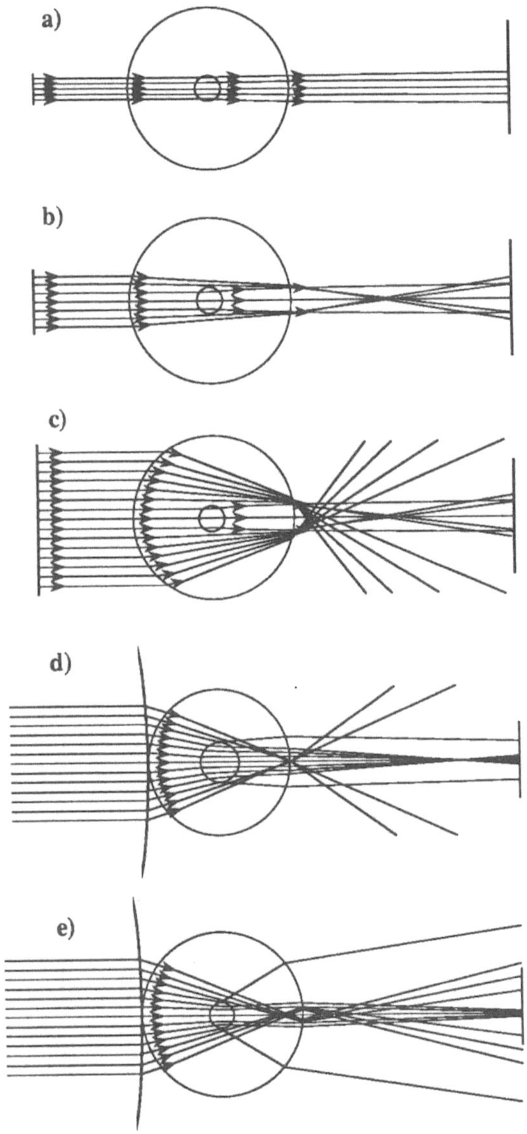

Fig. 4.7. Light refractions through cylindrical UV detection cells according to Ref. 102. Simulations for a cell with different aperture widths of 50 μm (a), 145 μm (b) and 350 μm (c) are shown in **a**-**c** for a capillary of 50 μm I.D. and 350 μm O.D.; simulations for a focusing lens are shown in (**d**) and (**e**) for capillaries of 75 μm I.D and 275 μm O.D (d) and 50 μm I.D. and 350 μm O.D. (Reprinted with permission of Elsevier Science Publishers)

length at smaller aperture widths an increase in sensitivity with decreasing aperture can be observed. On the other hand, the optical transmittance decreases rapidly if the aperture width is decreased, resulting in poor linearity and increased noise levels. The optical aperture can be used as a measure of the performance of cylindrical flow cells. It is defined as the ratio d_c/l_c, where d_c is the aperture width and l_c the optical path length [106]. For UV detectors with common cylindrical flow cells, the ideal optics have numbers for the optical aperture within the range 1/10 to 1/5. For values below 1/10 the available light energy rapidly decreases, resulting in increased noise and poor linearity. For values above 1/5 the path length becomes too short, resulting in poor sensitivity. Therefore, an optimum aperture width has to be found, where the highest signal-to-noise ratio can be achieved. Bruin et al. [102] have found that adjusting the aperture width to the inside diameter of the capillary results in the lowest LOD and the highest linear range. This result has the additional advantage that adjustment of the aperture width to the inside diameter is easy to do, because the inner wall is clearly visible under the microscope.

Several attempts have been made to extend the path length for UV detection in capillary separation techniques by changing the design of the detection cell. Chervet et al. [106] have shown the usefulness of a Z-shaped longitudinal capillary flow cell, which is constructed as part of the capillary by bending a small section into a Z-shape. The bending procedure can be found in Ref. 107. The flow cell (Fig. 4.8.) is prepared by sandwiching a shim with the bend capillary between two plastic disks of black polyethylene. The thickness of the shim and the sharpness of the capillary bends determine the total path length of the flow cell. For shims of 1 mm thickness, flow cells of ca. 3-mm path length can be obtained, providing that the bends are sharp. The shim has a centered hole of 300 μm that is adapted to match the o.d. of the capillary. Each plastic disk has a groove to fix the capillary with epoxy resin in order to obtain a stable flow cell. As they have stated, the bending of the capillary has virtually no effect on the

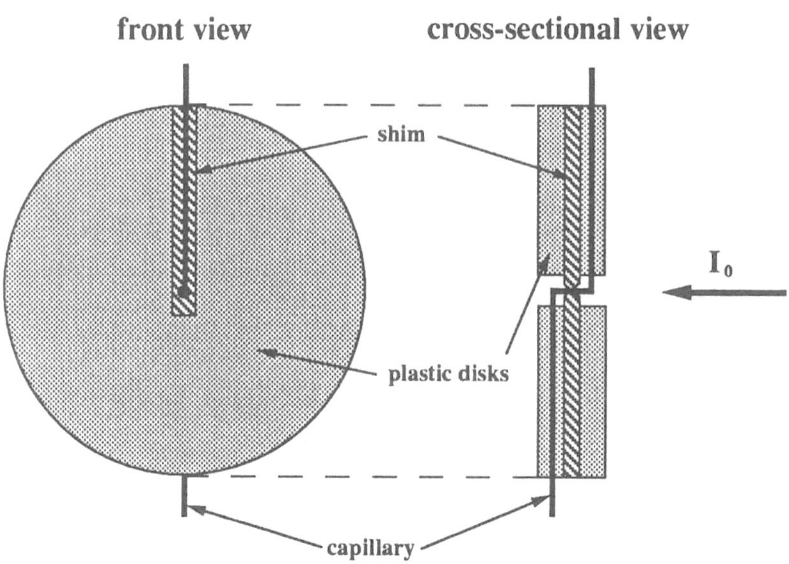

Fig. 4.8. Schematic diagram of a Z-shaped flow cell used in CE according to Ref. 106. (Reprinted with permission of Elsevier Science Publishers)

electrophoretic process. Since Z-shaped flow cells are more susceptible to "chemical noise", capillary washing and pretreatments are important for stabilizing the baseline. In general, extended washing with water before flushing with buffer solution improves the baseline considerably. They have found a sixfold improvement in signal-to-noise ratio and a loss in efficiency of 17 - 32% compared with normal on-column detection.

Tsuda et al. [108] have described the use of rectangular tubing with optical path lengths up to 1 mm. The rectangular cross-section allows a significant increase in the sensitivity of UV-VIS absorbance due to the higher effective path length. In addition, optical distortion and scatter is reduced by the flat capillary wall.

Another possibility to increase the effective path length across the capillary is the use of an on-column multireflection absorbance cell [109] as shown in Fig. 4.9. On a fused silica capillary of 75 μm i.d. and 364 μm o.d. an opening of about 1 cm is made by burning off the polyimide coating. A silver layer is deposited by redox reaction of $Ag(NH_3)_2^+$ and glucose. Black paint is applied on the silver layer to protect it from physical damage. The light windows made on the cell are separated by distance D_1 and D_2. The cell volume for D_2 is about 6.6 nL. With this arrangement, a detection limit of $3 \cdot 10^{-7}$ M (S/N=3) can be achieved for brilliant green.

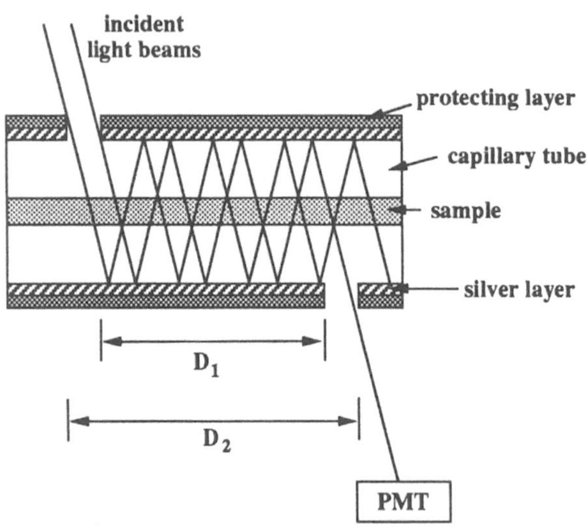

Fig. 4.9. Schematic diagram of a nanoliter-scale multireflection cell used for CE according to Ref. 109. (Reprinted with permission of the American Chemical Society)

Finally, the optical path length can be increased by directing the incident light beam along the axis of the capillary rather than perpendicular to the axis.

Such an axial detection system has been developed by Yeung and coworkers using a He-Ne laser as light source [110] as well as a conventional light source [111]. The light beams are focused into a 50-µm i.d. capillary. Light entering the capillary is transmitted by either partial or total internal reflection. A 60-fold improvement in the path length over conventional on-column UV detection was achieved. As a consequence, the limit of detection is raised by a factor of 15.

4.2.2 Fluorescence Detection

Fluorescence detection has emerged as one of the most sensitive detection modes used in CE, especially for trace analysis of derivatized amino acids, peptides and oligosaccharides. The high sensitivity of this technique is a result of the low background noise and the direct proportionality between excitation power and emission signal intensity. For fluorescein, a detection limit of approx. 10^{-8} M (S/N =2) has been found for conventional fluorescence detection in the static mode [112]. Using laser-induced fluorescence (LIF) detection, LOD's of as low as $2 \cdot 10^{-12}$ M (which corresponds to approx. 1000 analyte molecules injected onto the capillary) have been realized for fluorescein thiohydantoin derivatives of amino acids [113]. A fluorescence detector is comprised of an efficient excitation source, a detection cell and an arrangement to detect the fluorescence signal. Fluorescence detection in CE is mostly performed on-column by imaging the excitation beam into the capillary and collecting the emission at an angle perpendicular to the plane of the incident beam and the capillary on a photo-sensitive device such as a photodiode or a photomultipler tube (Fig. 4.10.).

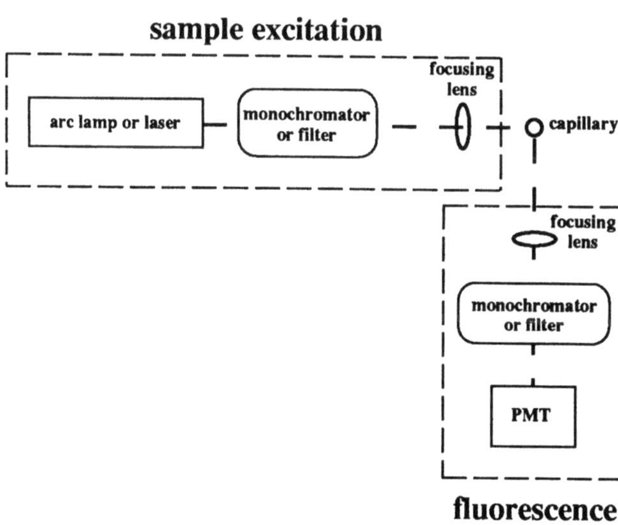

Fig. 4.10. Schematic layout of on-column fluorescence detection. (PMT: photo-multiplier tube)

4.2.4.1 Excitation Sources for Fluorescence Detection

The excitation source consists of the light source and the optics to select the spectral band and to focus the light onto the sample. The light source can be either a lamp or a laser. The most common lamps are arc lamps such as mercury-, xenon- and mercury-xenon-arc lamps. The characteristics of these lamps are dependent on the pressure of the gas and the metal steam, respectively. The light intensity is increased with rising pressure. High-pressure lamps are therefore favorized as excitation sources, although low-pressure lamps possess longer lifetimes. Arc-lamps can either have strong emission lines such as the mercury-arc lamp or a continuum background which allows flexibility in the choice of excitation wavelength to match most analytes. Xenon-arc lamps belong to this second type of lamp.

In contrast to arc lamps, lasers emit highly coherent light. The most commonly used laser sources are the helium-cadmium and the argon ion lasers. He-Cd lasers are relatively inexpensive and emit at 325 and 442 nm. Argon ion lasers can be adapted to different wavelengths in the green, blue and the UV-range of the spectrum (most often to 350 - 360, 476, 488 and 514 nm) and are available with powers ranging from a few mW to more than 10 W. The use of semiconductor lasers in the near-infrared and deep-red region with an output power of 3 - 40 mW has also been reported [114]. According to the authors, it is less expensive than a conventional light source, and the lifetime exceeds 10 000 h. While lasers provide greater sensitivities, their use is limited by the lack of spectral lines matching analyte absorbance bands. This problem can be circumvented by designing fluorophores to match the available laser lines or changing to more complex laser systems.

For a first approximation, a high laser power is desirable. Both the fluorescence and the background signals increase linearly with the laser power. Under shot-noise conditions, the noise in the background increases with the square root of laser power, and the detection limit increases with the inverse square root of laser power. Although lasers typically emit less total power than most commonly used arc lamps (50 - 200 W) nearly 100% of the laser power is available for excitation because of the extremely high spatial coherence of the laser beam. In contrast, not more than 15% of the emission of arc lamps can be used for excitation. For the same reason, lasers can be focused to a few micrometers spot size and, thus, allow a highly efficient excitation of the analyte. Therefore, they are more suitable for coupling into small capillaries than light produced by arc lamps. Capillaries having inner diameters of 10 µm or even less can be used in combination with LIF [115].

Too much laser power and irradiance, however, can lead to optical saturation or photodegradation of the analytes. Under saturation conditions, a significant fraction of the analyte is raised to the excited state. Saturation becomes important when the number of photons absorbed per second approaches the spontaneous emission rate. To eliminate saturation, the laser irradiance should be held to a value less than 10^5 W·cm^{-2} for highly fluorescent molecules. To minimize photodegradation, either a low-power beam must be employed or the illumination time of the analyte must be kept very short by using a high flow velocity. Even an argon ion laser beam of only 20 mW of power which is focused to a 1 μm spot will produce a peak irradiance of over 10^6 W·cm^{-2}. In practice, laser powers between 1 and 50 mW are commonly used.

4.2.4.2 Optical Layout of a Fluorescence Detector

The kind of excitation wavelength selection depends on the light source. Lasers usually do not need a wavelength selection because of the highly monochromatic laser beam. Nevertheless, some groups use band-pass filters to select only one of the different laser wavelengths. In the case of arc lamps, narrow band-pass interference filters are used to select the specific excitation wavelength. A grating monochromator is advantageous over filters for two reasons. First, significant spectral overlap in the transmittance of the filter results in a large background signal and thus poor detection limits. Secondly, the filter can only select one single excitation wavelength, whereas a monochromator allows selection of excitation wavelengths over a broad range. Green and Jorgenson [116] first described the use of a double monochromator as excitation wavelength selector in CE with fluorescence detection. In combination with an arc lamp, this monochromator makes it possible to select the wavelength in the range of 200 - 800 nm. Since it provides extremely low levels of stray light, it exhibits lower background levels. To prevent overheating of the monochromator the light first passes through a water-filled liquid filter to absorb a portion of the infrared radiation. Albin et al. [112] have incorporated a 75-W xenon-arc lamp fluorescence detector into a commercial available instrument (Model 270A Capillary Electrophoresis System, Applied Biosystems, San Jose, CA, USA) which is equipped with a UV absorbance detector. Excitation wavelengths in the range of 190 - 700 nm can be selected by a built-in monochromator.

To minimize scattered light, the capillary is mounted at Brewster's angle. Before the light beam enters the capillary, it has to be collimated and focused through an optical lens to a small spot. Conventionally, this is achieved by placing a plano-convex fused silica or sapphire lens at a distance of a few millimeters up to about 1 cm away from the capillary. Alternatively microscope objectives made of fused silica can be used [117]. According to the authors, these lenses are very well corrected for aberrations, are readily available and are easily manipulated and exchanged for other optics.

The fluorescence generated from the illuminated sample stream must be collected with high efficiency while scattered light reaching the detector must be minimized. The collection of emitted light is usually performed at a right angle to both the capillary and the excitation light path by using fused silica or sapphire lenses. To enhance the amount of emitted light, again, microscope objectives with high numerical apertures can be used [117]. The fraction of light collected by a lens is related to its numerical aperture and the refractive index of the surrounding medium so that lenses of very high numerical apertures are required for a high collection efficiency. The working distance which describes the distance from the exit of the lens to the sample is another characteristic which is of importance for the right choice of the objective. Only a few lenses are equipped with both long working distances and high numerical apertures. Therefore compromises have to be made when choosing the optimal objective.

Instead of using optical lenses, the emitted light can also be collected by one or two optical fibers. In the latter case, one fiber is installed on each side of the detection cell perpendicular to the incident light beam [112]. Light from these optical fibers is directed through interchangeable glass filters onto the photocell. According to the authors, this arrangement leads to a considerable reduction in scattered light. Optical fibers can also be used to transmit the light from the source to the detection cell as is done in the commercial available LIF detector for the P/ACE system from Beckman Instruments. This allows placement of the laser module in a remote location.

If a rectangular detection cell is used for fluorescence detection and the emitted light is collected at a right angle to the excitation beam, the amount of scattered light arising from total and multiple reflection is significantly reduced. In CE, however, a cylindrical flow cell is used. Four boundaries occur where reflection can take place: (i) air-to-silica, (ii) silica-to-liquid, (iii) liquid-to-silica and (iv) silica-to-air. Scattering by total and multiple reflection is greatest when light is incident on a boundary between high and low refractive indices. Two such boundaries, (ii) and (iv), cause internal reflection in a CE flow cell resulting in a high background noise. In contrast to the rectangular flow cell, the scattering of the excitation beam takes place in every direction [118]. For water, there are two major bands arising from Raman scattering, one of them at 585 nm. The excitation wavelength should be chosen so that the analyte emission wavelength lies between the Raman line at 585 nm and the excitation wavelength. To isolate spectrally the fluorescence from scattered laser light as well as from the Raman band of water at 585 nm, sharp-cut-off optical filters can be used.

The limits of detection of on-column fluorescence detectors as described above ranges from 10^{-6} - 10^{-9} M. The sensitivity of a fluorescence detector can be further improved by reducing the background signal levels. One of the most powerful approaches to reduce light scattering is the use of a post-column detection system based upon a quartz sheath-flow cuvette [117, 119]. In this design, the end of the capillary is inserted into a 250-µm square-flow chamber made from high-grade optical glass. A sheath stream, provided by a high-stability chromatographic pump, surrounds the sample stream as it exits the separation capillary. Since the sheath stream has the same composition as the carrier electrolyte, no light scattering occurs at the capillary - sample interface, minimizing the level of the background signal. With this arrangement detection limits in the zeptomole range have been obtained for fluorescein thiohydantoin derivatives of amino acids [113].

Kurosu et al. [118] have developed an immerse flow cell to reduce light scat-

cell. The space between the capillary and the flow cell is filled with a liquid having an appropriate refractive index to minimize the differences between the refractive indices of the boundary materials. Propanol has proved to be best suited for this purpose.

Another possibility to minimize scattered light is the use of a collinear arrangement instead of the orthogonal as described so far. Hernandez and coworkers have designed such a system using an epillumination fluorescence microscope. Originally equipped with a conventional light source [120], they have modified their arrangement by combining the epillumination fluorescence microscope with an argon ion laser [121, 122]. Detection limits of 10^{-13} M have been achieved for fluorescein thiocarbamyl-arginine. The CE instrument Iris 2000, which is available from Europhor Instruments, Toulouse, France, is equipped with this collinear LIF system. A schematic representation is given in Fig. 4.11. An air-cooled argon-ion tunable laser emitting at 476 or 488 nm with powers of 2 and 4 mW, respectively, is used as excitation source. The beam of coherent light is filtered through a 450 - 490 band-bass filter, reflected

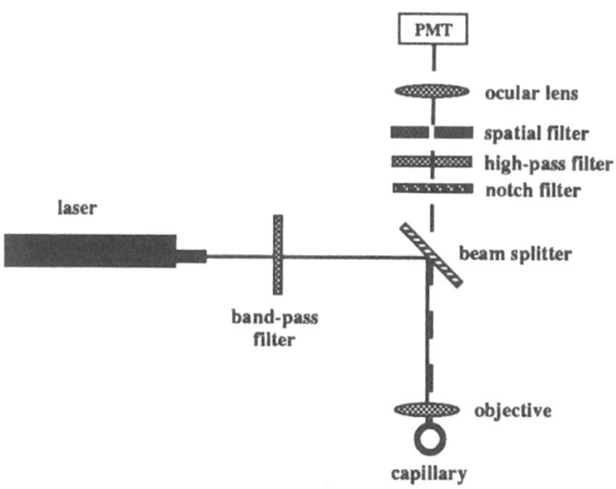

Fig. 4.11. Schematic representation of the collinear arrangement of fluorescence detection by means of an epillumination fluorescence microscope (with permission of Ref. 122)

by a 510-nm chromatic beam splitter and condensed onto the capillary by means of a 0.85-numerical aperture fluorite objective. After crossing the chromatic beam splitter, it is filtered through a notch filter centered at 492 nm (to suppress the 488 nm line that the capillary reflects toward the photocell), a 520-nm long-wave-pass filter and a spatial filter. A 10x glass ocular is placed after the spatial filter to focus the light onto the photomultiplier tube.

The collinear arrangement offers some advantages over the orthogonal. Because the microscope already has a highly precise XYZ displacement device, the alignment of the capillary can be guided visually. The directions of incident and scattered light are generally perpendicular to each other; thus the scattered light is observed in the same plane as the emitted light. Whereas it is difficult to use lenses of less than 1 mm working distance and higher numerical apertures in the orthogonal arrangement, the emitted light can easily be collected with such lenses in the collinear arrangement.

4.2.4.3 Derivatization with Fluorescent Tags

In on-column fluorescence detection the emitted fluorescence can arise from the intrinsic fluorescence of the sample or the fluorescence of labels attached to the sample components. Relatively few organic molecules exhibit intense native fluorescence. The emission characteristics of molecules are difficult to predict. Extended conjugation favors a high fluorescence quantum yield. Although most fluorescent compounds are aromatic, aromaticity is not a guarantee that a molecule will fluoresce. Furthermore, planarity and rigidity of the molecule decrease the quantum yield. Finally, fluorescence is very dependent on the sample matrix such as solvent properties, pH and sample contaminants. Only a few examples exist in literature of cases where native fluorescence can be exploited for detection. Peptides and proteins containing tryptophan can be detected by fluorescence without derivatization at an excitation wavelength of 280 nm and an emission wavelength of 305 nm [112].

In all other cases, analytes have to be labeled with a fluorescent tag. Many derivatization reagents can be used for the detection of amino acids, peptides, proteins, amino sugars and nucleic acids. Some have been especially developed to match the excitation wavelengths of the available lasers. The most important derivatization procedures are summarized in Table 4.2. Additionally, some reaction schemes are presented in Fig. 4.12.

CBQCA, NDA and OPA react with primary amines to form highly fluorescent isoindol derivatives. According to the Edman degradation, FITC reacts with primary amines like phenyl isothiocyanate under alkaline conditions to form fluorescein thiocarbamyl (FTC) derivatives. Hydrolysis of the FTC derivative with acid produces the corresponding thiohydantoin (FTH) derivative. The different reagents offer different advantages and disadvantages. FITC provides good sensitivity for primary and secondary amines, but the derivatization reaction is relatively slow and large artifact signals due to free label are observed in the electropherogram. Fluorescamine reacts very quickly with primary amines, and excess reagent is hydrolyzed to a nonfluorescent product. OPA also reacts in seconds with primary amines to form isoindole derivatives, and excess reagent is not fluorescent. The derivatives, however, are very unstable, and have to be analyzed directly after derivatization. NDA and CBQCA also react with primary amines to form isoindole derivatives, which are more stable than those obtained with OPA and show higher quantum efficiencies, even in the aqueous buffer solutions required for CE. The reaction, however, is slower than in the case of OPA. FMOC reacts with primary and secondary amines in less than 1 min. The fluorescent reagent can be extracted with pentane.

Table 4.2. Derivatization procedures for fluorescence detection in CE

Reagent	Abbreviation	Excitation Wavelength [nm]	Emission Wavelength [nm]	Applications	Ref.
3-(4-Carboxybenzoyl)-2-quinoline carboxaldehyde	CBQCA	442	550	amino acid and amino sugar analysis, peptide mapping	123,124 125
Dansyl chloride	Dns-Cl	325	600	amino acid analysis	126,127
4-Phenylspiro[furan-2(3H), 1'-phthalan]-3,3'-dione	Fluorescamine	390	450	amino acid analysis analysis of protein drug substances	112,128
9-Fluorenylmethyl-chloro-formate	FMOC	260	305	amino acid analysis	112
Fluorescein isothiocyanate	FITC	494	525	peptide, protein and DNA sequencing, amino acid analysis	113,119 121,129
JOE, TAMRA, FAM, ROX		514.5, 488	540,560,580,610	DNA sequencing	130
Naphthalene dialdehyde	NDA	440	490	amino acid analysis	131,132
o-Phthaldialdehyde	OPA	340	450	amino acid analysis	112, 133

a) CBQCA

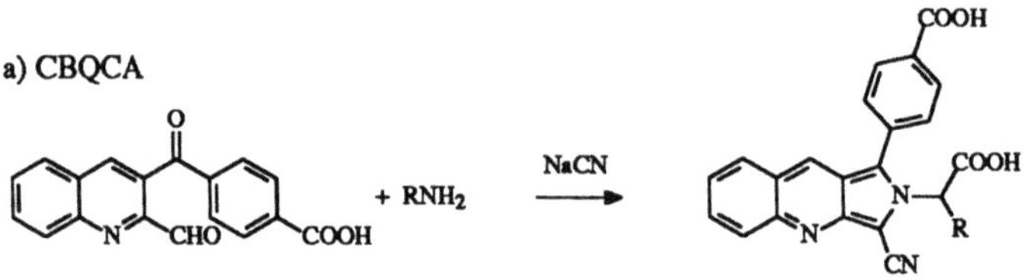

b) Dns-Cl

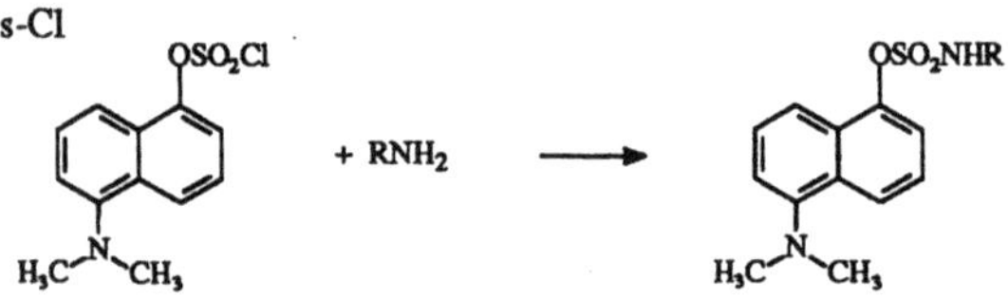

c) Fluorescamine

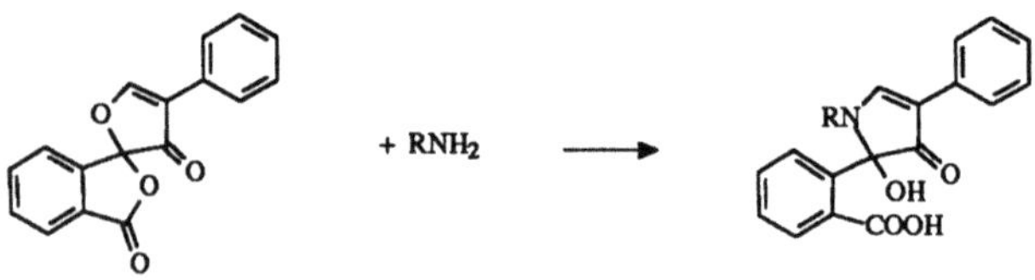

d) FMOC

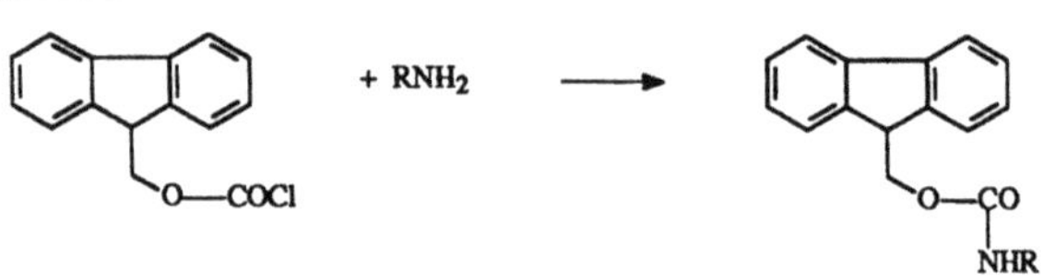

e) Fluorescein isothiocyanate

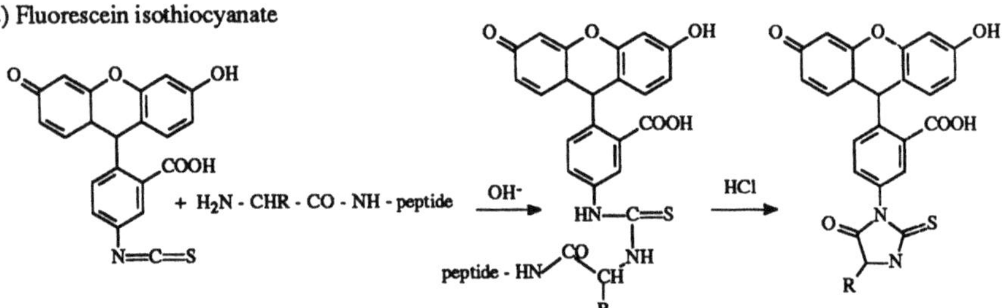

Fluoresceinthio-
hydantoin - AA

f) NDA

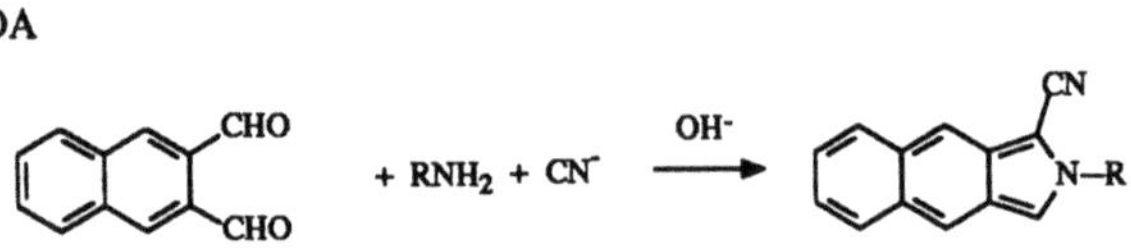

g) OPA

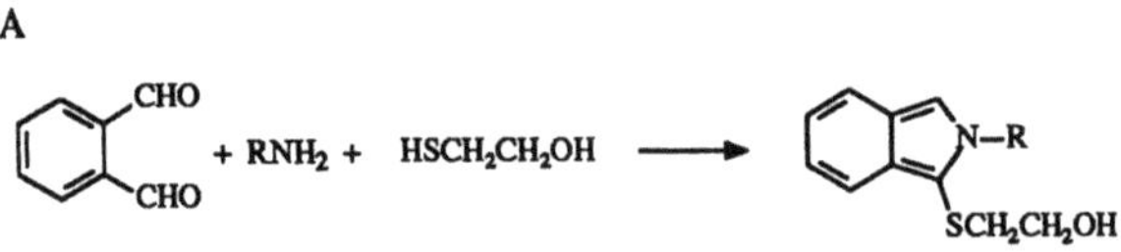

As one can see from Fig. 4.12., some of the labeled products are uncharged like NDA derivatives and some bring new charges into the molecule as in the case of fluorescamine and FITC. As a consequence, the electrophoretic behavior of the derivatives is different from that of the underivatized analytes. A separation scheme has to be developed for each derivatized analyte. Obviously, the kind of derivatization agent used is dependent on the available light source. Furthermore, the given applications are only a crude guideline, which means that one has to find the optimal tag for each sample to be analyzed. A detailed description of derivatization procedures can be found in Appendix 8.2. Further informations about the choice of derivatizing reagent in special cases can be found in chapter 7 under the substances of interest.

4.2.4.4 Pre- and Post-Column Derivatization

Derivatization can be performed in two ways. Pre-column derivatization is the most simple and most commonly used procedure. Because the sample is labeled prior to separation, no modification of the instrument is needed to incorporate the derivatization step. However, this procedure has some limitations. For example, OPA-labeled amino acids decompose over time with the rate of decomposition being amino acid-dependent. Additionally, the labeling of a given protein with more than one amine function produces a number of different labeled protein species, each with a different number of tags and thus each with a different electrophoretic mobility. Furthermore, pre-column derivatization must contend with impurities and product stability. Finally, variations in reaction times and the time between derivatization and analysis will impact the quantitative results.

Alternatively, the label can be attached to the separated components after the separation. Several designs of post-column derivatization reactors have been described. Rose and Jorgenson [134] have created a coaxial reactor consisting of two concentric fused silica capillaries allowing a sheath-flow of labeling reagent around the effluent stream (Fig. 4.13.). The reaction capillary is held in a stainless-steel tee. It has a 1-2 cm wide detection window formed by burning off the polyimide coating. The separation capillary, which has an outer diameter smaller than the inner diameter of the reaction capillary, passes through the tee and enters the reaction capillary such that the two capillaries are concentric to form the coaxial reactor. In their study, OPA has been used as labeling reagent in the separation of amine-containing compounds. The best results in terms of separation efficiency have been obtained by a combination of a 25-μm i.d./40-μm o.d. separation capillary with a 50-μm i.d. reaction capillary.

Rose [135] has reported a reactor that simplifies post-column detection by terminating the electrophoresis capillary into a static solution of *o*-phthalaldehyde reagent which acts as both the cathodic reservoir and a free-solution reactor. Zones from the capillary mix and react with the OPA reagent to produce a fluorophore that is detected just beyond the capillary tip. This system, however,

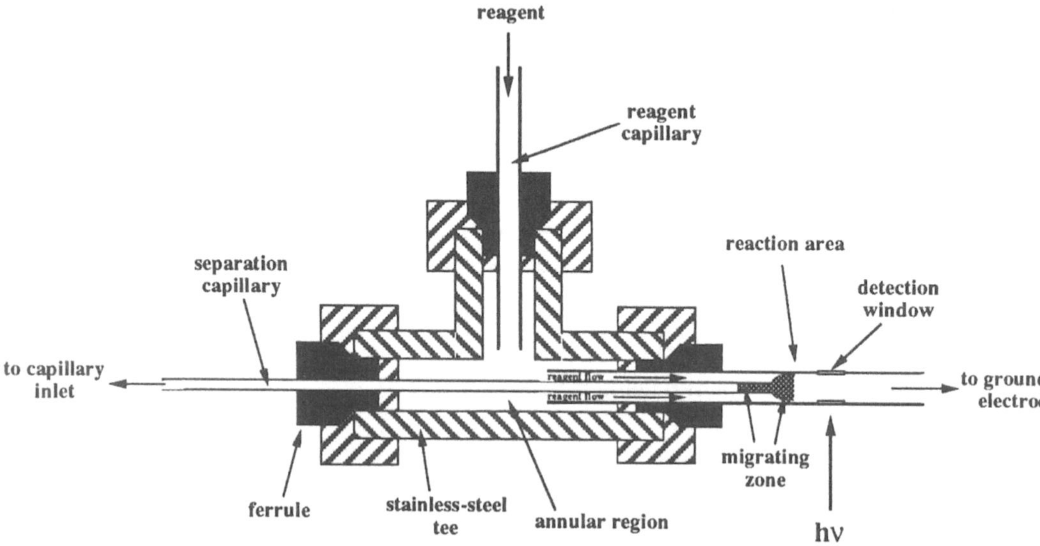

Fig. 4.13. Cross-sectional schematic of the post-column derivatization reactor of Rose and Jorgenson [134]. (Reprinted with permission of Elsevier Science Publishers)

shows significant zone broadening due to convective forces and high background fluorescence due to the large illumination volume. Tsuda et al. [136] have designed a post-column reactor consisting of two mixing parts, a four-way and a three-way connector, and three pumps. This arrangement builds a closed system to balance the pressure which is generated at the outlet side of the capillary due to the mixing of the reagents with the effluent.

Albin et al. [112] have incorporated a four-way polyacrylic tee as the derivatization reactor into a commercial available instrument (Model 270A Capillary Electrophoresis System, Applied Biosystems, San Jose, CA, USA). Via the four-way tee, the end of the separation capillary (50 μm i.d., 375 μm o.d.) is connected to a larger i.d. capillary with a 10 - 50-μm distance between them resulting in a "gap junction" (Fig. 4.14.). The larger i.d. capillary is connected with the detection cell. The labeling reagent is introduced through the third vent and exits to waste through the fourth vent of the tee. Solute zones migrating through the gap are confined by the electric field lines extending across the gap. The introduction of buffer fluid containing the derivatizing reagent takes place due to the EOF which is greater in the larger capillary than in the smaller separation capillary.

Post-column derivatization reagents should not contribute to the fluorescence background signal, produce strong fluorescent derivatives and react rapidly, and they should be stable. According to the authors, OPA in particular meets all of

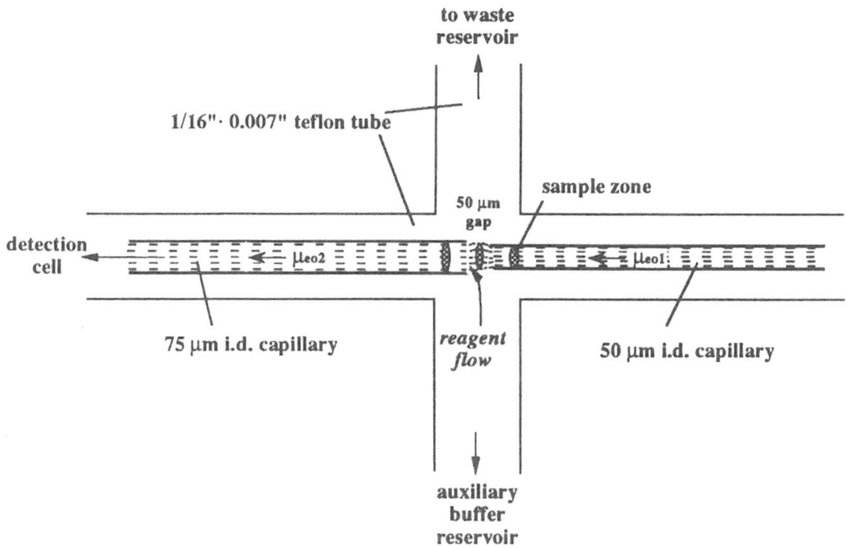

Fig. 4.14. Principle of function of the gap junction reactor built by two capillaries with different inner diameters in a four-way Teflon tee; $\mu_{eo2} > \mu_{eo1}$ (with permission of Ref. 112)

these criteria. Due to the fixed nature of reaction conditions, post-column derivatization provides reproducible labeling procedures. The main disadvantage of post-column derivatization, however, is that the CE instrument has to be modified to incorporate the derivatization reaction cell between the capillary and the detector.

4.2.5 Electrochemical Detection

Electrochemical methods which are used as detectors for CE include potentiometric, conductometric and amperometric detection. Potentiometric detection for zone electrophoresis was introduced in 1974 by Virtanen for 200-µm i.d. capillaries [4]. The electrode system consists of an Ag/AgCl-coated platinum wire electrode encased in a glass capillary. One end of the capillary is inserted into the cathodic buffer reservoir. Potential changes caused by eluting analyte zones are monitored via a Wheatstone bridge. Zones of small inorganic ions such as K^+, Na^+ and Li^+ have been detected with this system [4]. Potentiometric detection has not found widespread application in modern CE because of its low sensitivity. In this section, we will therefore focus on conductometric and amperometric detection.

4.2.5.1 Conductometric Detection

Conductometric detection or potential gradient detection was first used in CZE by Mikkers et al. [5] in 200-μm i.d. PTFE capillaries. It is accomplished by measuring the potential between two electrodes while passing through a small constant current. Analyte zones are detected because of their different conductivity to that of the background electrolyte. Since conductometric detection belongs to the unselective detection modes, it is universally applicable. It is, however, limited to systems showing sufficient conductivity difference between the electrolyte solution and the analyte zones. The most widespread application field of conductometric detection is the analysis of small inorganic ions and carboxylic acids. The analyte peaks can be "positive", if they project above the baseline, or "negative", if they project below it, depending on the mobility of the buffer compound with the same sign as the analytes. If the mobility of the buffer compound is higher than those of the analytes, "negative" solute peaks will result.

Unlike in optical detection modes, the response of the conductometric detector is directly related to the effective mobility of the species being detected [137]. Since the migration time of a species is also related to the mobility, the peak area correlates linearly with the migration time. This relationship allows the response from an internal standard to be used to calibrate the response of all analytes present in the mixture on an absolute basis.

One problem arising in conductometric detection is the fact that the concentration of the background electrolyte must be high relative to the concentration of the sample to minimize electrophoretic dispersion (see also Sect. 3.1.4). This condition results in a loss in sensitivity due to the elevated background conductivity. If, however, the ratio of the analyte to background electrolyte concentration is kept constant and the background electrolyte concentration is reduced, an increase in sensitivity while maintaining resolution is achieved [138]. A fourfold decrease in background electrolyte concentration as well as in sample concentration results in a 12-fold sensitivity increase.

Three possible designs for a conductometric detector for CE are on-column, off-column and end-column structures. On-column conductometric detection has the advantage that no loss in efficiency occurs. Foret et al. [104] have designed an on-column conductometric detection cell which is simultaneously used with an on-column UV absorbance detector. The cell consists of platinum wires protruding from the wall of the separation capillary and, thus, staying in direct contact with the electrolyte solution. The LOD for picric acid is 10^{-5} M (S/N=2), the linear range extends from 10^{-5} M to 10^{-3} M.

Another on-column conductometric cell has been constructed by Huang et al. [139] by fixing two 25-μm o.d. platinum electrodes in diametrically opposite holes in 50-μm or 75-μm i.d. fused silica capillaries. The 40-μm i.d. holes are made with a computer-controlled CO_2 laser. The electrodes are placed exactly opposite to each other in order to minimize background noise associated with the high electrical field used for separation. In order to hold the electrodes in place, polyethylene glycol (PEG), heated to liquid, is applied to the area surrounding the electrodes. After removing the solidified PEG from the outside surface of the capillary, an epoxy is used to seal permanently the electrodes in the capillary. Wires are soldered to the platinum electrodes, and the entire cell is sealed in a Plexiglas jacket. The detector end of the capillary is made the cathode. The detector measures a constant value for conductivity of the buffer, until an analyte zone migrates between the detector electrodes and changes the conductivity. The buffer conductivity and the changes are transmitted to a data acquisition system. A schematic diagram of the whole CE-conductometric cell system is shown in Fig. 4.15. A detection limit of 10^{-7} M (S/N=2) for Li^+ has been found with this system [140].

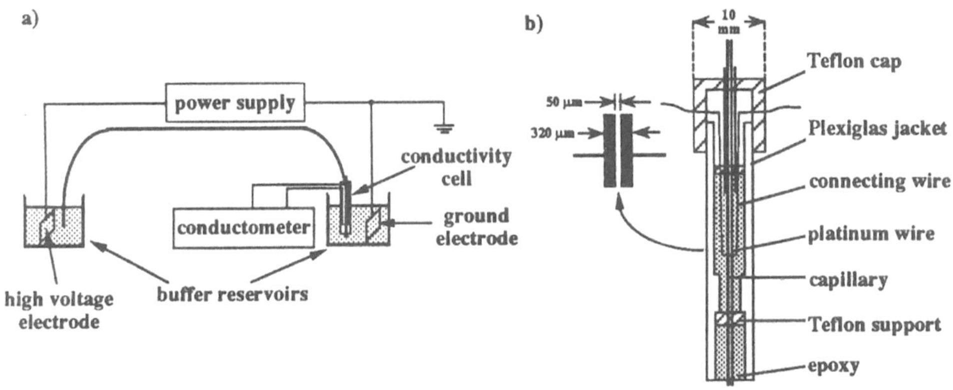

Fig. 4.15. (**a**) Schematic diagram of a CE-conductometric cell system according to Ref. 140. (**b**) Extended view of the conductometric cell (with permission of Elesevier Science Publishers)

Although on-column conductometric detection works well, the question arises of how to produce such structures reliably and inexpensively. Additionally, if the sensing electrode in the on-column mode is, for instance, 5 cm away from the capillary outlet, the sensing electrode is at 1500 V with respect to ground, assuming a potential of 300 $V \cdot cm^{-1}$. Therefore, a specially designed, isolated conductivity meter has to be used. Whereas systems for off-column conductometric detection are not described in recent literature, a so-called "end-column detector" has been developed [141, 142] which has the advantage that it can be constructed directly at the outlet of the capillary of a commercial CE separation system, for instance to the CE system P/ACE 2000 of Beckman Instruments. This arrangement allows for both UV absorbance and conductometric detection. The use of both detectors provides not only greater analysis power, particularly for ions that cannot be detected by UV absorbance, but also a calibration of the UV absorbance detector.

Figure 4.16. shows a CE device with end-column conductometric detection. At a distance of 7 mm from the outlet of a 50-µm or 75-µm i.d., 360-µm o.d.

fused silica capillary, a 40-μm hole is drilled through the wall using a CO_2 laser. The capillary serves as the separation capillary, and the hole provides the outlet for the eluent. A platinum or stainless steel 50-μm wire, which serves as the sensing electrode, is inserted along the length of the capillary and sealed with epoxy at the capillary outlet. A fine, insulated lead wire is connected to the sensing electrode. The conductivity is measured between the sensing and the ground electrode in the outlet buffer reservoir. If the detection cell is used in combination with the P/ACE unit, the capillary is prepared exactly in the same way as described here and then placed in the capillary cartridge. The efficiency loss of this end-column detector compared with on-column detection has been found to be about 25%. Because only a few volts difference occurs between the grounding and the sensing electrode, capacitors can be used to couple the detector to an AC conductance meter.

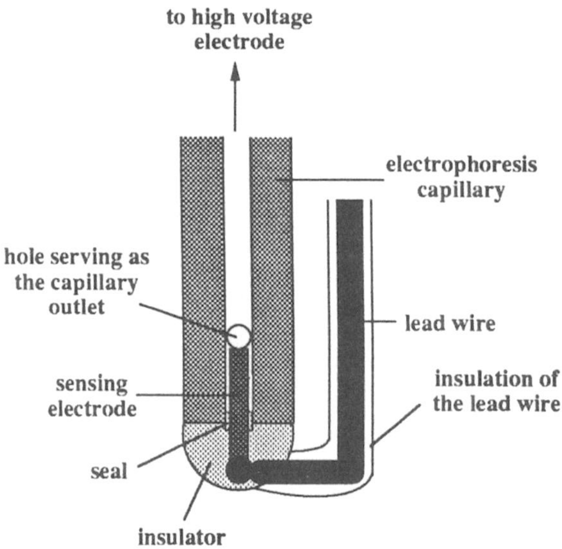

Fig. 4.16. Schematic drawing of the CE separation device with end-column conductometric detection according to Ref. 142. (Reprinted with permission of the American Chemical Society)

4.2.5.2 Amperometric Detection

Amperometric detection has been shown to be among the more sensitive detectors available for capillary electrophoresis. It affords sensitive detection of many biologically important molecules without derivatization. In addition to its extreme sensitivity, the amperometric detector is quite selective which means that only compounds electroactive at a given potential are detected. This feature is advantageous in those cases where trace components have to be detected in a complex matrix such as body fluids or plant materials. A large number of important compounds of biomedical and environmental interest are electroactive and can therefore be studied by CE combined with amperometric detection.

In amperometry, current is measured at a working electrode as analytes are oxidized or reduced. The working electrode is held at a fixed applied potential, relative to a reference electrode, and current is recorded as a function of time. The baseline of the electropherogram consists of the background current which depends on the buffer, on the potential and on the electrophoretic current. For analytical work, 3 V is available between those potentials at which the medium itself begins to oxidize and reduce. For a given working electrode and a given electrolyte system, the practical working range or the "potential window" is even lower and may be less than 1.5 V, typically between 0 and 1 V.

The electrochemical properties of a molecule are determined by the nature of the electroactive functional groups. Electroactive substances are either oxidizable, reducible, or both. Phenols, aromatic amines, quinones, imines and nitro compounds are electroactive. The applicability of amperometry to a given analysis problem depends on the voltammetric characteristics of the molecules of interest in a suitable electrolyte and at a suitable electrode surface. It is therefore important to evaluate the voltammetric behavior by cyclic voltammetry which is convenient for quickly assessing the thermodynamic and kinetic properties of the electrode reaction and the stability of the initial product. A cyclic voltammogram is obtained by scanning the voltage with a high velocity, for instance 100 $mV{\cdot}s^{-1}$, and measuring the resulting current. The potential at which the maximum current and minimum background signal is obtained serves as the optimum electrode potential for the CE experiments.

Often, the behavior of different molecules bearing the same electroactive function will be quite similar, and therefore voltammetric studies are not always necessary prior to setting up an electrochemical detector. When determining whether or not a particular compound can be successfully analyzed, it is not sufficient to know that the compound can react electrochemically. The type of electrode surface, nature of the solvent, and relative ease of oxidation or reduction must be carefully considered. The most important consideration for trace analysis is the relative contribution which can be expected from the background current under certain conditions. In general, detection limits are lower for more easily oxidizable (or reducible) substances, since these can be determined at potentials well inside the available potential window. Amperometric detection is mostly performed for the detection of oxidizable rather than reducing analytes. One reason for this fact is that disturbing reduction of dissolved oxygen, trace metal ions and hydrogen ions are difficult to eliminate.

Amperometric detection has been carried out in capillaries ranging from 75 µm i.d down to 2 µm i.d. The signal in amperometry is proportional to the efficiency of the redox reaction. Oxidation efficiencies in amperometry are typically less than 10%. In CE, however, higher efficiencies are observed because of the electrochemical cell geometry. As the capillary diameter decreases, the annular flow region between the surface of the working electrode and the capillary wall also decreases resulting in enhanced coulometric efficiency and, hence, increased sensitivity. A 5-µm o.d. carbon fiber, for instance, inserted into a 12.7-µm i.d. capillary produces an annular flow region thickness of about 3.8 µm

[143]. This leads to a 50% increase in electrochemical reaction of the analyte at the electrode surface. The greater coulometric efficiency achieved with small-bore capillaries allows for increased detection sensitivity and therefore lower detection limits. A detection limit of $8.5 \cdot 10^{-9}$ M corresponding to 6 amol of injected material has been realized for serotonin with the system described above.

As already mentioned in the previous section, problems arise when electrochemical detection in CE is used, due to electrical interference by the applied high voltage. Amperometric detection is even more sensitive to the electrophoretic current than conductometric detection. The current through the capillary is usually several orders of magnitude higher than the faradaic current measured at the sensing electrode. Only if the electrophoretic current is very small (1 - 15 nA), is electrical interference minimized. This can be realized by performing CE in narrow bores of 5 µm i.d. [141]. Such a CE system for amperometric detection consists of a 5-µm i.d. fused silica capillary and a 10-µm i.d. cylindrical carbon fiber microelectrode as working electrode. Since the electrode possesses a larger diameter than the capillary, it is aligned with the bore of the capillary and positioned up against but not into the capillary, thereby creating a thin-layer cell at the capillary outlet. Good oxidation efficiency is achieved because the diameter of the electrode is twice the i.d. of the electrophoresis capillary. A detection limit of 56 amol for catechol and a linear range of $5 \cdot 10^{-7}$ - $2.5 \cdot 10^{-5}$ M have been obtained with this end-column detection system.

If capillaries of greater diameters have to be used, the electrical circuit of the CE system has to be completed prior to the capillary outlet. To achieve this, Wallingford and Ewing [144] have developed a system utilizing a porous glass junction created in the capillary near the cathodic end (Fig. 4.17.). This junction

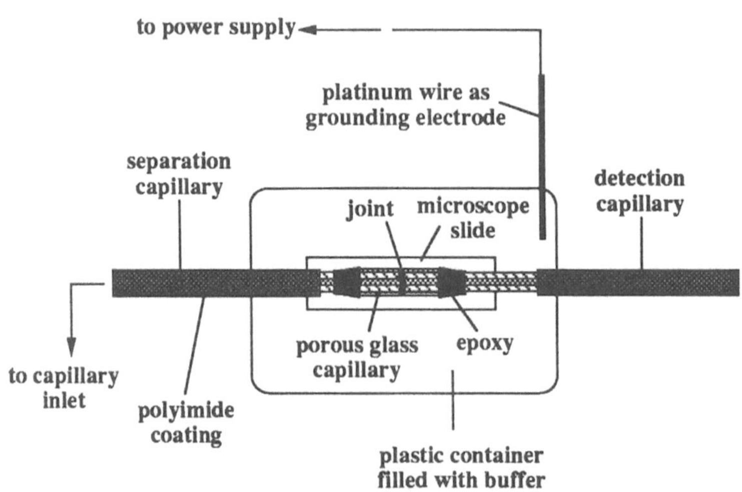

Fig. 4.17. Schematic representation of the porous glass junction according to Ref. 144. (Reprinted with permission of the American Chemical Society)

is prepared by breaking the capillary tubing into two segments and placing both inside a custom-made porous glass sleeve. The porous joint rather than the end of the capillary is submersed into the outlet buffer reservoir along with the cathode (ground electrode). The applied potential is dropped across the electrophoresis capillary prior to the porous junction, and the resulting EOF pumps the solution through the short section of detection capillary after the joint. The pores within the porous glass joint are large enough to allow permeation of small electrolyte ions, thereby allowing current to flow upon application of a potential to the system. Larger analyte and solvent molecules are excluded by their size from permeating through the pores. The porous glass capillaries are extremely fragile. Therefore the coupler assembly has to be immobilized on a microscope slide to allow subsequent handling and manipulation. As long as the coupler assembly is immersed in buffer, many runs can be performed without deterioration. Efficiency loss due to the use of the porous glass joint lies in the range of 0 - 46%. It depends on the length of the detection capillary and the precision of the bore alignment. The authors have found that detection capillaries of less than 2.5 cm in length produce minimal band broadening.

More recently, Huang and Zare [145] designed an on-column frit which also serves to isolate the final section of the capillary column from the applied electrical field (Fig. 4.18.). A 40-µm in diameter hole is made into the wall of a 75-µm i.d. fused silica capillary by means or a CO_2 laser, after the polyimide coating has been removed. In order to cover the hole, a tungsten wire is placed in the capillary. Then a 4:1 mixture of solder glass and powdered fused silica with particle size between 1 and 10 µm is added to amyl acetate to make a slurry, which is used to paste over the hole in the capillary. Once the hole structure has set, which is aided by gentle heating, the wire is removed. The whole structure is sintered by heating the mixture to about 1000 °C for 30 s. To reduce the fra-

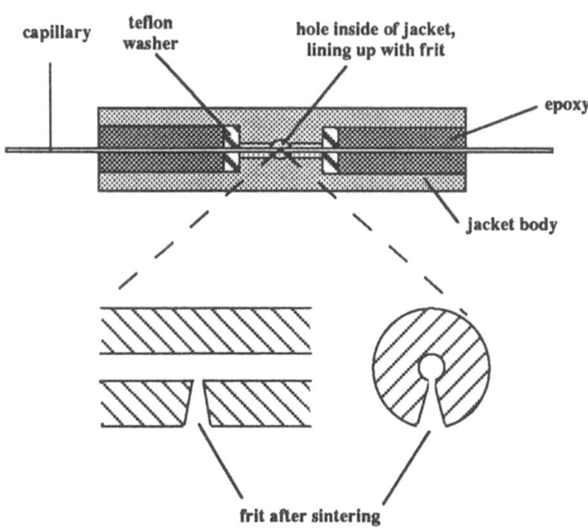

Fig. 4.18. On-column capillary frit according to Ref. 145. (Reprinted with permission of the American Chemical Society)

gility of the frit, it is placed in a protective jacket. A hole in the jacket allows it to be filled with electrolyte so that the frit structure becomes part of the electrical circuit. Once an acceptable frit structure has been fabricated, no degradation in performance is observed during several months of operation. The frit is placed at a distance of 0.5 - 1.5 cm from the outlet.

Another structure intended to separate the detection zone from the high voltage zone of the capillary has been developed by O'Shea et al. [146]. Polyimide coating is scored with a capillary cutter ca. 1.5 cm from the end of a fused silica capillary (50 µm i.d., 360 µm o.d.) . A 1-cm length of Nafion tubing (0.33 mm i.d., 0.51 mm o.d.) is carefully threaded over the score mark. Both ends of the Nafion tubing are sealed to the capillary using an epoxy resin and cured overnight. After curing, the capillary is fractured by gentle pressure to either end of the Nafion tubing. The Nafion tube holds the capillary joint securely in place and insures correct alignment. For additional support, the joint is epoxied to a small section of glass. According to the authors, the Nafion joint is easily constructed with a 100% success rate and is extremely durable with no adverse effects on the CE separation.

In all these off-column detection modes, detection is performed at the end of the detection capillary and is carried out by using a two-electrode amperometric format. The low currents measured require that the detection end of the system is housed in a Faraday cage in order to minimize the effects of external noise sources. Typically, cylindrical carbon fiber electrodes of 5 - 10 µm diameter protruding 0.2 - 0.5 mm from drawn glass capillaries are used. They are inserted into the end of the detection capillary by means of a micromanipulator and a mi-

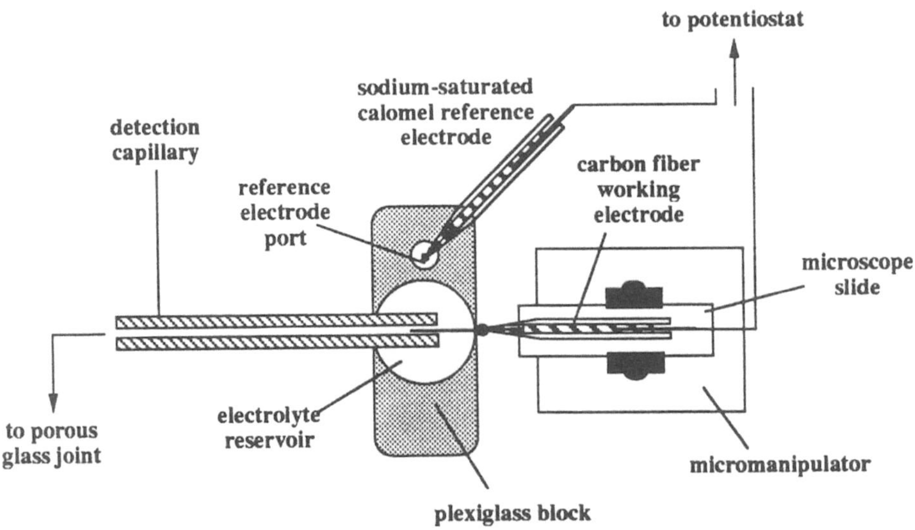

Fig. 4.19. Top view of the amperometric detection system according to Ref. 147. (Reprinted with permission of the American Chemical Society)

croscope. A top view of the detection system is shown in Fig. 4.19. The end of the detection capillary is positioned into the center of a ≈ 0.65-cm diameter cell in a Plexiglas block which is filled with an electrolyte solution. The end of the microelectrode is manipulated through the opposite slot in the Plexiglas block and into the end of the detection capillary with a micromanipulator while viewing under a microscope. A sodium-saturated calomel reference electrode is placed into a second reservoir in the Plexiglas block which is connected to the cell. A potential is applied between the working and the reference electrode with a mercury battery and a voltage divider.

CE with amperometric detection is especially useful in those cases where extreme low sample volumes are available. One impressive example which has been reported by Olefirowicz and Ewing is the analysis of single cell cytoplasm [96, 148]. Because of the high sensitivity even in small-bore capillaries, sample volumes as low as 270 fL can be injected. The smallest i.d. of the electrophoresis capillary that can be used with amperometric detection is limited by the size of the detection electrode. To perform amperometric detection in 2 - 5-µm-i.d. capillaries, electrochemically and flame-etched carbon fiber electrodes have to be employed. Cylindrical carbon fiber electrodes can be electrochemically etched from 5 µm to 2 µm diameter by cycling the electrode potential between ±1.75 V in a 3 M KOH solution. Flame etched electrodes are constructed by placing 5-µm diameter carbon fiber electrodes into a methane-air flame for a few seconds. The electrodes have a conical tip less than 1 µm in diameter. Consequently, electrochemically etched carbon fibers are used in combination with 5 µm i.d. capillaries, whereas one has to employ flame etched carbon fibers for 2-µm i.d. capillaries. For a 50-pL injection volume of catechol, a LOD of $6 \cdot 10^{-8}$ M has been achieved in a 50-µm i.d. capillary. This corresponds to 3.3 amol of catechol.

Carbon fiber electrodes are only useful in those cases where easily oxidized analytes such as catechols and indoles are to be detected. Engstrom-Silverman and Ewing [149] have reported the use of a copper wire amperometric detector inserted in the end of the detection capillary of a two-segmented capillary system. Amperometric detection on copper/copper oxide electrodes is based on complexation between solutes and Cu^{2+} ions. These ions are present in the porous passive bilayer produced on the copper surface at mildly positive potentials in weakly acidic or alkaline buffer solutions. The amperometric response of copper electrodes results from the interaction of complexing agents (the analytes) with Cu^{2+} ions contained in the porous outer layer. This leads to an enhanced solubility of the film and results in an increase in the otherwise steady-state anodic current. Using this system, nonelectroactive native amino acids and dipeptides have been detected.

Special care has to be taken, if amperometric detection is used in the presence of surfactants as in MEKC [147]. Surfactant molecules or micelles adsorb onto the surface of the carbon fiber electrode resulting in an inhibition of the oxidation of hydrophobic analytes by a selectively permeable barrier. Thus, detection limits in MEKC are higher than in CZE.

4.2.6 Indirect Detection

4.2.6.1 General Aspects

So far, indirect detection has been reported for UV-VIS absorbance, fluorescence and amperometry. There are several reasons for the investigation of different indirect detection modes for CE:

> Indirect detection can be used as a universal detection mode for analytes that cannot be visualized without the need for time consuming pre-column derivatization or experimentally complicated post-column derivatization procedures.

> There is no universal detector available working for capillaries smaller than 25 µm in i.d. and for analyte concentrations below 10^{-6} M.

> Indirect detection can be performed using the same instrumentation as for the corresponding direct detection mode. In an ideal case, all analytes can be detected with the same instrumentation simply by varying the electrolyte composition.

There are, however, some restrictions that have to be taken into account when working out an indirect detection scheme for a special analytical problem. For indirect fluorescence and indirect absorbance detection, the major factors determining the detection limits are the concentration of visualization agent, C_a, dynamic reserve, D_r, and displacement ratio, R_d. The limit of detection can be estimated from the following relationship [150]:

$$LOD = \frac{C_a}{R_d \cdot D_r} \qquad (4\text{-}10)$$

The visualization agent is identical with the coion of the background electrolyte having the same charge as the analyte. D_r describes the ability to measure a small change on top of a large signal and is equal to the signal-to-noise ratio of the background signal. Typically, the value of D_r lies between 100 and 1000. The displacement ratio is defined as the number of molecules of the visualization agent displaced by each analyte molecule. Consequently, a value of $R_d = 1$ is desirable. Because all analytes separated by CZE are charged, displacement is guaranteed in order to maintain local charge neutrality. The exact value for R_d can be derived from the Kohlrausch function and the effective mobilities of the analyte and the coion. R_d can deviate considerably from 1 and will only become 1 for analytes having the same mobility as the coion.

From Eq. 4-10 it becomes obvious that the LOD can be improved by keeping C_a as low as possible and D_r as high as possible. The three parameters, however, are not independent of each other. By lowering C_a, D_r will also be decreased. Equilibrium and surface effects can further reduce the sensitivity. Therefore, one has not only to optimize the detector to provide a large dynamic reserve, but also the separation process with respect to the concentration and mobility of the coion. Figure 4.20. shows a typical electropherogram as obtained by indirect detection in comparison with a conventional electropherogram. Substance 1 has a higher mobility than the coion of the background electrolyte resulting in a fronting peak, if the concentration of substance 1 is higher than 0.01 times the concentration of the coion. If the analyte ion has a lower mobility than the coion such as substance 3, a tailing peak is observed if the concentration of substance 2 exceeds 0.01 times the concentration of the coion. If the analyte and the coion have similar mobilities as in the case of substance 2, electrophoretic dispersion becomes negligible (see also Sect. 3.1.4).

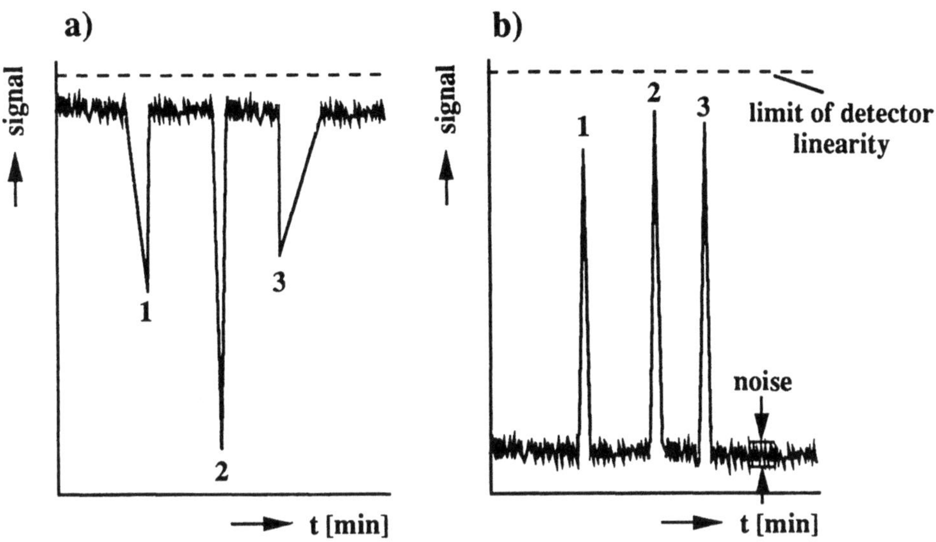

Fig. 4.20. Comparison of indirect (a) and direct (b) detection in CE. For details see text

Foret et al. [151] were the first to investigate the relationship between electrophoretic dispersion and indirect detection. Although this work deals with indirect UV detection, the rules regarding the choice of the electrolyte system are also valid for other indirect detection modes. Electrophoretic dispersion is normally suppressed by keeping the concentration of the analyte two orders of magnitude lower than that of the coion of the background electrolyte. If indirect detection is used and the concentration of the analyte ion is 100 times lower than that of the coion, the decrease of the buffer absorbance or the fluorescence signal due to migration of an analyte zone through the detection cell can be too low for practical use (i.e. < 0.001 AU). Therefore, it is better to suppress electrophoretic dispersion by choosing a coion with a mobility close to those of the sample components. In this case, the concentration of the sample component in the migrating zone can be high, while electrophoretic dispersion is still negligible. This method of minimizing electrophoretic dispersion has the additional advantage that the LOD is increased due to an increase in R_d according to Eq. 4-10.

Another important parameter which influences the sensitivity in indirect detection is the pH value of the background electrolyte. For strong electrolytes like inorganic cations and anions, the charge displacement ratio is independent of the pH. If, however, weak electrolytes have to be analyzed by indirect detection, the pH chosen must be low or high enough in order to guarantee a substantial amount of analyte in the ionized form. An extreme example is the analysis of sugars via CZE/indirect UV detection [152]. Sugars are only very weak acids. Therefore, the pH of the background electrolyte has to be adjusted to 12 to ionize the sugar molecules. At such a high pH, however, the concentration of OH^- is no longer negligible relative to the concentration of the chromophore. This results in a decrease in the charge displacement ratio, which can be described by the following formula:

$$R_d = \frac{\alpha \cdot [\text{sugar}]}{[C_a] + [OH^-]} \qquad (4\text{-}11)$$

where $\alpha \cdot$ [sugar] is the amount of ionized sugar molecules. At constant sugar and chromophore concentration, α and $[OH^-]$ are competing functions of pH. Thus, R_d goes through a maximum when plotted versus pH. The pH at the maximum is the most sensitive pH for detection.

4.2.6.2 Indirect Absorbance Detection

Indirect absorbance or UV detection was first reported in 1987 [153]. Today, it has become the detection method of choice for the analysis of small ions by CZE (see also Sect. 7.1). This is due to the fact that this group of substances possesses only very low absorptivities and cannot be detected by direct absorbance detection. Thus, indirect UV detection offers an exciting alternative to conductometric detection. LOD's of 10^{-5} M can be achieved. Nevertheless, the linear range of indirect UV detection is generally not higher than two orders of magnitude. Additionally, one has to be aware that the indirect detection system is much more unstable than direct UV detection. Large baseline drifts and disturbances can occur due to the low electrolyte concentration.

The following rules can be regarded as a guideline for the successful performance of indirect UV detection:

> Choose an absorbance of the visualizing coion which is close to the upper limit of detector linearity (≈ 0.1 AU).

> The mobility of the coion should be close to the mobilities of the pair of substances in the sample which are most difficult to separate.

- The concentration of the background electrolyte should be as low as possible. Use therefore low-concentration background electrolytes containing a coion with a high absorptivity at a given detection wavelength.
- In order to minimize Joule heating select a counterion of low mobility.
- Because significant UV absorbance of an analyte ion decreases the dynamic reserve significantly, the coion should provide a broad UV absorbance in order to be able to choose a detection wavelength where all sample components possess a very low absorptivity.

These rules show clearly that the choice of visualizing agent is dependent on the sample composition. Jandik and Jones have evaluated the use of different electrolyte systems for indirect UV absorbance of inorganic anions [154]. They have found chromate to be best suited for the analysis of highly mobile inorganic anions, because it provides suitable UV absorbance in a wide range of wavelengths, while at the same time matching the mobility of F^-, CO_3^{2-}, Cl^-, NO_2^-, NO_3^-, Br^-, PO_4^{3-} and SO_4^{2-} more closely than, for instance, benzoic acid used in previous investigations. Less mobile anions like carboxylic acids or alkylsulphonates, however, produce highly asymmetric peaks when using chromate because of the higher mobility difference. Aromatic carboxylates, e.g. *o*-phthalate, salicylate, benzoate and p-hydroxybenzoate have proven to be useful for indirect detection of these less mobile anions. In Table 4.3., some selected visualization agents which have been used so far for indirect UV detection are summarized. For further informations about the separation of small ions see chapter 7.1.

4.2.6.3 Indirect Fluorescence Detection

Although LIF is among the most sensitive detector for CE, the technique requires derivatization for most analytes, which is difficult to perform for ultramicro sample amounts at low concentrations. An alternative approach to derivatization of non-fluorescent analytes is indirect fluorescence detection. A charged fluorophore is used as the coion of the background electrolyte having the same charge as the analyte. So far, salicylate [162, 163] and coumarin [164] have been investigated for negatively charged analytes and quinine [165] for positively charged analytes. Typically, fluorophore concentrations in the range of 0.1 - 2.0 mM are used. One problem arising with indirect fluorescence detection is the need for a very stable laser to increase the stability of the background fluorescence signal and, thus, the dynamic reserve. For this purpose, the laser has to be stabilized with an external power stabilizer [163]. LOD's in the 10^{-7} M range, corresponding to $5 \cdot 10^{-17}$ mol injected onto 20 μm columns have been realized.

Indirect fluorescence detection can also be employed in combination with micellar electrokinetic chromatography (MEKC) [166]. Although the aforementioned charge displacement mode cannot be used to detect neutral analytes, the

Table 4.3. Some selected visualization agents for indirect absorbance detection

Visualization Agent	λ [nm]	Carrier Electrolyte	Application	Ref.
benzoate	254	20 mM benzoic acid, titrated with histidine to pH 6.2; + 0.1% Triton-X 100 as EOF modifier	model anions (inorganic anions and carboxylates), $c = 2 \cdot 10^{-5}$ M	151
benzyltrimethyl-ammonium	254	6 mM benzyltrimethylammonium chloride, 10 mM crown ether (18C6H_4), 5 mM Tris, titrated with citric acid to pH 2.5	chiral separation of aliphatic amino acids	155
chromate	254	5 mM sodium chromate, titrated with H_2SO_4 to pH 8.0; + 0.5 mM NICE-Pak OFM Anion-BT as EOF modifier	highly mobile inorganic anions, c = 1 ppm	156
creatinine	220	30 mM creatinine, 30 mM acetic acid, pH 4.8 in a polyacrylamide coated capillary	metal cations, c = 1mM	157
creatinine	220	30 mM creatinine, 30 mM acetic acid, pH 4.8, 4 mM hydroxy isobutyric acid in a polyacryl-amide coated capillary	rare earth metals, c = 0.5 mM	157
imidazole	214	3 - 5 mM imidazole, pH 4.0 - 6.0	alkali and alkaline earth metal cat-ions, amines and amino alcohols	158
phthalate	254	5 mM phthalate, pH 5.6; + 0.5 mM OFM Anion-BT as EOF modifier	short chain carboxylates	159
salicylate	234	5 - 10 mM sodium salicylate, titrated with NaOH to pH 11.0	amino acids, c = 0.05 - 1 mM	160
sorbate	254	2 - 20 mM sorbic acid, adjusted to pH 12.1 with 0.25 mM NaOH	monosaccharides, c = 1.5 - 12.5 mM	152
sorbate	254	7 mM sorbic acid, titrated with histidine to pH 6.2	model anions (inorganic anions and carboxylates), $c = 2 \cdot 10^{-5}$ M	151
veronal (5,5-diethyl-barbituric acid)	240	6 - 12 mM veronal, pH 8.6	C_2 - C_{12} sodium alkylsulphate surfactants, c = 0.1 mM	161

analyte perturbs the partitioning of the fluorophore between the background electrolyte and the micelles. This results in an only partial displacement of the fluorophore in the micelle by the analyte. On the other hand, fluorescence intensities of most fluorophores are significantly enhanced in micellar solutions. Therefore, it is possible to detect neutral analytes separated by MEKC with sensitivities only a little bit lower than for CZE. The separation buffer consists of sodium dodecyl sulphate (SDS) as the micelle forming surfactant and quinine sulphate as the fluorophore. The enhanced fluorescence of quinine in presence of the micelles leads to a large background signal. When an analyte molecule interacts with the micelle, the fluorophore-micelle complex is perturbed, resulting in a decrease in the fluorescence signal due to a combination of displacement and reduction of the quantum efficiency. For a separation of a mixture of alcohols, a LOD of 10^{-7} M has been obtained.

4.2.6.4 Indirect Amperometric Detection

Indirect amperometric detection has been introduced by Olefirowicz and Ewing [167]. A cationic electrophore, 3,4-dihydroybenzylamine (DHBA), is added to the electrophoretic buffer. A constant background current is produced by continuous oxidization of DHBA at the carbon fiber working electrode, which is held at +0.7 V versus a sodium-saturated calomel electrode. Zones of non-electroactive cations displace the cationic DHBA during migration to preserve electroneutrality in the zone. Thus, a decrease in the current is observed as a negative peak. Indirect amperometric detection of cationic amino acids and dipeptides has been accomplished with 26-µm i.d. and 9-µm i.d. capillaries. Additionally, it is shown that simultaneous detection of both electroactive and non-electroactive species is possible by indirect amperometry.

4.2.7 Other Optical Laser-Induced Detection Modes

Besides LIF, several other optical detection modes using lasers as light source can be found in literature. Among these are Raman spectroscopic detection [168, 169], laser-induced capillary vibration detection [170], refractive index (RI) detection [171 - 176] and thermooptical absorbance detection [98, 177, 178]. For Raman spectroscopic and capillary vibration detection, the reader is asked to refer to the cited literature. In the following, we will give a short description of refractive index and thermooptical absorbance detection.

4.2.7.1 Refractive Index Detection

In chromatography, refractive index (RI) detection is commonly used, when the substance of interest neither absorbs nor fluoresces. The measured quantity is a change in the refractive index n of the solution, defined as:

$$n = \frac{\sin \phi_1}{\sin \phi_2} \tag{4-12}$$

where ϕ_1 and ϕ_2 are the angles of the incidence and departure of a beam crossing the interface between two different media. The RI of a solution is particularly sensitive to the presence of solutes, the basis for its frequent use in determining liquid purity.

The operational principle of RI detectors developed so far for detection in CE is based on the interference pattern arising from side-illuminated fused silica capillaries. When coherent light strikes the capillary transversally, the light is scattered over 360° in the plane perpendicular to the capillary axis to form a characteristic pattern of many light and dark fringes. This effect, which is the primary limitation in LIF detection because of the high background signals produced, can be used for RI measurements, because the position of some of the fringes changes with the refractive index of the solution in the capillary. Thus, a sensitive RI detector can be built up by monitoring the position of a predominant fringe. For this purpose, a small area photodiode is located at the fringe boundary. As the RI of the solution changes, the fringe changes its annular position and sweeps over the photodiode, generating a voltage change which is proportional to the RI of the solution.

The interference pattern is produced by four types of scattered rays which depend on the incident position and angle of the laser beam, namely rays that

- > are reflected at the outer capillary surface,
- > enter the capillary, propagate through the wall, and emerge without intercepting the inner bore,
- > are reflected at the inner capillary - liquid interface or
- > propagate through both the capillary wall and bore.

Consequently, the resulting interference pattern is rather complex and it is difficult to select a sensitive fringe with a high contrast. Furthermore, only a very small fraction of the total optical power is available in every fringe. In order to select the fringe which shifts the most for a given change of the refractive index Δn and to direct the optical power into the fringes, the fringe pattern is simplified by immersing the capillary into a transparent liquid, the refractive-index matching fluid (RIMF), which has the same RI as the capillary wall [171]. In this way, the reflection and refractions at the external capillary wall are eliminated.

The limiting factor of performance is the thermal stability of the system. At room temperature, thermal coefficients of RI are in the order of 10^{-4} °C·n^{-1}. Thus, temperature control plays an important role when combining RI detection with CE because of the Joule heat generated during the separation process. To reduce thermal fluctuations in the background signal, Morris and coworkers [172, 179] have developed the so-called analyte velocity modulation method. The detection frequency region is shifted from DC to a frequency in the 100 - 400 Hz region by imposing an AC voltage on the driving DC voltage. The obtained response is a derivative of the conventional electropherogram. Synchronous demodulation of the detected signal ensures that all signals that are not within a narrow bandpass around the modulation frequency are rejected resulting in a further refinement of the system. Analyte velocity modulation does not only reduce thermal fluctuations but also other background fluctuations, laser beam drift and small capillary position changes. Since it produces stable background derivative responses by electrical rather than optical implementation, it should also be useful for other shot-noise or thermal-noise limited detectors.

Another approach to minimize thermal fluctuations has been made by Pawliszyn [173]. He has built up a concentration gradient differential detector based on Schlieren optics which is rather insensitive to thermal fluctuations inside the capillary. It has shown good sensitivity for CE modes exhibiting self concentration and focusing effects such as capillary ITP and capillary IEF [174 - 176]. Finally, the use of RI-matching [171] fluids mentioned above can also improve temperature stability because the thermal conductances of liquids are 20 times larger than that of air and, thus, the surrounding matching fluid acts as a heat sink.

4.2.7.2 Thermooptical Absorbance Detection

Thermooptical absorbance techniques can be employed for the determination of small absorbances. Nonradiative relaxation following absorbance of a laser beam produces a temperature rise within the sample, which is proportional to both the laser power and the absorbance of the sample. Since the RI of most substances changes with temperature, a modulated RI field will be generated within the sample. Consequently, the temperature rise can be detected as a change in the RI of the heated sample. The detection system consists of a probe laser delivering the beam to the sample and a pump laser producing the interference pattern for RI measurement. By using a 5-mW KrF excimer laser operating at 248 nm and a 3-mW helium-cadmium pump laser operating at 442 nm, 20 phenylthiohydantoin-amino acids have been separated and detected by thermooptical absorbance detection. A LOD of $9 \cdot 10^{-7}$ M and a mass LOD of 0.5 fmol has been found for PTH-glycine [177].

4.2.8 Radiometric Detection

Radiometric detection has been payed little attention in CE so far [180 - 183]. An on-line radioisotope detector, however, has several advantages which make it an interesting detection system for CE:

> high sensitivity,
> unrivaled selectivity because only radiolabeled samples are detected,
> the radiolabeled molecule possesses the same chemical properties as the unlabeled molecule,
> direct calibration to provide a measurement of absolute concenteration of the labeled species.

Detection of radiolabeled molecules is based on scintillation counting of the CE eluent. Pentoney et al. [180] have reported the design of three on-line radiometric detectors applicable to both high-energy ß-emitters and γ-emitters:

> a CdTe semiconductor device responding directly to γ rays and ß particles that pass through the fused silica wall of the CE capillary,
> a parabolic plastic scintillator surrounding the detection region of the capillary, and
> disk fashioned from plastic scintillator material positioned between two room temperature PMT's operating in the coincidence counting mode.

The three detectors were characterized for the analysis of ^{32}P-labeled species. The limit of detection is strongly dependent on the separation conditions. For CE separations performed at a relatively high constant voltage, the detection limit lies in the low nanocurie range, corresponding to an analyte concentration of 10^{-9} M of injected sample. The LOD can be extended to 10^{-10} M by reducing the flow rate to increase the residence time of the labeled sample within the detection volume. Flow programming is accomplished by manually reducing the separation voltage as the solute zones elute. Further sensitivity improvement is achieved by freezing the contents of the capillary after separation and exposing the frozen capillary to film.

Altria et al. [183] have reported on the construction and evaluation of a gamma-ray detector for CE which can be used for the detection of radiopharmaceuticals containing ^{99}Tc. The light emitted as the labeled sample zone traverses the detection volume is measured by passing of the capillary through a solid block of scintillator material. The linear range of detection is only given in Bequerel per cm^3 and is 10 - 500 Bq·cm^{-3}.

4.2.9 Comparison of the Presented Detection Modes for CE

Table 4.4. gives a comprehensive comparison of the most important detection modes presented in this section.

Table 4.4. Comparison of different detection modes for capillary electrophoresis

Detection Mode	Approx. Dynamic Range [M] (S/N = 2)	Approx. Mass LOD [mol]	Applications	Advantages	Disadvantages
UV-VIS Absorbance	10^{-6} - 10^{-3}	10^{-15}	peptides, proteins, nucleic acids, drugs, small ions	easy to use and relative universal	relative low sensitivity
Fluorescence	10^{-8} - 10^{-5}	10^{-17}	amino acids, peptides, proteins, nucleic acids	higher sensitivity than UV detection, rather selective	not universally usable
Laser-Induced Fluorescence	10^{-12} - 10^{-9}	10^{-21}	trace analysis of amino acids, peptides, proteins, nucleic acids	very high sensitivity, rather selective	expensive, not universally usable
Conductometry	10^{-6} - 10^{-3}	10^{-16}	ion analysis	peak area correlates linear-ly with migration time	relative low sensitivity not universally usable
Amperometry	10^{-8} - 10^{-5}	10^{-20}	trace analysis of electroactive compounds in complex matrices such as body fluids	very high sensitivity and selectivity	limited to electroactive compounds, difficult to establish
Indirect UV-VIS Absorbance	10^{-5} - 10^{-3}	10^{-14}	ion analysis, carbohydrates	universal	rel. low sensitivity, re-stric-tions in buffer choice
Indirect Amperometry	10^{-8} - 10^{-5}	10^{-20}	simult. det. of elec-troactive and non-electroactive compounds	high sensitivity	difficult to establish
Indirect Fluorescence	10^{-7} - 10^{-5}	10^{-17}	detection of non-fluorescent compounds	universal and rather high sensitivity	restrictions in the choice of buffers

4.3 Capillary Column

The core of each CE system is the capillary. Although there are columns made by Pyrex borosilicate glass or Teflon, fused silica is by far the most frequently used material. This is owing to the intrinsic properties of fused silica like the superior transparency for UV light, the high thermal conductance and the feasibility of manufacturing capillaries with diameters of a few micrometers. Fused silica capillaries are available from a number of suppliers. Most manufacturers of CE instruments offer tailor-made columns of bare fused silica, with internal coating (Sect. 5.1.2) or gel-filled columns (Sect. 5.2). More inexpensive fused silica capillaries are available in bulk from e.g. Polymicro Technologies, Phoenix, Arizona or SGE Inc., Austin, Texas.

The internal diameter of the capillary may be chosen in the range of 10 to 100 µm at an outer diameter ranging from about 190 to 366 µm. The real inner diameter of a capillary usually deviates from the declared value. This is shown in Table 4.5. for several capillaries with a quoted i.d. of 75 µm. While small variations of ± 3 nm should be tolerable, the real i.d. of 53 µm of the Chrompack probably originates from a mix-up. The thickness of the fused silica wall varies with the supplier. Wall thickness and quality of the silica influence the absorptivity for UV light. According to Engelhardt [184] the Chrompack capillary shows the highest transparency for UV light between 190 and 210 nm of those types presented in Table 4.5. The highest absorbance was found with the capillary from Polymicro.

Table 4.5. Comparison of fused silica capillaries from different suppliers [184]

	Chrompack	Polymicro	SGE	Siemens
declared i.d. [µm]	75	75	75	75
measured i.d. [µm]	53	74	74	71
measured o.d. [µm]	232	366	190	260
coating [µm]	16	17	9	28
wall [µm]	74	129	49	67

Figure 4.21. shows a cross-sectional view of a capillary. Because bare fused silica capillaries are extremely fragile, they are externally coated with a polyimide polymer to improve the flexibility and to make handling easier. However, approximately 0.5 cm of the polyimide coating has to be removed to allow light passage for detection. Preparation of the capillary for use can be done by the following procedure, which is recommended because of its simplicity:

To obtain a proper cut, scratch the silica (through the coating) several millimeters from the end of the bulk ware with e.g. a sapphire cutter (from i.e. J&W Scientific or Supelco SA).
Carefully bend the capillary at the scratch until it breaks.
Measure the desired capillary length from the bulk ware (take into account the length to the detector and to the buffer reservoirs) and cut again.
Look at the cuts of both ends with a magnifying viewer and make sure that they are proper (see Fig. 4.22.).
Remove the polyimide coating at a distance of 1 - 2 mm from both ends of the capillary by burning with a butane lighter (from e.g. Supelco SA).
Measure the desired length to the detector and remove the polyimide coating again by burning with a butane lighter.
Clean all burned regions with a tissue moistened with methanol or acetone (avoid bending the fragile part of the capillary).
Insert the capillary carefully into the cartridge as described by the instrument manufacturer.
Rinse the column first with 1 M sodium hydroxide (10 min), then with 0.01 M sodium hydroxide (30 min), finally with running buffer (30 min).
The system is now ready for operation.

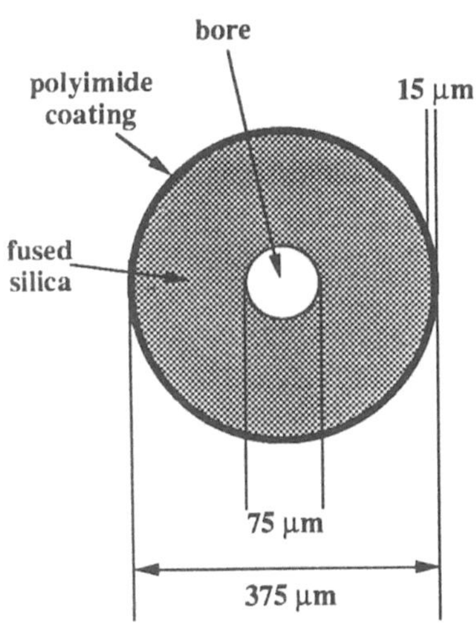

Fig. 4.21. Cross-sectional view of a fused silica capillary

For obtaining an accurate sample plug during injection, it is important to have a clean cut as shown in Fig. 4.22. A bad cut leads to unreproducible sample injection and decreases the separation efficiency by distortion of the sample plug.

Besides the procedure given above which leaves the glass brittle, there are a number of alternatives described for removing the polyimide coating. McCormick and Zagursky [185] have developed a mechanical stripping device for pro-

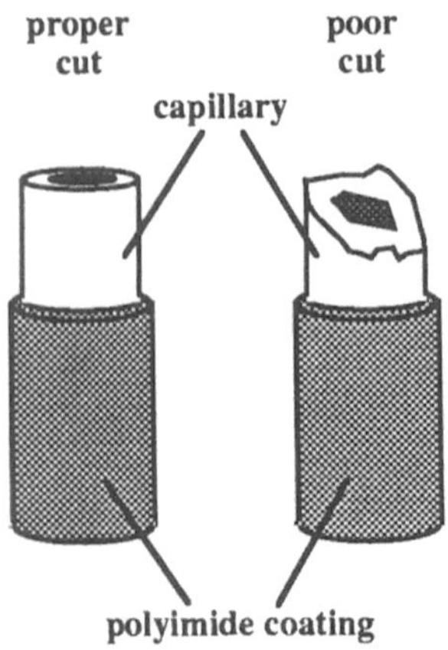

Fig. 4.22. Proper and bad cut of the capillary

ducing windows on a fused silica capillary. This procedure offers some advantages over the conventional technique. Capillaries are not subjected to extreme heat. Sharp, well-defined windows are produced even on chemically derivatized or gel-filled columns without destroying the surface modification or the gel inside the capillary. Another methodology has been described by Schomburg and coworkers [186]. They use a filament which is electrically heated by a low voltage/high current transformer. The polyimide is burnt off at the contact point with the filament. Thus, very small windows can be produced. The temperature of the filament is easily adjusted so that only the coating is removed. Concentrated sulphuric acid at 100 °C or strong bases also remove the polyimide in a few seconds. These techniques, however, partially leave brittle glass and must be used with care.

An interesting alternative to the polyimide coating is presented with a new UV transparent column coating by Supelco [187]. As claimed by the manufacturer this capillary allows analysis in the UV range without the need of removing the coating. At the same time the coating provides flexibility and durability similar to that of polyimide-coated fused silica capillaries.

Before using a capillary for the first time with a special separation buffer a conditioning procedure should be carried out to ensure that the surface of the tube is always in the same shape. To minimize conditioning, we strongly recommend the use of an individual capillary only in a narrow pH range, i.e. one capillary for pH 2 - 4, a second one for pH 4 - 6 and so on. Regeneration is necessary if migration times vary from run to run. For pH values below 3, it is suggested that the capillary should be rinsed with the running buffer for 10 minutes before each run. It should be noted that, particularly in the acidic pH range, migration times occasionally become constant only after several hours of use. Variation of migration times are frequently observed in a pH region of 3 to 7 where small pH changes have a big influence on the electroosmotic flow. In this range and also at higher pH values the following procedure should be carried out before each run to give reproducible results:

> Rinse with 0.1 M sodium hydroxide (5 min).
> Rinse with bistilled water (2 min).
> Rinse with running buffer (10 min).

For storing a fused silica capillary, the following operation should prevent damage of the capillary during storage:

> Rinse the capillary with bistilled water to remove the running buffer (5 min).
> Blow nitrogen or air through the capillary (2 min).
> Remove the capillary from the instrument.

Hint: Storing capillaries filled with buffer solutions should be avoided. The buffer solvent will evaporate leaving solid salt crystals which clog the capillary. Additionally, buffer solutions with low or high pH values will destroy the column by dissolving the fused silica.

As an alternative to the circular columns, rectangular capillaries have been introduced to CE by Tsuda et al. [108]. They can be purchased from Wilmad Glass as microvials in the dimensions 20 x 200, 30 x 300, 50 x 500, 500 x 1000, 50 x 50 and 100 x 100 μm^2. All are borosilicate glass and are not available with any protective coatings. A significant disadvantage are their thin walls, equal to the width of the narrowest dimension; for instance 20 μm for 20 x 200 μm^2 capillaries. Thus, special care must be taken not to break these capillaries. Some of the advantages of these systems are their efficient heat dissipation due to the large height-to-width ratio and, hence, their high surface-to-volume ratio and their high detection sensitivity for optical on-column detection modes. According to the authors, one of the most exciting possibilities of these flat separation channels is the ability to perform two-dimensional separations, with one force being applied across the separation channel, and with the sample zones detected by the use of a multi-channel array detector.

To simplify column thermostation and to permit a compact design of the instrumentation the flexible polyimide coated fused silica capillaries can be conveniently coiled into relatively small coil diameters. Column coiling, however, can effect the efficiency of the separation process. Karger's group [188] investigated the influence of coiling on the performance in CE using open tubes and gel-filled capillaries. In open tubes of 75 μm i.d., no significant effect of column coiling is observed, even at $2 \cdot 10^6$ plates·m^{-1}, probably because of the small i.d. and radial diffusion in free solution. On the other hand, with gel-filled capillaries of 10^7 plates·m^{-1}, the influence of coiling on efficiency is significant. Plate numbers decrease by a factor of 3 or more per coil. For relatively rigid gels (see Sect. 5.2) the observed loss of efficiency is attributed mainly to the inability of the polymer network to permit diffusional relaxation of the coiling effects. In medium concentration polymer networks, coiling causes a change of the network under the influence of mechanical stress, resulting in anisotropy across the capillary. Furthermore, the changes in structure and efficiency are time dependent and, in general, reversible.

4.4 Sample Collection

When CE analysis shows that a sample is impure, the problem arises how to purify it, if no other method can provide the separation efficiency required. For this purpose it can be very helpful, if the CE equipment could also be used for the micropreparative isolation of small sample amounts. In order to yield sufficient material for subsequent analysis, collection of several subsequently runs is required. Therefore, separations must be highly reproducible to assure minimal cross-contamination of neighboring peaks.

CE as it is normally practiced has both the inlet and the outlet of the capillary immersed in buffer reservoirs to complete a closed circuit. Thus, sample zones are directly discharged into the outlet reservoir, which hinders sample collection. One approach to overcome this problem has been made by Rose and Jorgenson [189] by using a programmable fraction collector consisting of the collector tray and three digital linear actuators which allow precise movement of the capillary. Ten collection cones with ≈ 25 µL volume each and a rectangular slot buffer reservoir with 4 mL are machined into the tray. Platinum wire electrodes dip into each collection cone and the buffer reservoir. The cones and the reservoir are filled with buffer to allow current to flow during the application of high voltage. In order to reduce the amount of time electrophoresis is interrupted, a tapered glass capillary filled with buffer and grounded via a platinum and a copper wire is mounted on one of the digital actuators. If a fraction has to be collected, the capillary is lifted up into the tapered glass capillary and is then moved over the cone, which is filled with 5 - 10 µL of buffer, and lowered into the buffer in the cone. the species to be collected migrates from the end of the capillary into the buffer in the cone. After collection of the samples species, the capillary can be moved in the original position or to the second cone of the collection tray.

Instead of an additional collection tray, an autosampler can be used in the way, that it holds both the samples for injection and the vials for sample collection. This approach is realized in the commercial CE equipment 270A-HT of Applied Biosystems. Samples and fractions can be cooled to prevent degradation and to minimize evaporation.

Since the capillary outlet must have contact to the electrolyte in the fraction collector to complete a circuit the collected samples are diluted by the factor of ≈ 10^2 - 10^3. Additionally, electrochemical reactions of the sample can arise in the small amount of electrolyte in the fraction collector. Another approach to sample collection eliminating these problems is to complete the electrical circuit in the capillary prior to its outlet. The techniques to achieve this have been already presented for the most part in Sect. 4.2.5.2 about amperometric detection, namely the porous glass junction of Wallingford and Ewing [143], the on-column frit structure of Huang and Zare [145] and the on-column Nafion joint of O'Shea et al. [146]. A similar device was presented by Fujimoto et al. [190] who surrounded a fractured capillary with polyacrylamide gel. All these designs do not only facilitate the use of detection schemes in which the detector is at ground potential, such as electrochemical detection, but also allow the collection of sample with no internal dilution.

Especially for high-speed multicomponent separations where the peak widths are only a few seconds wide, reproducible collection is difficult to perform. By electric field programming [191], it is possible to simplify collection while maintaining the high resolution of the separation. In this approach, the electric field is kept at high potential, until just before the species reached the end of the capillary. The field is then cut off, and the capillary and the electrode is placed into a microvial containing 1.5 µl of water. The field is then applied again at ≈1/10 of the original field strength and the collection is performed.

Cheng and coworkers described membrane fraction collection for CE [192] using a membrane assembly at the exit of the capillary to complete the electrical circuit. This membrane assembly consists of a poly(vinylidene difluoride) membrane, two layers of 3MM Chrom filter-paper as the buffer reservoir, sandwiched between the membrane and a stainless-steel plate serving as the ground electrode. Proteins are collected on the rotating membrane and identified by Coomassie Brilliant Blue staining. The collected protein samples can be sequenced trough direct protein sequence analysis.

4.5 Commercial Instruments

In parallel to the rapid development of capillary electrophoresis started in the early 1980s, a number of instruments have been developed and are now commercially available. All instruments on the market allow the application of the most important modes of CE such as CZE, CGE or MEKC.

Although a comparison of instruments is bound to be incomplete and might even be obsolete at the date of publishing of the book, we try in this section to give an overview of the individual features of a number of instruments. Table 4.6. summarizes their most important characteristics.

Most instruments are equipped with an on-column UV absorbance detector with nanoliter cell volume. Owing to the extremely short path length, sensitivity is limited. The almost universal application range of UV absorbance at wavelengths down to 190 nm make this detector most popular in CE. Three manufacturers offer fluorescence detectors with two of them being laser induced fluorescence detectors. High costs are attributes of these devices. However, fluores-

Table 4.6. Survey of a number of commercial CE instruments

Instrument	Detection	Injection	Autosampler	Column Thermo-statization	Sample Cooling Device	Remark
Applied Biosystems "270A"	UV-VIS, 190-800 nm	vacuum, electrokinetic	8 positions	forced air	no	fraction collector (optional)
Beckman Instruments "P/ACE 2010"	UV (filter) LIF (optional)	pressure, electrokinetic	34 positions	liquid, ambient - 40 °C	optional	
Bio-Rad Laboratories "HPE100"	UV-VIS, 190-800 nm	manual/ electrokinetic	no	no	no	
Bio-Rad Laboratories "BioFocus 3000"	UV-VIS, 190-800 nm fast scanning	pressure, electrokinetic	32 positions	forced air	yes	
Dionex "CES I"	UV, 190-360 nm Fluorescence	gravity, pressure, electrokinetic	39 positions	forced air	no	fraction collector
Europhore "IRIS 2000"	LIF	hydrodynamic, electrokinetic	24 positions	forced air, 15 - 60°C	yes	on-line degassing of buffer
Hewlett Packard "HP^{3D} Capillary Electrophoresis System"	UV-VIS, 190-600 nm diode array detector	pressure, electrokinetic	48 positions	forced air, 10°Cbelow air - 60°C	yes, 10 - 40°C (external water bath)	fraction collector, extended light path detection cell
ISCO "Model 3140"	UV, 190-360 nm	vacuum, electrokinetic	40 positions	forced air, 15 - 40°C	no	

Table 4.6. (continued)

Instrument	Detection	Injection	Autosampler	Column Thermo-statization	Sample Cooling Device	Remark
ISCO "Model 3850"	UV, 190-360 nm	manual split-flow sample injection	no	fan	no	
Kontron Instr. "Eureka 2000"	DAD, 190-800 nm	gravity, pressure, electrokinetic	20 positions	forced air, 5 - 50 °C	no	
LauerLabs "Prince"	UV-VIS, MS, Fluorescence, etc.	pressure, electrokinetic	optional	forced air	optional	modular system
SpectraPhysics "SpectraPHORESIS 500"	UV-VIS, 190-800 nm single wavelength selection	vacuum, electrokinetic	80 positions	forced air, 15 - 60 °C + N_2 to 5°C	no	on-line degassing of cathode buffer reservoir
SpectraPhysics "SpectraPHORESIS 1000"	UV-VIS, 190-800 nm multi wavelength selection	vacuum, electrokinetic	80 positions	forced air, 15 - 60 °C + N_2 to 5°C	no	on-line degassing of cathode buffer reservoir
Stagroma "Model 100"	UV-VIS, 190-800 nm single or multi wavelength selection	manual/ electrokinetic	no	fan	no	modular system
Waters "Quanta 4000"	UV-VIS, 190-800 nm	gravity, electrokinetic	13 positions	fan	no	fraction collector

cence detectors are more than three orders of magnitude more sensitive than UV, albeit limited to fluorescing solutes. Moreover, LIFs are limited to only one or a few wavelengths, which make derivatization with a suitable fluorophore necessary. This detection system may be extremely useful for the analysis of drugs or metabolites in biological samples like blood serum or urine where ultimate sensitivity is required. Electrochemical detectors based on amperometry or conductometry are not yet commercially available, although they are in the pipeline of several manufacturers. The coupling of mass spectrometry to CE provides mass sensitive detection and, ultimately, structural information, ideally as CE-MS-MS.

Except for the low-cost versions, all instruments allow both hydrodynamic (gravity, pressure and/or vacuum) and electrokinetic sample introduction. Although strong efforts are made with respect to the reproducibility of the injected volume, standard deviations are typically found in the range of about 2%. Thus, quantitative work by the external standard method is not feasible for many appli–cations especially for pharmaceutical analysis. Internal standardization is the only way out until injection reproducibility is improved.

Depending on the fields of application, the demands on the equipment may vary significantly. While, for industrial purposes, features like an autosampler allowing overnight sequence analysis may be most important, for university purposes flexibility and low costs might be decisive for the selection of a particular instrument.

With increasing acceptance of CE as a routine analysis technique commercial equipment will be further improved and additional features will be developed. The worldwide market for CE is predicted to reach nearly $ 100 million in 1994 [193]. Ideally, CE will supplement HPLC as one of the two most versatile separation techniques.

5 Techniques

In his foreword to the book "Praxis der elektrophoretischen Trennmethoden" Milan Bier wrote: Despite its admirable age electrophoresis is still a vivid science and a fruitful ground for innovative research. Its application range covers small ions up to living cells, their organelles and even intact chromosomes. Actually, some of the most important electrophoretic applications were developed during the last two decades and others will follow in the near future [194]. In particular the development of CE and its various modes has verified Bier's prognosis in an excellent way.

Besides ongoing efforts in the area of CZE in the presence of electroosmotic flow a broad variety of other CE techniques matching the needs of special analytical problems have been developed in the last decade. As a continuation of his work on free-zone electrophoresis, Hjertén introduced CZE in coated capillaries [29], where absorption as well as electroosmosis are eliminated. Especially for the separation of proteins which tend to absorb strongly on glass surfaces, this method has proved itself to be very useful. In 1984, Terabe and coworkers [195] presented micellar electrokinetic chromatography (MEKC) to facilitate the separation of uncharged compounds by CE. In 1983, Hjertén [196] used polyacrylamide gel-filled capillaries for the first time. Beginning in 1987, Karger and coworkers have introduced capillary gel electrophoresis (CGE) for the separation of proteins [197] and nucleic acids [198] to combine the very high resolution of gel electrophoresis with the simple instrumental set-up of CZE, predestined for automation and on-line detection of the separated compounds.

In the following section the different techniques of CE, namely capillary zone electrophoresis, capillary gel electrophoresis and micellar electrokinetic chromatography are presented and some of their unique features are discussed. Afterwards some techniques related to CE as capillary isotachophoresis, capillary isoelectric focusing, electrochromatography and hyphenated techniques, are shortly described. In the last section of this chapter special techniques such as affinity electrophoresis and sample stacking are presented.

5.1 Capillary Zone Electrophoresis

In analogy to chromatography the zonal mode of operation is by far the most important technique in capillary electrophoresis. This is probably because of its simplicity and separation power. The principles of CZE are stressed in detail in chapter 2.

5.1.1 General Aspects

To carry out CZE, the capillary, uniformly filled with a buffer solution, is dipped into buffer reservoirs at both ends of the tube. For sample injection one buffer reservoir, usually at the anodic side of the capillary, is replaced by the sample vial. A few nanoliters of the sample solution are introduced into the capillary by either pressure, gravity, vacuum suction or by applying voltage. After injection is completed the sample vial is again replaced by the buffer reservoir and voltage is applied across the capillary. Under the influence of the electric field the analytes migrate with different velocities to the corresponding electrodes, cations to the cathode, anions to the anode. If an appropriate electroosmotic flow exists, both cations and anions move through the detector which records the individual zones. Separation is based on differences in the electrophoretic mobility of the analytes. By varying the buffer pH, the ionic strength and the buffer composition, the electrophoretic mobilities can be manipulated.

The use of bare fused silica capillaries for CZE may cause some problems. Variations of the buffer pH do not only modify the electrophoretic mobilities of the analytes but simultaneously change the electroosmotic flow. Similar interactions are found for the ionic strength. These coupled effects make method optimization difficult. For this reason it is often desirable to control or even suppress the electroosmotic flow without altering the electrophoretic migration of the analytes at the same time. Elimination of EOF is also useful to enable capillary isotachophoresis and capillary isoelectric focusing. In addition, adsorption processes between analytes and the charged silica surface sometimes cause peak distortion and reduce the efficiency of the separation (see Sect. 3.1.2). Deactivation of the active sites of the fused silica by chemical derivatization (coating) have shown to suppress both adsorption and electroosmotic flow.

5.1.2 Capillary Coating

For many applications in CE a shielding of the analytes from the active sites of the fused silica is necessary. Especially for proteins which tend to adsorb strongly to the silica a coating of the capillary is essential. There are several ways to coat a fused silica tube. As already discussed in Sects. 2.5 and 3.1.2 the capillary surface can be coated dynamically by adding additives such as surfactants, zwitterionic salts or hydrophilic linear polymers to the buffer system. This procedure is advantageous because of its simplicity and low costs but it suffers from several drawbacks. First, reproducible dynamic coating is difficult to achieve and changes in the buffer composition alter the coating conditions. Secondly, disturbing interactions with the analytes may occur. It is, for instance, well known that several proteins precipitate in the presence of ionic surfactants.

An alternative strategy to reduce adsorption is to bond chemically a polymer to the capillary surface or to modify the active sites of the silica by derivatization. If a polymer is used for the coating, it is anchored to the silica by reaction of only a part of the silanol groups with a reagent. The long polymer chains then shield the remaining silanol groups. In contrast, if the surface is derivatized, the coating is only effective if all active silanol groups react with the reagent. A chemical capillary coating should satisfy the following requirements:

> effective in suppressing adsorption
> allow a constant electroosmotic flow over a wide pH range
> reproducible in preparation
> stable for a long time
> stable over a wide pH range.

Several suppliers of CE instruments offer coated capillaries with various coatings. Supelco Inc. were the first to offer bonded capillary electrophoresis columns under the tradename CElect. They are available as a neutral hydrophilic phase (CElect-P), a weakly hydrophobic C1 phase (CElect-H), a moderately hydrophobic C8 phase (CElect-H1) and a highly hydrophobic C18 phase (CElect-H2). As claimed by the manufacturers the electroosmotic flow is reduced with both columns by 33 - 43% compared to the bare silica. Residual EOF remains fairly stable in the large pH range of 3 - 10. Especially the moderate and the highly hydrophobic coatings reduce protein-silanol interaction such that separations can be obtained at pH values that are not possible in bare fused silica columns. Applied Biosystems offers a material which forms a noncovalent coating on the surface of a fused silica capillary by combination of ionic and hydrophobic interaction under the tradename Micro-Coat. Since the reagent is cationic, it produces a positively charged coating that reverses the direction of EOF and reduces wall interactions of proteins at pH values below their isoelectric points. Bio-Rad Laboratories offers 25 µm and 50 µm i.d. capillaries coated with linear polyacrylamide which eliminates EOF and reduces adsorption of biomolecules. Isco Inc. markets 3 different coated capillaries, all 80 cm in length and 75 µm i.d.: a bonded C18 phase (CE-100-C18), one with glycerol groups covalently attached to the capillary by an octyl linker (CE-200) and one with a sulfonic acid group also linked to the wall by an octyl linker (CE-300). Whereas the first two coatings are recommended for use in peptide and protein analysis, the latter should be used for nucleotide separations. Various other approaches have been employed to modify the capillary surface for CE, most of which were especially developed for protein analysis. A listing of the most commonly used coating procedures including the described commercial available coatings are given in Table 5.1.

Hjertén [29] developed a theory which explains the reduction of the electroosmotic flow by coating the capillary surface. The electroosmotic mobility de-

Table 5.1. Coating procedures of fused silica capillaries for CE

Coating	Remark	Ref.
polyacrylamide via Si - O - C bond	simple procedure; effective suppression of EOF	29
polyacrylamide via Si - C bond	difficult and time-consuming, but stable over pH 2 - 10.5; effective suppression of EOF	199 200
nonionic surfactant via octadecyl-silane (C18 phase)	simple; stable EOF over a wide pH range	201
diol-epoxy polymer	reduction of protein adsorption at sufficient EOF; useful pH range: 5 - 10; lifetime > 120 h	202
polyethylene glycol via 3-aminopropyltriethoxysilane	reduction of EOF related to the molecular weight of PEG	203
polyethylene glycol via γ-glycidoxy-propyltrimethoxysilane	good stability over a pH range of 3 - 5	204
protein coating	simple procedure; EOF is a function of protein disssoziation; adsorption is reduced	30
aryl pentafluoro group	works at neutral pH and moderate ionic strength	205
polyethylene glycol/polyethylene glycol diglycidyl ether via γ-glycid-oxypropyltrimethoxysilane	works well at pH 4.0 - 7.5; EOF reduction as a function of chain length	206
maltose coating	shields well the silica up to pH 7; an antimicrobial agent must be added to the separation buffer	207
polyethyleneimine coating	hydrophilic, positively charged; causes reversal of EOF; stable from pH 2-12; simple procedure	208

pends on the viscosity of the solution within the diffuse double layer according to the following equation, which is derived from Eq. 2-26:

$$\mu_{eo} = \frac{\varepsilon}{4\pi} \cdot \int_0^{\zeta} \frac{1}{\eta} \cdot dE \qquad (5\text{-}1)$$

If the viscosity in the double layer close to the wall approaches infinity the integral and, consequently, the electroosmotic mobility will approach zero. As a result, the coating of the inner capillary surface by a polymer solution of high viscosity will eliminate electroosmosis. Any neutral polymer which is soluble in water can be used, e.g. methyl cellulose or non-cross-linked polyacrylamide. The same principle holds for dynamic coating, where the water soluble polymer is dissolved in the buffer. The polymers tend to adhere to the capillary wall and thereby create dynamically a thin surface layer with high viscosity.

Before the coating solution is introduced into the capillary, the fused silica wall should be pretreated to ensure reproducible conditions at the surface. Additionally, the treatment with strong bases etches the capillary surface resulting in a higher number of reactive silanol groups. For this purpose the capillary is rinsed with 1 M KOH or NaOH, followed by distilled water and, eventually, HCl. Before the rinsing procedure the capillary can be further activated by heating it in excess of 100 °C, generally for several hours. Detailed descriptions of the most important coating procedures are given in the following sections.

5.1.2.1 Polyacrylamide Coating via Siloxane Bond [29]

This method is based on the use of a bifunctional compound, in which one group reacts specifically with the silica wall and the other with a monomer taking part in a polymerization process. Besides γ-methacryloxypropyltrimethoxysilane (MPS, e.g. from Pharmacia, Sweden, or as Bind Silane from Aldrich, Wisconsin), other bifunctional compounds such as vinyltriacetoxysilane, vinyltri(β-methoxyethoxy)silane, vinyltrichlorosilane and methylvinyltrichlorosilane can be used. While reaction with the silanol groups is accomplished by the methoxy, acetoxy, methoxyethoxy or chloro functional group, the acryl or vinyl group is involved in the polymerization step. The reaction scheme is shown in Fig. 5.1. The following procedure can be used:

> Rinse the capillary with 1 M NaOH and then with distilled water both for 30 min.

> Mix 4 µL MPS with 1 mL of 6 M acetic acid, pH 3.5, and introduce the solution into the capillary.

> Allow the reaction to take place for 1 hour at room temperature before removing the silane solution from the tube.

> After washing with distilled water, fill the capillary with deaerated 3 - 4% (w/v) acrylamide solution containing 1 mg/mL *N,N,N',N'*-tetramethylene ethylenediamine (TEMED) and 1 mg/mL $K_2S_2O_8$.

> Remove the excess polyacrylamide after 30 min of reaction and rinse the capillary with water.

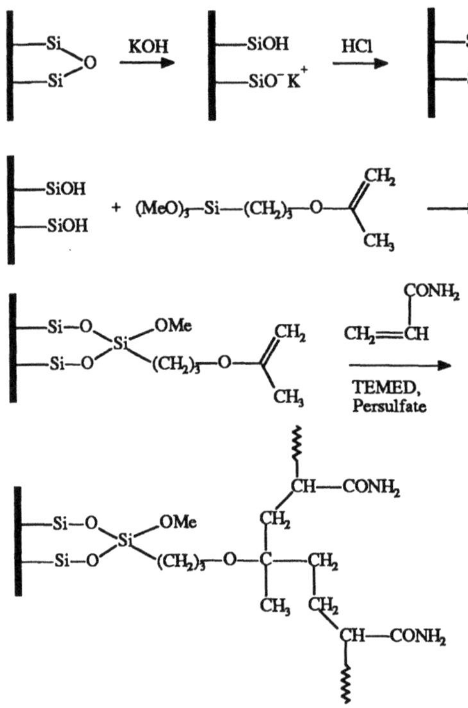

Fig. 5.1. Reaction scheme for the coating of silica with polyacrylamide by using a bifunctional agent

5.1.2.2 Polyacrylamide Coating via Si-C Bond [199, 200]

This approach utilizes a Grignard reaction to form a hydrolytically stable Si-C bond which is more stable than the Si-O-Si-O-C bond described above. Figure 5.2. depicts the reaction scheme for the multi-step process. Capillaries coated by this procedure can be used over a pH range of 2 - 10.5, without noticeable degradation of the coating. The following method is described by the authors:

> First rinse the capillary with 1 M NaOH for 30 min, followed by distilled water for 30 min.

> Dry the capillary overnight with a N_2 stream at 110 °C.

> Fill the capillary with thionyl chloride (in a pressurizing chamber which is flushed with N_2).

> After the capillary is filled with thionyl chloride, seal one end of the capillary by using a small propane torch.

> Quickly attach the open end of the tube to a vacuum pump and evacuate for approx. 20 min to achieve a vacuum of 8 Pa or less. Throughout the evacuation process maintain the capillary at 60 °C by keeping it in a heating bath.

- Seal the capillary with a propane torch near the connection to the vacuum pump and heat it for 12 hours at 70 °C.
- By using a dry syringe dissolve 1 mL vinyl magnesium bromide in 5 mL dry THF placed in a 10 mL vial which is fitted with a rubber septum.
- Break off one end of the sealed capillary while it is immersed in a dry THF solution. Place this open end immediately into the THF - vinyl magnesium bromide solution. Break off the other end of the tube and connect it to the vacuum line to suck the solution into the tube.
- After several minutes of suction seal the tube near the vial septum with a propane torch. Maintain the capillary at 50 °C for 30 min. Seal the other end of the capillary near the vacuum line and place it in a heating bath for 12 h.
- Break off both ends of the capillary and rinse first with THF and than with bidistilled water for several minutes. Fill a deaerated solution of 0.3 mL of 10% acrylamide, 0.7 mL of water, 1 μL of TEMED and 10 μL of 10% $(NH_4)_2S_2O_8$ into the tube. After 30 min of reaction rinse the capillary with water to remove excess acrylamide.

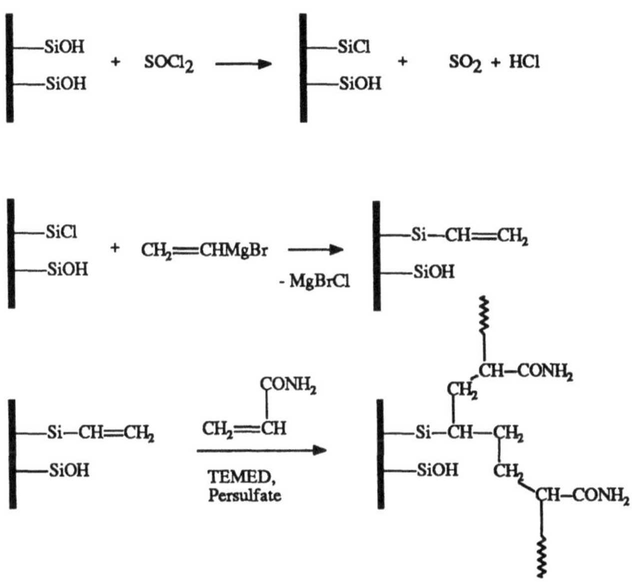

Fig. 5.2. Reaction scheme for the preparation of vinyl-bound polyacrylamide coating

5.1.2.3 Nonionic Surfactant Coating via Octadecylsilane [201]

By this coating the silica surface is derivatized with octadecyltrichlorosilane to a C18 phase followed by a deposition of a layer of nonionic surfactant, e.g. Tween or Brij series surfactant, from a micellar aqueous solution. The reaction is shown in Fig. 5.3. This (dynamic) coating shows sufficient layer thickness

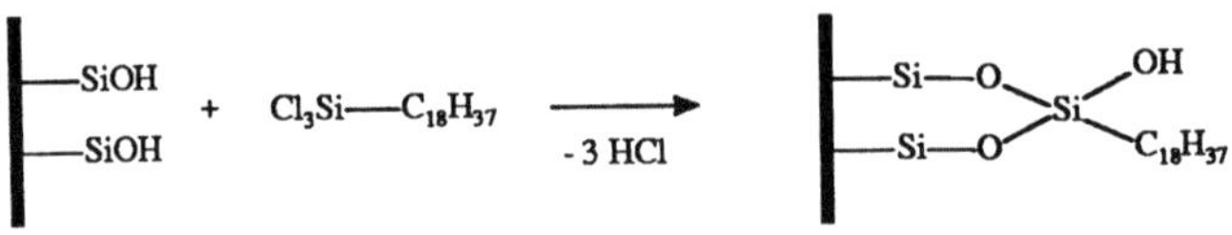

Fig. 5.3. Reaction scheme of the derivatization of silica with trichlorooctadecylsilane

and reduces adsorption of hydrophobic proteins and the EOF. It is stable over a pH range of 4 - 11 with a relatively constant electroosmotic flow. Performance parameters for five selected surfactants obtained by the separation of myoglobin and lysozyme are summarized in Table 5.2.

Table 5.2. Performance parameters for some surfactants adsorbed on alkylsilane coated capillary according to Ref. 201

Surfactant	$\mu_{eo} \cdot 10^8$ $m^2V^{-1}s^{-1}$	$N_{myoglobin}$ (L_T = 80 cm)
TWEEN 20	2.03	170 000
TWEEN 40	2.48	135 000
TWEEN 80	2.27	150 000
BRIJ 35	1.50	240 000
BRIJ 78	1.26	115 000

The following method can be used to prepare a stable coating:

> Treat the capillary with 1 M NaOH for 15 min followed by washing with deionized water for 15 min. Evaporate residual water from the capillary by heating at 100 °C under a N_2 stream.

> Pull a solution of octadecyltrichlorosilane with 5% methylene chloride through the capillary by syringe.

> Place the capillary in an oil bath at 90 °C for 3 h, with new solution used every 5 min.

> Remove excess octadecyltrichlorosilane from the capillary by forcing N_2 through the tube. Wash with several tube volumes of methanol followed by washing with deionized water.

> Dissolve 0.5 % (w/v) of TWEEN 20 or BRIJ 35 in double-deionized water and pull the solution through the capillary for 2 hours to complete coating. Now rinse with running buffer to remove residual surfactant.

5.1.2.4 Diol-Epoxy Coating [202]

The diol-epoxy coating deactivates the surface silanols of fused silica capillaries. The coating consists of a crosslinked diol covalently bonded to the silica with

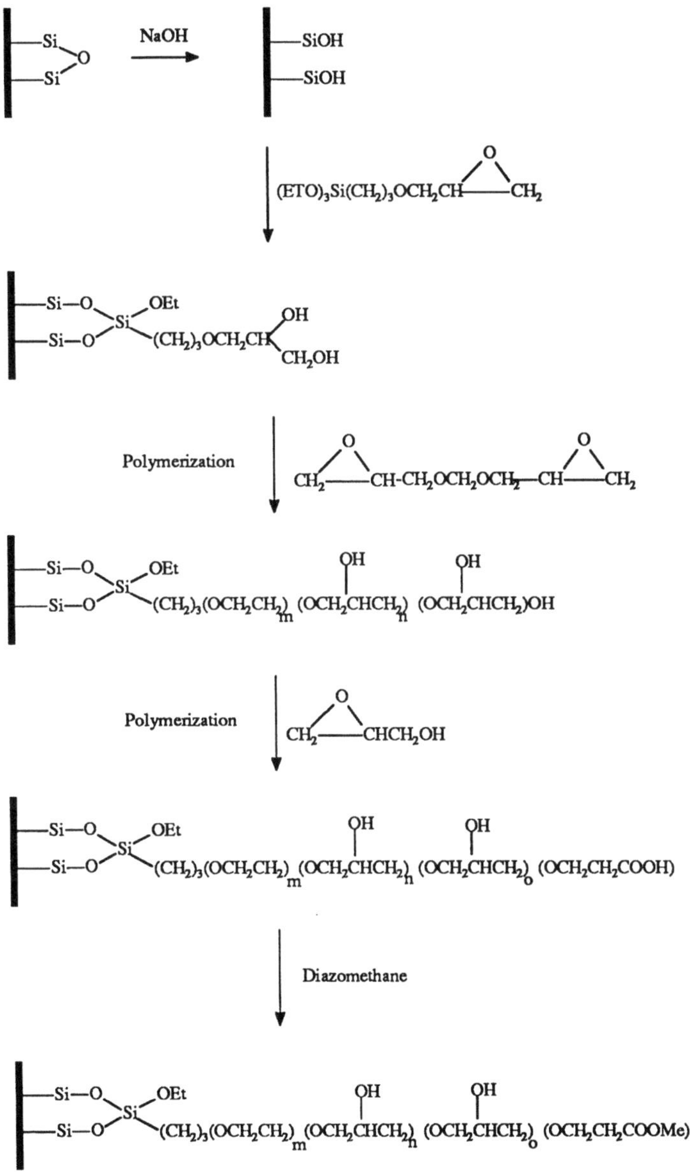

Fig. 5.4. Reaction scheme for the preparation of diol based epoxide coating

oxiranes, e.g. ethylene glycol diglycidyl ether and glycidol. The polymer eliminates negative charged silanols and shields residual surface silanols, thus limiting interactions of proteins with the surface. The coating is of sufficient hydrophilicity to allow enough electroosmotic flow for transporting positive and negative species through the detector. The coating procedure is difficult and contains some hazardous steps, e.g. the use of diazomethane. Therefore, this derivatization process should only be performed by experienced workers. The reaction scheme is given in Fig. 5.4.

5.1.2.5 Polyethylene Glycol Coating [203, 204]

The covalent coating of the inner capillary surface with polyethylene glycol (PEG) reduces the electroosmotic flow in relation to the molecular weight of the PEG (Table 5.3.). Whereas coatings of PEG 5000 and PEG 20 000 virtually eliminate electroosmosis in the pH range 3.5 to 7.8, PEG 400 and PEG 1900 reduce electroosmosis by about 50%. Two different coating techniques are described in the literature.

Coating	$\mu_{eo} \cdot 10^4$ [$cm^2 \cdot s^{-1} \cdot V^{-1}$]
none	5.0
aminopropyl	-3.2
PEG 400	3.1
PEG 1 900	2.5
PEG 5 000	0.3
PEG 20 000	0.3

Table 5.3. Electroosmosis in polymer coated capillaries in 7.5 mM NaCl solution, pH 5.8, according to Ref. 203

1.2.5.1 PEG Coating via 3-Aminopropyltriethoxysilane [203]

PEG coatings of average molecular weights 1900, 5000 or 20 000 are covalently bonded to the silica surface in two steps. If the capillaries are stored filled with distilled water, the PEG coatings are stable for at least 6 months.

> Rinse the capillary for 1 h with alcoholic NaOH followed by distilled water and aqua regia. Rinse over night with distilled water. Dry the tube for 4 h at 65 °C and 676 Pa.

> Place the capillary in a glass pressure vessel and cover it with 20% (w/v) solution of 3-aminopropyltriethoxysilane (e.g. Pierce) in toluene and apply a vacuum of ≈ 130 Pa to remove trapped air.

> Seal the vessel and heat it in an oil bath at 100 °C for 24 h. Subsequently, wash the capillary several times first with acetone, than with water.

> Repeat the entire procedure once more. Rinse the tube with acetone and dry under vacuum.

> Place the tube in a pressure vessel and cover it with an aqueous solution of 20% PEG (w/v) activated with cyanuric chloride. Apply a vacuum of 130 Pa.
> Seal the vessel and heat it in an oil bath at 100 °C for 24 h. Wash the capillary several times with water and repeat the second step once more.

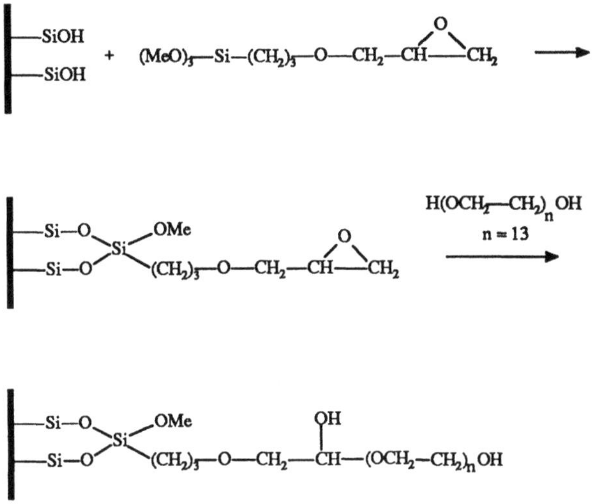

Fig. 5.5. Scheme of the deactivation of fused silica by derivatization with PEG 600

5.1.2.5.2 PEG Coating via γ-Glycidoxypropyltrimethoxysilane [204]

This surface modification significantly decreases adsorption and reduces the electroosmotic flow. Symmetric peaks are obtained for a number of proteins in the pH range 3 - 5. However, at higher pH values noticeable peak deformation occurs. The reaction scheme for the deactivation of the fused silica is shown in Fig. 5.5. The coating procedure described as follows is fast and simple to perform.

> Etch the capillary with 1 M KOH for 3 hours followed by rinsing with bidistilled water. Flush with 1 M HCl solution to remove potassium ions from the wall and to produce free silanol groups.
> Dry the capillary at 200 °C for 3 h with a gentle stream of helium.
> Coat the dried capillary with γ-glycidoxypropyltrimethoxysilane (e.g. from Serva) dissolved in dry toluene (10%, v/v) at 110 °C for 3 h. Remove the residual reagent by flushing with toluene.
> Couple PEG 600 (e.g. from Merck) to the epoxide by flushing a solution of 20% PEG 600 and 2% boron trifluoride etherate in dioxane for 1 h at 100 °C. Finally, rinse the tube with bidistilled water.

5.1.2.6 Protein Coating [30]

Surface deactivation with amphoteric compounds like proteins allow control over both direction and magnitude of EOF by adjusting the pH of the buffer. Depending on the pI of the protein bound to the wall the electroosmotic flow is directed toward the anode for pH < pI, but toward the cathode for pH > pI. EOF is zero for pH is equal to pI. Since bound proteins possess positively and negatively charged groups as well as hydrophilic and hydrophobic domains, complete elimination of adsorption seem to be impossible.

> Etch the capillary with 1 M KOH for 3 hours followed by rinsing with bidistilled water. Flush with 1 M HCl solution to remove potassium ions from the wall and to produce free silanol groups.

> Flush the capillary with a 10% (v/v) solution of 3-aminopropyltriethoxysilane in dry toluene for 3 h (110 °C). Rinse with dry toluene and dry the capillary by flushing with helium overnight.

> Pump a solution of 5 % glutardialdehyde in 100 mM phosphate buffer, pH 7.0, through the column at a rate of 1 - 2 column volumes per minute for 30 min. Allow the filled column to react for 4 h before removing residual glutardialdehyde with buffer.

> Pump a solution of 5 $mg \cdot mL^{-1}$ protein, e.g. α-lactalbumin, in phosphate buffer, pH 7.0, through the capillary and allow to react overnight. Finally, wash out the protein solution with buffer alone and store the capillary in a refrigerator before use.

5.1.2.7 Polyethyleneimine Coating [208]

This coating produces a hydrophilic, positively charged surface. Thus the direction of the electroosmotic flow is reversed. Polyethyleneimine (PEI) 200 (MW = 20 000) is physically adsorbed to the inner surface of the fused silica capillary and subsequently cross-linked into a stable layer. The final coating has proven to be stable over a pH range of 2 to 12. The positive electroosmotic flow declines 50% from pH 3-7 and remains constant in the pH range 8-12. Proteins which are positively charged at pH 7 were resolved fast and efficiently with good recovery. Coating procedure:

> Treat the capillary first with 1.0 M NaOH for 15 min followed by 15 min with deionized water.

> Dry the tube by flushing with N_2 at 80 °C for 1 h.

> Pull a methanolic solution of PEI 200 through the capillary with a syringe and allow adsorption for 8 h. Remove excess solution by pushing N_2 through the tube at 80 °C for 4 h.

Pull a 70% solution of ethylene glycol diglycidylether (EDGE, Aldrich Chemicals) in triethylamine into the capillary and allow to react for 1 h. Push the solution out with a nitrogen stream for 3 h.
Finally, heat the tube at 80 °C for 30 min.

5.2 Capillary Gel Electrophoresis

5.2.1 Principles of Capillary Gel Electrophoresis

Gels were introduced in electrophoresis originally as anticonvective media. Due to the "anticonvective wall effect" of capillaries with small inner diameters, the addition of gels to avoid convectional mixing of analyte zones is not necessary. So, why are gels nevertheless very interesting media for their use in CE? As we have seen, electrophoretic separations are based on different effective mobilities of the analytes. If, however, the charge densities (charge to mass ratios) of the analytes are rather similar, their separation becomes difficult or even impossible. In these cases, separation can be performed on the basis of the different molecular sizes of the analytes. Molecular sieving materials like gel matrices having controlled pore sizes can be produced for this purpose. Separation results from differences in the abilities of the different sized analytes to migrate through the gel matrix.

The most important areas of conventional slab gel electrophoresis are nucleic acid, carbohydrate and protein analysis. While proteins can be separated, in general, on the basis of their charge and size, nucleic acids such as synthetic oligonucleotides, DNA restriction fragments and higher DNA strains as well as charged oligo- and polysaccharides possess very similar charge densities limiting high-resolution separations in open-tube CZE. Thus, especially in these two latter fields, capillary gel electrophoresis (CGE) has become the mostly intensive investigated technique. In the meantime, only few efforts have been made in CGE of proteins, probably mainly due severe unforeseen problems in this field.

Two theories describe the migration of a macromolecule through a polymer network: the Ogston sieving model and the reptation model [209]. The real migration mechanism lies somewhere between these two models which have to be considered as borderline cases. In the Ogston model the gel matrix is considered as a molecular sieve consisting of a random network of interconnected pores having an average pore size ξ. The migrating solutes behave as undeformable spherical particles. According to this model, the smaller molecules migrate faster because they have access to a larger fraction of the available pores. Assuming that the mobility in the gel matrix is only a function of the polymer concentration the following equations describe the migration of a solute through the gel network:

$$\mu_{i,g} = \mu_i \cdot \exp[-T \cdot b(R_g + r)^2] \tag{5-2}$$

$$\mu_{i,g} = \mu_i \cdot \exp\left[-\frac{1}{4}\pi\left(\frac{R_g + r}{\xi}\right)^2\right] \tag{5-3}$$

$\mu_{i,g}$ electrophoretic mobility of the analyte i in the gel matrix
μ_i effective electrophoretic mobility of the analyte i (mobility in free solution)
T polymer concentration
b constant
R_g radius of gyration of the macromolecule
r average pore radius of the gel network

The term b $(R_g + r)^2$ is called the retardation coefficient k_R. Eqs. 5-2 and 5-3 hold strictly only for $\mu(E \to 0)$. Thus, the mobilities of the analytes have to be extrapolated to E = 0. A plot of log $\mu_{i,g}$ versus the polymer concentration gives

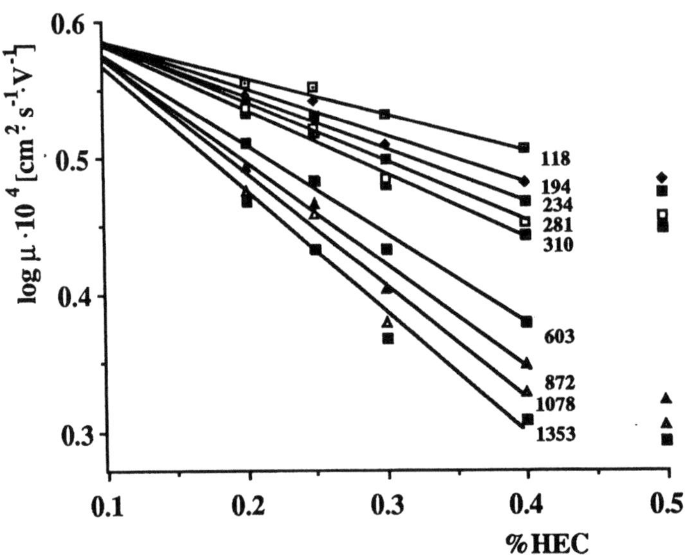

Fig. 5.6. Ferguson plots for a commercially prepared restriction digest of the ΦX 174 plasmid as a sample mixture of DNA fragments [209]. It consists of fragments with 118, 194, 234, 281, 310, 603, 872, 1078 and 1353 base pairs (bp). Experimental conditions: fused silica capillary, 50 cm x 50 µm i.d., hydrodynamic injection for 2 s, field strength 300 V·cm^{-1}, temperature 30 °C, UV detection at 260 nm, electrolyte system 89 mM TRIS - 89 mM boric acid with 5 mM EDTA. 0.2%, 0.25%, 0.3%, 0.4% and 0.5% hydroxyethylcellulose (HEC) is added to the electrolyte solution as the gel matrix. All mobilities are extrapolated to E = 0. (Reprinted with permission of Elsevier Science Publishers).

a linear relationship with a slope equal to k_R and an intercept equal to log μ_i. This so-called Ferguson plot can be used to characterize the size selectivity of a given gel matrix. Fig. 5.6. shows the Ferguson plots for 9 different sized DNA fragments separated by CGE. For the fragments consisting of 118, 194, 234, 281 and 310 base pairs the intercepts are identical, implying a value of $3.87 \cdot 10^{-4}$ $cm^2 \cdot V^{-1} \cdot s^{-1}$ for the mobility of these five fragments in free solutions. The curves are linear up to 0.4% HEC indicating that these DNA fragments migrate as globules with smaller radii than the average pore size through the gel matrix. For the larger fragments consisting of more than 310 base pairs, the linearity degrades significantly even for lower HEC concentrations. This is due to the gradual transition from the Ogston regime to the reptation regime for these larger fragments. For a flexible macromolecule moving through the gel network, the transition from the Ogston sieving model to the reptation model takes place when R_g is 1.4 ξ [209].

The reptation model describes the migration of a macromolecule through a highly crosslinked gel, assuming that the length of the flexible macromolecule chain is very large when compared to the distance between neighboring links in the fixed network. The molecule "snakes" through the tubes of the gel matrix head first. The mobility is now proportional to the reciprocal of the chain length or, for nucleic acids, to the base number N of the analyte [210]:

$$\mu_{i,g} \sim \frac{1}{N} \tag{5-4}$$

As the molecular size of the DNA fragments increases further, relation 5-4 becomes also invalid because of the deformation of the DNA coil caused by the electric field. In CGE, relation 5-4 is valid up to chain lengths of ca. 1000 base pairs, depending on the electric field applied. For a more detailed description of the migration behavior of nucleic acids in gel electrophoresis see Ref. 210.

Gel electrophoresis is usually carried out either in non-denaturing or in denaturing gels. In the latter case, denaturing agents are added to the separation buffer that distroy both intermolecular and intramolecular interactions. These interactions determine the ternary and quaternary structure of biopolymers such as nucleic acids and proteins. The most commonly used denaturing agent is urea which is added to the buffer in a very high concentration of 5 - 8 M (see Sect. 3.3.7). A special case of denaturing gel electrophoresis is polyacrylamide gel electrophoresis of proteins in the presence of SDS (SDS-PAGE). In this procedure, proteins are fully denaturated, and disulfide bonds are cleaved by heat in the presence of excess SDS and a reducing agent, e.g. mercaptoethanol. The remaining polypeptide chains bind SDS in a constant weight ratio (1.4 g of SDS per g of protein) to yield detergent-protein complexes of constant charge density and, thus, similar electrophoretic mobilities in free solution. Separation is then based only on the size or molecular weight (MW) differences in a polyacrylamide (PAA) gel. SDS-PAGE in slab gels is a standard method for purity control and MW estimation of proteins. For this purpose, a mixture of standard proteins of known MW is separated by SDS-PAGE. A linear relationship exists between their mobility and the logarithms of their MW's. By comparing the mobilities of unknown proteins under standard electrophoretic conditions with the calibration chart plotting the log of the molecular weight of standard materials versus their mobilities the MW of the protein can be determined. It has been shown that SDS-PAGE can be carried out, in principle, also in CGE [197].

In general, all buffer systems which are employed for CZE can also be used in CGE. By far the most widely used buffer solutions are TRIS - borate buffers (TBE) based on a mixture of TRIS, 50 and 100 mM, and boric acid, 50, 100 and 250 mM. The pH lies in the range of 7.6 - 8.5, depending on the concentration ratio of TRIS and boric acid. Sometimes it is adjusted to a specific pH value by adding NaOH. These buffers have been shown to work well for nucleic acid separation.

Thermal effects are less important in CGE than expected [24]. Efficiencies in dependence on the applied field strength are similar to open tubes. On the other hand, stability of the corresponding gel limits the maximal field strength that can be applied. Working at field strengths higher than 300 $V \cdot cm^{-1}$ increases the frequency of capillary breakdown (see also Sect. 5.2.2.2). In order to increase the lifetime of the gel-filled capillary, it is therefore recommended to work at moderate field strengths, at the expense of analysis time.

Depending on the characteristics and the size of the analytes, a great number of different gel matrices has been used so far. One can differ between chemical and physical gels. Chemical gels are prepared by polymerization of acrylamide in the presence of crosslinking agents to give a polyacrylamide gel matrix. In these crosslinked gels the network structure is formed by covalent bonds between the polymer chains. In physical gels the network structure is formed by physical interactions, e.g. van der Waals forces and hydrogen bonds. These physical interactions are in dynamic equilibrium, providing a more flexible structure than chemical gels. The most popular used physical gels are made of agarose. Another class of physical gels are entangled polymer solutions which are also referred as liquid gels, syrupy or viscous solutions. They consist of long chains of linear or branched polymers forming a gel or polymer network by physical interactions. If the concentration of these polymers is low, they are also described as nongel sieving media.

5.2.2 Crosslinked Polyacrylamide Gels (Chemical Gels)

5.2.2.1 General Aspects

Gels of relatively small pore sizes are commonly prepared by copolymerization of acrylamide as monomer and *N,N'*-methylenebisacrylamide (BIS) as the crosslinking agent. Crosslinked PAA gels consist of long chains of PAA crosslinked at intervals to form a random 3-dimensional network (Fig. 5.7.). The average

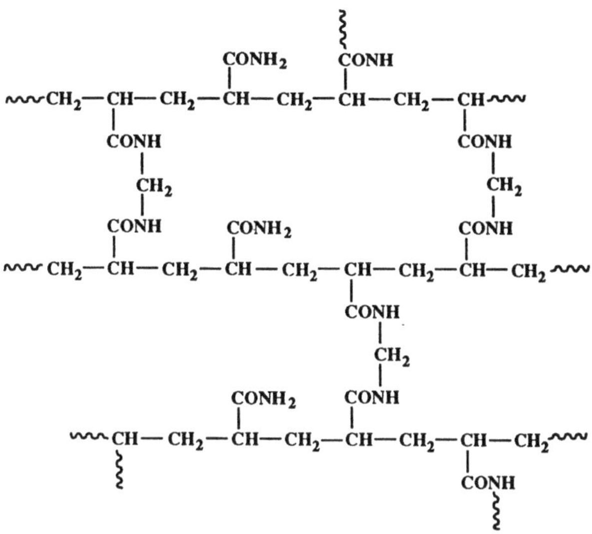

Fig. 5.7. Chemical structure of crosslinked polyacrylamide gel

pore size of these crosslinked polyacrylamide gels lies in the range of 2 - 8 nm [211] and can be varied by changing the amounts of monomer and crosslinking agent. The total monomer concentration and the concentration of crosslinking agent are expressed as % T and % C, respectively, as follows [212]:

$$\% \mathrm{T} = \frac{\text{grams of acrylamide + grams of crosslinker}}{\text{100 mL of solvent}} \tag{5-5}$$

$$\% \mathrm{C} = \frac{\text{grams of crosslinker} \cdot 100}{\text{grams of crosslinker + grams of acrylamide}} \tag{5-6}$$

Commonly used gel compositions for the separation of single-stranded oligonucleotides are, for instance, 2.5% T, 3.3% C or 4% T, 3.3% C [213, 214] and 3% T, 5% C [191]. While smaller pore sizes, e.g. 6% T, 5% C, possess a greater size selectivity for the shorter oligonucleotide chains, a gel with medium pore size, e.g. 3% T, 5% C, allows one to separate a broader MW range of oligonucleotides. To compensate for the lower resolving power of the larger pore size for short nucleotide chains, the effective column length has to be increased. Thus, good separation are achieved, but at the expense of longer analysis times. A very impressive example for the high resolution of capillaries with crosslinked PAA gels is illustrated in Fig. 5.8. showing the complete separation of all degradation products from a 431-mer poly(uridine 5'-phosphate) in 140 min. If the number of bases (or base pairs in the case of double-stranded DNA fragments) becomes too high, poor resolution and peak distortion occour with these gel compositions. An increase of the pore size can be achieved by reducing the amount of crosslinking agent for a given % T. A ten-fold reduction of the crosslinker percentage resulting in a gel composition of 3% T, 0.5% C has been found to give excellent resolution of double-stranded DNA restriction fragments having 75 - 12 000 base pairs [216]. Obviously, no single gel composition is optimal for the resolution of all mixtures of nucleic acids. If the pore size is too low the analytes are totally excluded from the capillary. If, on the other hand, the pores are too large, the analytes move freely through the capillary and no molecular sieving occurs. It is therefore necessary to select appropriate gel concentrations for particular samples.

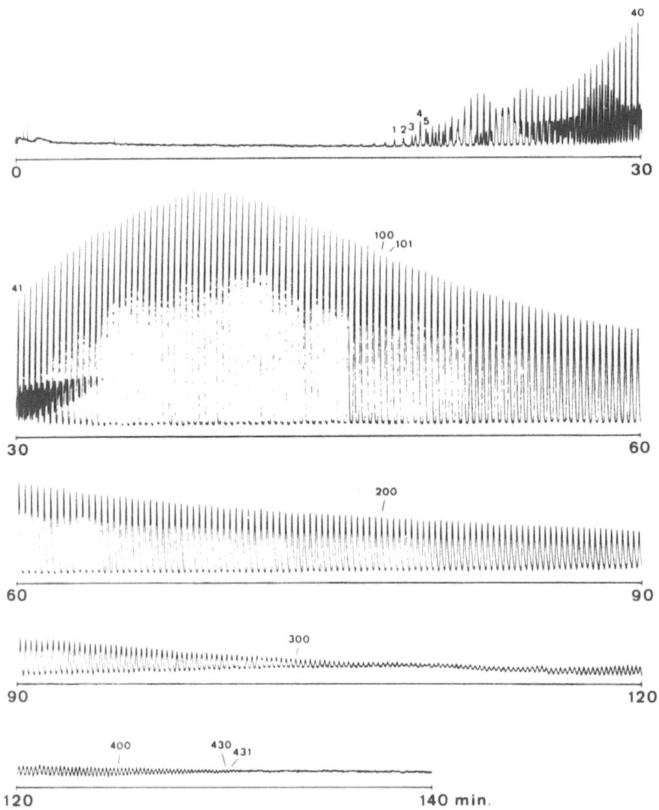

Fig. 5.8. CGE of poly(uridine 5'-phosphate). The numbers in the electropherogram indicate the number of nucleotide units. Experimental conditions: fused silica capillary, 45 cm (L_D) x 100 μm i.d., electrokinetic injection for 2 s at a voltage of 5 kV, field strength 300 V·cm^{-1}, UV detection at 260 nm, electrolyte system 100 mM TRIS - 250 mM boric acid - 7 M urea, polyacrylamide gel, 6% T, 5% C. (From Ref. 215 with permission of Hüthig Publishers)

In CGE with crosslinked PAA, the observed (cathodic) EOF is either very small, e.g. in uncoated capillaries, or virtually absent, e.g. in coated capillaries where the gel is covalently bound to the capillary surface. If an anodic separation is performed in an uncoated PAA gel-filled capillary, the EOF is opposite to the direction of migration of the analytes and leads to extrusion of the gel at the injection end of the capillary.

Sample injection can only be performed electrokinetically. Typically, sample concentrations are 20 - 200 µg/L. The samples migrate into the capillary according to their mobility. In the case of nucleic acids, the injected amount is independently of their different size and sequence. This lack of discrimination between analytes during electrokinetic injection is in contrast to the discrimination found in CZE and is based on their equal effective mobilities in free solution.

One problem arising with gel-filled capillaries is the low UV transmittance of PAA (and also other gel media) resulting in low sensitivity if UV absorbance detection is employed. The UV transmittance of PAA (6% T, 5% C) gel-filled capillaries at 260 nm, which is the typical wavelength for nucleic acid analysis, is 15% lower than that of a water-filled capillary [217]. If additives like urea or PEG are used, the transmittance is even lower. As an alternative, on-column LID [217] as well as off-column LIF with the sheath flow cuvette [129] has been described. Because the analytes are not forced through the capillary by the EOF, the electrophoretic velocity of the analytes has to be high enough to inject the eluted analyte zone into the sheath stream. The mass LOD is about 10^{-20} mol for a 20-nucleotide long strain.

5.2.2.2 Preparation of Crosslinked PAA Gel-Filled Capillaries

Capillaries filled with crosslinked PAA gels are currently available from Applied Biosystems (Micro-Gel 100, 50 cm x 50 µm), Beckman Instruments (eCAP U100P, 60 x 100 µm) and J&W Scientific (µPAGE, 75 cm x 75 µm, filled with either 5% T - 5% C or with 3% T - 3% C in 100 mM TRIS - borate - 7 M urea, pH 8.3). They are, however, only suitable for special applications and their lifetime is limited. The best approach is to prepare your own gel-filled capillaries for your special needs. Although several groups are involved in the preparation of reproducible bubble-free crosslinked gels, a simple and reliable method for the production of gel-filled capillaries with long lifetimes matching the special needs of the customer has not been developed so far. The main problem which has to be overcome is the formation of vaccuum bubbles during polymerization inside the capillary resulting from shrinkage. Also during transport and use of the gel-filled capillary (especially during injection), bubble formation can take place. Bubbles in the gel lead to diminished resolution, decreasing current during operation and changing separation patterns. Moreover, the lifetime of a gel-filled capillary is very difficult to predict. The "natural" life span of a PAA gel depends on the extent of hydrolysis of the neutral polymer to form polyacrylate. As soon as the hydrolytic process has reached a certain level, the electroosmotic flow is reactivated and the charged polyacrylate chains migrate slowly out of the column. Hydrolysis depends on the pH of the electrolyte buffer and the storage temperature. In addition, the higher the electric field strengths, the higher is the propability that capillary breakdown occurs during use. The reported numbers of injection that can be carried out with one capillary vary from 3 to over 200 runs. In the following we will give an overview of the different approaches to prevent bubble formation during preparation of gel-filled capillaries and describe procedures of which we think they are best suited to be carried out in your own laboratory.

Independently of the method used, the preparation of PAA gel-filled capillaries includes the following steps:

- preparation of a stock solution of acrylamide and the crosslinking agent in the buffer used for the separation
- pretreatment of the silica surface
- introduction of the adequately diluted stock solution of the monomers into the capillary, addition of polymerization initiator, catalyst and, eventually, stabilizing media
- in situ polymerization inside the capillary
- pre-electrophoresis

Depending on the polymerization reaction, either a radical initiator or a light sensitive initiator is added to the stock solution. The most commonly used radical initiator is $(NH_4)_2S_2O_8$ in combination with *N,N,N',N'*-tetramethylene ethylenediamine (TEMED) [212]. Alternatively, riboflavin is used as a light sensitive initiator [218, 219]. The advantages of riboflavin over the radical initiation procedure are that only very low concentrations (5 ppm or less) have to be added in contrast to peroxodisulphate which must be present at concentrations 100 times higher and that initiation of the polymerization procedure is well controlled. In addition, riboflavin does not appear to oxidize or denature samples entrapped in the gel matrix as it has been reported for peroxodisulphate. Nevertheless, the riboflavin as initiator involves the use of UV transparent capillaries. The polymerization of acrylamide can also be accomplished by γ-radiation from a ^{60}Co-source [220]. The authors point out that this approach is a simple method for the production of gel-filled capillaries, but it is rather unlikely, that a conventional analytical laboratory is equipped with a ^{60}Co-source. Additionally, the power of the ^{60}Co-source is not given in the publication.

In order to prevent extrusion from the capillary due to electroosmosis, the crosslinked gel can be chemically attached to the silica wall by a bifunctional agent [29]. For this purpose the capillary is treated with the bifunctional agent prior to the introduction of the monomer solution. Since the gel is fixed at the capillary wall, the shrinking during polymerization can lead to vacuum bubbles which appear equidistantly in the capillary. The bubble formation can be avoided by interpositioning a non-crosslinked monolayer of PAA between the gel and the bifunctional agent [215]. Bubble formation is prevented because the structure of the PAA monolayer does not impede the shrinking of the gel. The latter procedure is identical to that described in Sect. 5.1.2.1.

Another possibility of avoiding bubble formation due to shrinking in spite of the presence of the bifunctional layer is the addition of a hydrophilic polymer such as PEG to the polymerization mixture [212]. Righetti et al. [211] call these gel mixtures 'laterally aggregated' PAA gels. They are of special interest because the pore size of the PAA gel of regular %C is increased significantly in the presence of such a preformed hydrophilic polymer. This is due to aggregation of the nascent polymer chains to thick gel fibers (Fig. 5.9.). Since hydrophilic polymers coordinate large amounts of water aroung their coils, the solvent phase is perturbated. Thus, the growing PAA strings are forced to seek hydrogen bonding among themselves, rather with the surrounding solvent. Once such large bundles are formed, they are stabilized in an irreversible structure by the crosslinks. The diameters of these fibers exhibit values of several hundred nanometers. The completion of transition from small to large pore sizes depends on the chain length of the polymer. While, for instance, only 1.5% PEG 20 000 is needed for the complete transition, ≈ 10% of PEG 2000 is required for the same transition. Consequently, if PEG is added to the gel matrix to avoid gel shrinkage, one has to be aware that the sieving properties are totally different if compared with those of the pure PAA matrix.

a) b)

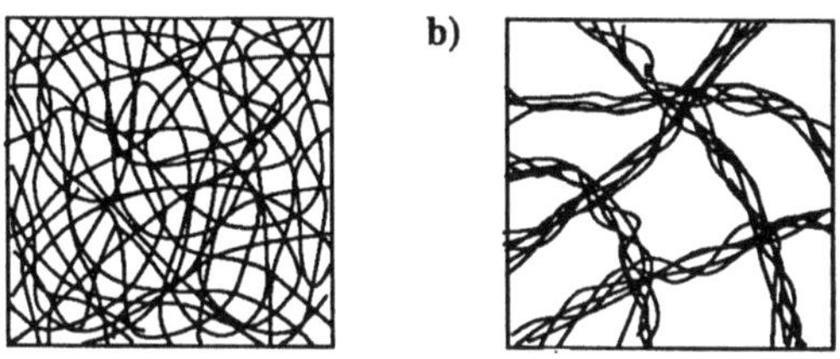

Fig. 5.9. Hypothetical model for 'laterally aggregated' gels according to Ref. 211. (**a**) Homogeneous network of crosslinked PAA and (**b**) heterogeneous network of PAA in the presence of PEG 10000

Shrinking can also be avoided by gradually polymerizing the acrylamide in only small portions of the capillary instead of simultaneous polymerization along its entire length. This can be done by exposing only a small cross-section of a capillary, which is filled with the monomer solution and a light sensitive initiator, to an irradiated area at a speed of 1 cm·min^{-1} [219]. Doing so, volumetric losses due to shrinkage can be compensated from the monomer solution which is always in front of the polymerized gel. Another possibility for gradual polymerization is the "isotachophoretic polymerization" technique developed by Novotny and coworkers [221], which is described below. Finally, Drossman et al. [222] reported the preparation of gel-filled capillaries under high pressure. The capillary with the solution to be polymerized is placed in a steel tube, the pressure is raised to 400 bar and maintained, until polymerization is completed.

Under these conditions, the density of the unpolymerized solution is comparable to that of the gel resulting in bubble free gels.

Nevertheless, it is also possible to prepare capillaries without any surface pretreatment taking the small EOF into account which extrudes a small part of the gel out of the capillary [223, 215]. The advantage of using capillaries without any bifunctional agent lies in the fact, that much less bubble formation is observed during their preparation because the gel is not fixed at the capillary surface and, thus, shrinking during polymerization is not hindered. However, the separation performance is reduced with these capillaries and their lifetime is not very long, especially if field strengths > 200 $V \cdot cm^{-1}$ are required. Moreover, the stability of these capillaries is strongly influenced by the gel concentration (the lifetime of a 3% T capillary is only half of that of a 5% T capillary) and the performance varies with the kind of sample solution [223]. Whereas capillaries with gels chemically bound to the surface can be used for more than 100 runs, only 5 - 50 separations can be performed successfully with the capillaries used in Ref. 223. Their lifetime, however, can be enhanced by occasionally cutting off the first few millimeters of the capillary at the cathodic end [215].

Another critical point which has to be considered very carefully in order to obtain reproducible gels is the introduction of the monomer solution into the capillary. In the simpliest case, the solution is introduced by a syringe. Other possibilities are the use of N_2 pressure or a vacuum line. Schomburg's group [224] employs a standard filling device as used in the production of capillary columns in GC.

Pre-electrophoresis before the first sample injection seems to be necessary to remove impurities and to ensure constant run conditions. Huang et al. [214] found, that, during a pre-electrophoretic run at constant voltage, the current steadily decreases, until after about 2 hours depending on the experimetal conditions, it reaches a stable value. Furthermore, the baseline UV absorbance, initially stable, decreases sharply and finally stabilizes at a new lower level. This suggests that a mixture of ions from the polymerization that occur along the entire length of a fresh gel-filled capillary are being eluted during this pre-run.

The following procedures for the production of crosslinked PAA gel-filled capillaries are suggested by the authors. They can be applied to 75-µm and 100-µm i.d. capillaries. It should be mentioned that, in order to check the quality of the prepared capillaries, a microscope is indispensable.

5.2.2.2.1 Radical Polymerization According to Karger and Cohen [212]

This procedure can be carried out with or without adding PEG to the polymerizing solution. According to the authors the addition of PEG allows an easier and more reproducible preparation, since the capillaries can be filled more readily with the polymerization mixture without bubble formation, and the polymerization reaction occurs smoothly. Furthermore, the capillaries containing PEG have longer shelf lives and better stability in use than capillaries without PEG.

Finally, higher voltages can be applied without bubble formation permitting higher resolution in shorter analysis times. Nevertheless, sensitivity in UV absorbance detection is reduced when using PEG because of additional loss in light transmittance. Moreover, the pore diameters should be much larger than those of the pure PAA gels [211]. If the capillaries are prepared correctly, over 100 injections can be carried out, when working at accommodate field strengths.

> (1) Remove the polyimide coating from a 1 cm section at one end of a 40 - 60 cm long, 75 or 100 μm i.d. fused silica tube.

> (2) Heat the empty capillary over night at ≈ 120 °C. Bring it again to room temperature for the following procedure and flush it with either dry NH_3 gas or fill it with 1 M NaOH for ≈ 2 hours.

> (3) Install a sheathing of small i.d. Teflon tube at one end of the capillary and fill it with 100 μL of a 50% solution of 3-methacryloxypropyltrimethoxysilane in methanol by connecting the Teflon tube to a syringe filled with the bifunctional agent.

> (4) Remove the syringe and connect both ends of the capillary via the Teflon tubing, which is also filled with bifunctional agent. Leave the capillary overnight (or for at least 3 h) at room temperature.

> (5) Remove the Teflon tubing from one end of the capillary and flush successively with 250 μL each of methanol and water to remove unreacted agent. Occasionally, cut the capillary in the middle resulting in two capillaries of 20 - 30 cm in length. Install another sheathing of Teflon on the free end of the capillary.

> (6) If PEG is to be used, dissolve 5% (w/v) PEG 8000 or above in triply distilled water which has been cooled to 10 °C. Stirr the suspension while temperature is raised slowly to room temperature. A clear transparent PEG solution with no precipitate should result. Alternatively, you can use the procedure suggested by Righetti et al.[211] who use 2.5% of PEG 10 000.

> (7) Prepare the buffer solution by dissolving 1.1 g of TRIS in 100 mL of 7 M urea solution, adding 0.01 g of EDTA and 0.1 g SDS. If PEG is used, dissolve the urea in the PEG solution instead of in bidistilled water. Adjust the pH to 8.6 by the addition of NaH_2PO_4. You can prepare an adequate buffer solution of your choice.

> (8) Dissolve 29 g of acrylamide and 1 g of BIS in 100 mL of buffer solution, giving a solution of 30% T and 3.3% C. Again, you are free to choose another crosslinking concentration in this stock solution. Store the solution at 4 °C.

> (9) Dissolve 0.2 g $(NH_4)_2S_2O_8$ in 2 mL of the buffer solution.

> (10) Filter the 3 solutions separately through 0.2 μm filters and degas them for 2 hours by treating them with ultrasound while applying a vacuum of 2.6 - 4 kPa.

> (11) Dilute the monomer solution with buffer solution to obtain the desired concentrations, for instance, diluting 2 mL with 8 mL of buffer, gives a final solution with 6% T and 3.3% C.

> (12) Add 3.0 µL of TEMED and 5 µL of the $(NH_4)_2S_2O_8$ solution to 1 mL of the solution prepared one step before. If you are working with another gel concentration determine the optimal concentration of initiator and catalyst experimentally at the desired %T and %C by varying the amount of TEMED and peroxodisulphate added to the mixture. The polymerization should be essentially complete in 45 - 60 minutes.

> (13) Connect a syringe filled with the reaction mixture with the Teflon tube at one end of the capillary and force the mixture very carefully through the capillary, until no more bubbles are observed exciting the capillary at the other end.

> (14) Remove the syringe carefully and dip both ends of the capillary in the running buffer reservoirs; keep the capillary there, while polymerization occurs.

> (15) Monitor the reaction separately in an aliquot of the reaction mixture by observing the loss of UV absorbance due to the vinyl groups at 260 nm. When the test solution indicates that the polymerization reaction is essentially complete (after ≈ 45 - 60 min), the reaction is allowed to proceed for ca. 2 hours more.

> (16) Remove the capillary ends from the buffer reservoirs and cut at least one end cleanly (see below). If the polymerization is incomplete in the first centimeters of both capillary ends, remove this area at each side.

> (17) Place the column in your electrophoretic device and apply a low electric field of ≈ 100 - 150 $V{\cdot}cm^{-1}$ to pre-electrophorese for ca. 1 hour. If the baseline is very noisy or no current is obtained, the capillary is improperly prepared and has to be rejected.

> (18) If the capillaries are not to be used immediately, they can be stored in a refrigerator by closing both ends with small rubber septa.

5.2.2.2.2 Photopolymerization According to Poppe and Coworkers [219]

Two methods are described, one in which the polymerization is performed at lower temperature and one in which the polymerization gradually occurs by pulling the capillary out of a dark box at room temperature. According to the authors, a success rate of about 80% has been achieved routinely with the first method. For the second method, the measure of success is somewhat lower. 10 - 20 injections can be carried out with capillaries made with the first method. Then, it is necessary to cut off a few millimeters from the top of the capillary in order to restore the separation performance. By using the second method for gel preparation, at least 50 - 60 injections can be done without any bubble formation at the injection end.

> Prepare a UV transparent capillary as described in steps (1) - (5) of Sect. 5.2.2.2.1. Alternatively to step (2), flush the tube with 1) 1 M KOH for 1 h, 2) water for 30 min, 3) 30 mM HCl for 30 min and 4) water for 30 min.

> Prepare the buffer solution containing 100 mM boric acid, 100 mM TRIS, 2 mM EDTA and 7 M urea. Adjust the pH to 8.7 by the addition of NaH_2PO_4. Alternatively, you can prepare an adequate buffer solution of your choice.

> Prepare the gel-forming solution according to steps (8), (10) and (11) of Sect. 5.2.2.2.1.
> Prepare an aqueous saturated solution of riboflavin (0.008% (w/v)) and mix 1 mL of gel-forming solution with the saturated riboflavin solution.
> Connect a syringe filled with the reaction mixture with the Teflon tube at one end of the capillary and force the mixture very carefully through the capillary, until no more bubbles are observed exciting the capillary at the other end.
> Remove the syringe carefully and seal both ends of the capillary with rubber septa.
> Place the capillary into a 3 L glass beaker containing an ice-water mixture. Irradiate the capillary overnight with a UV lamp through the bottom of the beaker. Alternatively, for gradually polymerization, place the capillary in a dark box and pull it out into the irradiated area at a speed of 1 $cm \cdot min^{-1}$.
> Cut the capillary to the desired length and pre-electrophorese.

5.2.2.2.3 Isotachophoretic Polymerization According to Novotny and Coworkers [221]

This method involves the principle of isotachophoretic polymerization, in which the polymerization initiator is introduced by electromigration to promote gradual polymerization of the acrylamide in a sequential manner along the length of the capillary. Modification of the inner surface of the capillary is essential to prevent EOF during the process. The capillary which is filled with a solution of acrylamide, BIS and triethanolamine hydrochloride as catalyst is inserted into the electrode reservoirs containing $(NH_4)_2S_2O_8$ at the cathode and triethanolamine hydrochloride at the anode. When voltage is applied, persulfate ions enter the capillary isotachophoretically behind Cl^- as the leading ion and, thus, initiates polymerization gradually. The speed of polymerization is primarily determined by the applied voltage. According to the authors, this procedure is superior to other methods for polymerizing gels inside capillaries of 50 µm i.d. because of the reduced bubble formation in the gels during polymerization.

> Remove the polyimide coating from a 1-cm section at one end of a 40 - 60 cm long, 50-µm i.d. fused silica tube.
> Coat the inner capillary wall with linear PAA as described in Sect. 5.1.2.1.
> Prepare a solution of 5.8% acrylamide (T = 6%), 0.18% BIS (C = 3%) and 100 mM triethanolamine hydrochloride, filter the solution through a 0.2 µm filter and degas it for 2 hours by treating them with ultrasound while applying a vacuum of 2.6 - 4 kPa.
> Fill the deaerated solution into the capillary and place one end into the cathode reservoir containing 10% $(NH_4)_2S_2O_8$. Place the other end into the anode reservoir containing 25% triethanolamine hydrochloride.
> Apply an electric field of 4 $V \cdot cm^{-1}$ for 8 - 12 h.
> Replace the electrode reservoirs by vials containing the background electrolyte, e.g. 100 mM TRIS, 200 mM MES, 1% SDS.

> Equilibrate the capillary with the background electrolyte by applying a voltage of 500 V for ca. 4 h. Increase the voltage stepwise so that the Joule heat generated does not exceed 0.5 $mW \cdot cm^{-1}$. The equilibration procedure is considered to be complete, when the current stabilizes at the maximum applied voltage or when all moving boundaries have passed through the detection cell and the detector output has been stabilized.

Hint: To achieve the highest resolution, it is necessary that at least the front end of the capillary is cleanly and squarely cut perpendicular to the axis of the capillary. Otherwise the surface of the polymer gel exposed at the end of the capillary is uneven, making it impossible to inject a narrow band of sample. To ensure a clean cut, form a tight sheath of Teflon around the end of the capillary and cut through the sheath, the capillary and the gel with a microtome leaving a smooth surface of gel material exposed at the end of the capillary. Alternatively, use a sapphire cleaver to score the capillary carefully at right angles to its axis and break it cleanly by bending.

5.2.3 Physical Gels

In order to avoid gel shrinkage, bubble formation and matrix collapse, a new way has been achieved in CGE involving the use of physical gels such as agarose and so-called entangled polymer solutions. The main difference between these "gels" and the crosslinked PAA gels is that the pores are created by physical interactions rather than chemically crosslinking. The composition of physical gels is by far more versatile than that of the chemical gels and can be comprised of a large number of different polymers.

The mechanism of separations in entangled polymer solutions has been investigated by Grossman and Soane [209]. An important difference exists between dilute and concentrated polymer solutions. Whereas in dilute polymer solutions the polymer chains are hydrodynamically isolated from each other, in concentrated solutions the chains begin to overlap and interact. The polymer volume fraction Φ where the chains begin to interact with one another is called the overlap threshold, Φ^*. Above the concentration of the overlap threshold the polymer solution is said to be entangled. As a consequence, sieving of macromolecular solutes takes place. Experimentally, Φ^* can be determined by plotting the logarithm of the specific viscosity versus the polymer volume fraction. For $\Phi < \Phi^*$, the slope of the curve is ≈ 1.0. For $\Phi > \Phi^*$, the slope increases. Φ^* is the point where the two curves cross each other.

5.2.3.1 Agarose Gels

Agarose gels are easily and rapidly prepared without catalysts and initiators. Agarose is obtained from algae and consists of a polysaccharide network of 1,3-

linked β-D-galactopyranose and 1,4-linked 3,6-anhydro-α-L-galactose. In natural agar, some of the sugar residues are replaced by sulfate, methoxy, pyruvate and carboxy groups. It is known from conventional slab gel electrophoresis on agarose gels, that the acidic groups cause electroosmosis toward the cathode. Agarose especially purified for use in electrophoresis contains only low concentrations of acidic groups and is commercially available from many suppliers, e.g. Serva and Sigma. Agarose gels exhibit average pore diameters of several hundred nanometers up to several µm and possess high mechanical strength even at low concentrations. Furthermore the gels are biologically inert and stable in the pH range of 4 - 9. The gelling temperature of agarose is ca. 35 - 40 °C, the melting point ca. 65 - 70 °C. The mechanism of gelation primarily involves the formation of double helices which rearrange to form bundles.

So far, only few applications of agarose filled capillaries are reported. Schomburg's group [224] have successfully separated medium sized DNA fragments and unsaturated sulfonated disaccharides. Bocek and Chrambach [225] have also shown the separation of DNA fragments by using liquified agarose solutions, maintained above their gelling temperature during separation.

Capillaries filled with agarose gels of 0.3 - 5% by weight which are stable for a few days of operation and can be prepared according to the following procedure without the need of a special pretreatment of the capillary wall [224]:

> Mix the required amount of agarose with 10 mL of buffer in a reaction vessel which can be tightly closed during subsequent heat treatment in order to prevent water evaporation and, thus avoiding changes in the agarose concentration.

> To improve the stability of the gel, a small amount of a polyalcohol, such as sorbitol, can be added.

> Place the suspension in a water bath at a temperature of ca. 100 °C for 15 min.

> Transfer the liquified agarose sol to an ultrasonic bath and degas it at 60-70 °C.

> Rinse the capillary with the buffer used for the production of the gel.

> Introduce the agarose gel obtained into the capillary taking care that the temperature is higher than the melting point (≈ 65 °C) to prevent gelling during filling. For this purpose you can either use your hydrodynamical injection system or a filling station as it is normally used for coating capillaries for gas chromatography.

> Allow the gel to form in the capillary at room temperature for 1-2 h.

> Place the two ends of the capillary into the electrode vials containing the separation buffer.

> If the gel is not used immediately, it can be stored a few days under refrigeration after gelling.

Alternatively, capillaries containing liquified agarose above their gelling temperature can be prepared as follows [225]:

> Coat the capillary with linear PAA according to the procedure described in Sect. 5.1.2.1.

> Prepare the agarose solution by suspending the appropriate amount of agarose in the separation buffer, e.g. 89 mM TRIS - 89 mM boric acid, 2.5 mM EDTA, and bring the suspension to a boil on a magnetic stirrer hot plate.
> Reweight the solution and replenish with water to compensate for the loss during boiling and store the solution at 50 °C in a thermostated oven.
> Fill the agarose - buffer solutions into the electrode vials just prior to analysis and keep the temperature at 40 °C during the hydrodynamic filling procedure and the subsequent electrophoresis.

5.2.3.2 Linear Polyacrylamide Gels

Linear non-crosslinked or liquid PAA was introduced as a sieving medium for electrophoresis by Bode as early as in 1977 [226, 227] and has been employed for the separation of proteins in slab gels. The minimum monomer concentration required to achieve sufficient anticonvective stabilization in slab gels is approximately 10% T. The anticonvective properties of the narrow capillaries in CGE allow the use of a much broader concentration range of linear PAA from 3 to 14% T. The viscous character of such compositions range from nearly liquid-like at 3% to that of a gelatinous material at 14%. Chiari et al. [228] investigated the viscosity of different concentrated PAA solutions between 3% and 8% T. The values range from ≈0.05 Pas to 100 Pas. The viscosity of solutions below 3% T could not be determined because it is too low. On the other hand, viscositities above 8% T are too high for dynamic measurement and the solutions behave essentially as a solid. Gels of 3 and 4% T show Newtonian behavior, whereas the higher concentrations are non-Newtonian. They conclude, as a consequence of the high viscosities of gels > 4%, that it seems to be impossible to pump these gels in or out of a capillary. A pressure of $8 \cdot 10^{12}$ Pa should be needed, for instance, to fill the capillary with a gel of 8% T. Thus, only low concentration gels can be simply manipulated by preparing them outside the capillary and exchange them easily after each run by refilling the capillary with a new gel solution. The majority of users, however, employ gels of higher percentages of T - at least 6%.

The size selectivity is again a function of polymer concentration. The sieving mechanism in linear PAA has been suggested to be similar to that occurring in crosslinked gels [226, 227]. According to the model of Bode, the molecules migrate through "dynamic pores" which are formed by the fluctuating polymer chain network. These dynamic pores also exist in crosslinked PAA gels, but the enlargment of the pores is not as flexible as in linear PAA gels. Although linear gels were originally introduced in CGE because of their believed larger pore sizes, it has been shown, that they are also well suited for the separation of oligonucleotides [228, 216]. Heiger et al. [216] have investigated the size selectivity of linear PAA gels as a function of polymer concentration for the separation of DNA fragments ranging from 72 - 1353 base pairs. Their results indicate that the sieving capability and with this the analysis time increases with the percentage of T (6%, 9% and 12%). Thus, for a given field strength and base pair size range, the column length can be shortened with increasing polymer concentration, resulting in a decrease in the analysis time from 30 to 12 minutes. We can conclude that linear PAA gels have the same application range as crosslinked gels.

As already mentioned, the gels above 3 - 4% T have to be prepared again in situ by filling the monomer solution into the capillary. Even though the filling solution is highly viscous, EOF would slowly pump out the gel, if the capillary surface was not coated. If using linear PAA as the sieving medium the coating procedure is simplified in the way, that, once a capillary, pretreated with bifunctional agent, is filled with the monomer solution, surface coating with linear PAA strings and gel formation occurs simultaneously. Some researchers, however, work with untreated capilarries taking the small EOF into account. Chiari et al. [228] have found that it is impossible to drive the conversion of acrylamide to linear PAA to better than 80 - 85% for a 10% T mixture because of the extreme viscosity of a physical gel, as opposed to a chemically crosslinked gel. This means that, under these conditions, the concentration of unreacted monomers is as high as 300 mM. Two problems arise from these large amounts of ungrafted acrylamide left over: the high UV absorbance of acrylamide (> 20 mAU) and its toxitity and potential reactivity toward macromolecules. The authors propose a chemical scavenging method to reduce the amount of free acrylamide: after polymerization, cysteine is drven into the capillary from the cathode and allowed to react with acrylamide to give a charged acrylamido adduct, that can be driven out electrophoretically. Their whole procedure for in situ polymerization of linear PAA giving capillaries, stable for two weeks of operation, is as follows:

> Treat the capillary with the bifunctional agent according to Sect. 5.1.2.1.

> Prepare a 100 mM TRIS - borate buffer by dissolving the appropriate amounts of TRIS and boric acid in distilled water and adjust the pH to 8.6 by the addition of NaH_2PO_4. You can prepare an adequate buffer solution of your choice.

> Dissolve 0.5 µL TEMED and 0.5 µL of a 40% $(NH_4)_2S_2O_8$ solution per mL of buffer and filter and degas the solution.

> Prepare a 10% T (0% C) acrylamide solution by dissolving the appropriate amount in the gelling solution. Again, you are free to choose another crosslinking concentration in this stock solution.

> Fill the capillary with the gelling solution and allow to proceed polymerization for 2 h at room temperature (to simultaneously coat the capillary wall and prepare the sieving medium). The ends of the capillaries are placed into electrode vials containing the 100 mM TRIS - borate buffer.

> After the polymerization step, replace the electrode solution with a solution containing 200 mM TRIS - borate, pH 9.0, and 100 mM cysteine. After the addition of cysteine, re-adjust the pH to 9.0 by adding TRIS.

> Allow the solution to migrate into the capillary toward the anode by applying an electric field of 20 V·cm^{-1} for 10 h. The current should rise slightly from ≈ 3.5 to ≈ 5 μA.

> Replace the electrode vials against vials containing 200 mM TRIS - borate, pH 9.0, and apply an electric field of 12 V·cm^{-1} for 4 h. The current should remain constant, but after 50 min or so the rear boundary of the reaction products emerges from the capillary resulting in a sharp drop of the UV absorbance.

> Equilibrate the capillary prior to CGE with the separation buffer by pre-electrophoresing for 3 h at 12 V·cm^{-1}.

5.2.3.3 Molecular Sieving in Entangled Polymer Solutions of Low Viscosity

Besides linear PAA, there are a number of other hydrophilic polymers, which can be used as molecular sieving media above their overlap threshold Φ*. Because of the low concentrations needed, these polymers possess only low to moderate viscosities allowing easy handling of these sieving media. In contrast to the linear PAA, these polymer solutions do not have to be prepared by in situ polymerization. The polymer is simply dissolved in the electrolyte solution. They can easily be filled into and pumped out of the capillary. Another advantage is their low UV transmittance (see below).

Zhu et al. [86] have first suggested the use of hydroxymethyl cellulose (HMC) as a sieving medium for the separation of DNA fragments and referred this technique to as nongel sieving. Methyl cellulose and hydroxypropylmethyl cellulose are other examples for linear polymers derived from cellulose. Interestingly, the overlap threshold concentration is rather low for those cellulose derivatives in comparison to the concentrations needed to prepare a linear PAA gel with the same size selectivity. The overlap threshold for hydroxyethyl cellulose, for instance, is only 0.3% [209]. The mentioned cellulose derivatives with molecular weights of approximately 900 kDa have been found to provide a very good sieving effect for DNA fragments if they are used at concentrations of about 0.5%. Similar separation patterns are received for DNA size standards (88 - 1746 base pairs) as for a linear PAA gel of 8% T by using TRIS - borate, pH 8.0, as the separation buffer.

Cellulose derivatives seem to be not suited for the separation of proteins. The group of Karger [229] showed recently, that SDS gel electrophoresis of proteins can be performed in polymer networks of dextran and polyethylene glycol. The relatively low viscosity of the buffer media results in a significant increase in column lifetime because of the simple replacement of the polymer network. If the dextran or PEG solution is replaced after each run, the column is still usable after 300 injections. In addition, UV detection of the protein bands at 214 nm becomes possible leading to a significant increase of sensitivity compared to gel electrophoresis in PAA gels where proteins have to be detected at 280 nm. Polymer network formulations based on these so-called UV transparent polymer networks are commercial available from Beckman Instruments.

If electroosmosis should be prevented, the capillaries have to be coated with linear PAA as described in Sect. 5.1.2.1. By using dextran at a concentration of 10% (w/v) the separation of six protein standards is comparable to that obtained in conventional PAA gels. Size selectivity is dependent on the molecular weight of the dextran. The higher the molecular weight, the higher is the obtained selectivity. Thus, dextran of MW 2 000 000 gives better results than dextrans of 72 000 or 500 000. Whereas dextran is a branched polymer, PEG consists of linear polymers. As in the case of cellulose derivatives which are also linear polymers, lower concentrations are employed. By using a 3% PEG (MW 100 000) polymer network a rapid separation of the standard SDS-protein complexes is achieved.

Molecular sieving media are commercial available from Bio-Rad Laboratories and Beckman Instruments as readymade polymer solutions. Bio-Rad offers a PCR analysis buffer containing the entangled polymer in TRIS-borate-EDTA buffer for the separation of PCR fragments as large as 600 base pairs. Beckman's eCAP SDS 200 polymer network sieving system can be used for protein separation and mass determination in the range of 29 - 205 kDa.

5.3 Micellar Electrokinetic Chromatography

Micellar electrokinetic chromatography (MEKC) allows the resolution of even uncharged molecules, thus extending the application range of CE. Pioneering work in this technique was done by Terabe and coworkers [195, 230, 231]. Since their initial work, MEKC has developed into a routine separation technique not only for uncharged compounds but also for a large number of ionic compounds to improve the selectivity. The purpose of this section is to describe the basic separation principles of MEKC and the influence of different separation parameters on the performance of MEKC.

5.3.1 Principles of MEKC

Micelles are molecule aggregates of surfactants that are compounds with amphiphilic properties. Amphiphilic molecules contain both hydrophilic and hydrophobic regions in their structure. Depending on the hydrophilic functional group, surfactants are classified as

- anionic, e.g. sulfonate or carboxylate groups
- cationic, e.g. alkylammonium groups
- zwitterionic, e.g. sulfobetains groups
- nonionic, e.g. polyethoxy groups

The hydrophobic region is usually a straight or branched alkyl chain or a steroidal skeleton.

Table 5.4. Critical micellar concentration (CMC) and average aggregation number (AN) of surfactants in water at 25 °C [232]

Surfactant	CMC [M]	AN
anionic		
lithium dodecyl sulphate	$8.77 \cdot 10^{-3}$	
sodium dodecyl sulphate	$8.10 \cdot 10^{-3}$	62
sodium tetradecyl sulphate	$2.20 \cdot 10^{-3}$	138
sodium dodecanate	$2.40 \cdot 10^{-2}$	56
sodium cholate	$1.40 \cdot 10^{-2}$	3
sodium deoxycholate	$5.00 \cdot 10^{-3}$	4 - 10
sodium taurodeoxycholate	$3.00 \cdot 10^{-3}$	8
cationic		
cetyltrimethylammonium chloride	$1.3 \cdot 10^{-3}$	
cetyltrimethylammonium bromide	$9.2 \cdot 10^{-4}$	23
dodecylammonium chloride	$1.5 \cdot 10^{-2}$	55
zwitter-ionic		
N-dodecyl-*N*,*N*-dimethylammonio-3-propane sulfonate (Sulfobetain SB 12)	$3.3 \cdot 10^{-3}$	55
3-(3-cholamidopropyl)dimethylammonio-3-propane sulfonate (CHAPS)	$4.2 - 6.3 \cdot 10^{-3}$	9 - 10
non-ionic		
octylglucoside	$2.5 \cdot 10^{-2}$	27
digitonine	$6.7 - 7.3 \cdot 10^{-4}$	60
n-dodecylglucoside	$1.9 \cdot 10^{-4}$	
n-dodecyl-β-D-maltoside	$1.9 \cdot 10^{-4}$	98
dodecyl-(polyethyleneglycol[23])-ether (BRIJ 35)	$9.0 \cdot 10^{-5}$	40
polyoxyethylene[20]-sorbitane monooleate (TWEEN 80)	$1.0 \cdot 10^{-5}$	
polyoxyethylene[20]-sorbitane monolaurate (TWEEN 20)	$5.9 \cdot 10^{-5}$	

At low concentrations in aqueous media the surfactant molecules are in a molecular-disperse stage, where they may be associated as dimers, trimers or oligomers depending on the type of surfactant. If the concentration exceeds the so-called critical micellar concentration (CMC) the molecules aggregate to form spherical micelles. This spontaneous aggregation is caused by increasing hydrophobic interactions of the surfactants at higher concentrations. It should be noted that hydrophobic interactions should not be understood as attraction forces, they are moreover pushed together by the polar medium to diminish the degree of order of water (entropy increase). The CMC value is depending on the type of surfactant and on external factors like temperature, ionic strength and pH of the medium. Moreover, chaotropic ions and organic solvents destabilize micellar systems by breaking water structures and decrease the polarity of the medium, respectively. The average number of molecules per micelle is termed the aggregation number (AN). Table 5.4. summarizes the CMC values and average aggregation number of some often used surfactants.

In aqueous solutions the structure of the micelle is commonly spherical but may change with higher surfactant concentrations. Within the spherical structure the hydrophobic moieties of the surfactants are oriented toward the center of the spheres forming non-polar "droplets" with molecular dimensions. The surface of the micelles is formed by the hydrophilic groups which are in contact to the aqueous phase. The diameter of typical micelles lies in the range of 30 - 50 Å. Macroscopically, micellar solutions behave like ordinary molecular-disperse solutions. For instance, micelles pass normal filters and do not show light scattering effects. Micellar systems are dynamic systems which are always in equilibrium with surfactant monomers in the solution. A schematic representation of an ionic micelle is shown in Fig. 5.10.

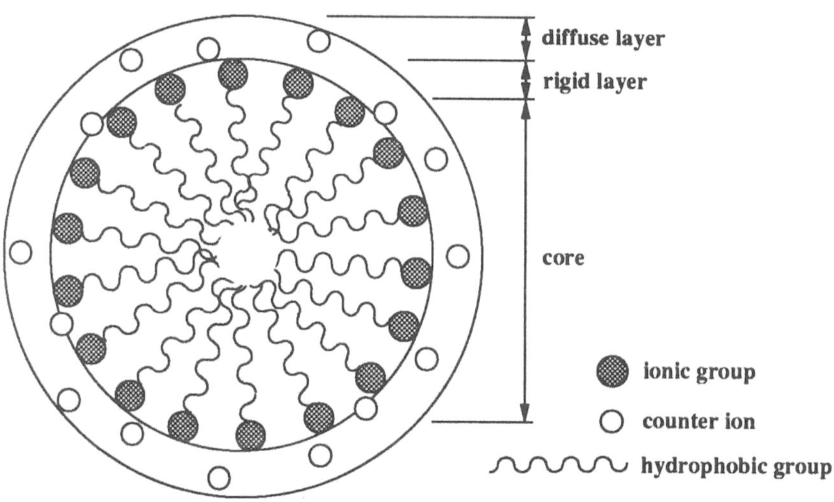

Fig. 5.10. Simplified schematic representation of an ionic micelle with its most important regions

As the surfactant concentration is further increased in the aqueous phase, the number of micelles as well as the micelle shape, size and conformation may change significantly. At very high concentrations, large aggregates, lyotropic liquid-crystalline phases and finally solid gels can be formed.

Micelles of ionic surfactants migrate electrophoretically due to their surface charge, e.g. negative sodium lauryl sulphate (SDS) micelles to the anode and positive cetyltrimethylammonium bromide (CTAB) micelles toward the cathode. This is schematically illustrated in Fig. 5.11. for an anionic micelle. Because in most cases the electroosmotic velocity is higher than the electrophoretic velocity of the micelle, the net velocity even of negatively charged micelles is toward the anode. If a solute is introduced into a micellar system, it will partition between the hydrophobic micellar phase and the aqueous phase with a particular partition coefficient P depending on the polarity of the analyte. Based on differential solubilization in the micellar phase, the partition between the slow moving micelles and the fast moving aqueous phase causes differential retention and resolution of the solutes. Thus, with respect to the separation mechanism MEKC is analogous to reversed phase (RP) liquid chromatography.

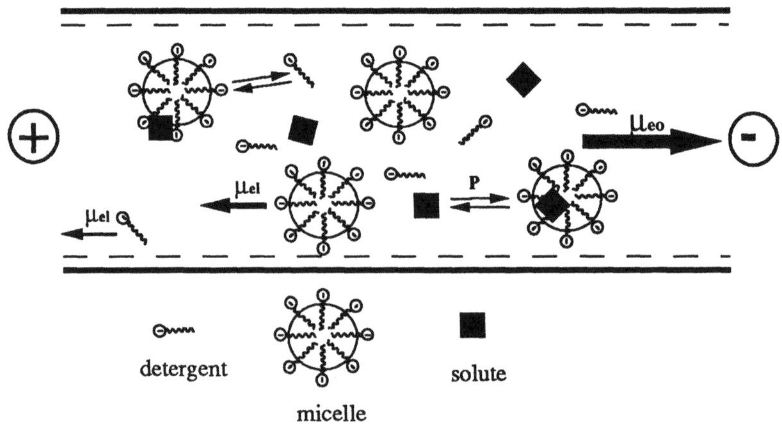

Fig. 5.11. Schematic representation of the principle of MEKC

However, in MEKC chromatography is carried out in a homogeneous solution which enables a rapid establishment of partition equilibrium between the micelle (the so-called pseudostationary phase) and the aqueous phase. This and the low axial dispersion are the origins of the high efficiency. Another significant difference to reversed phase liquid chromatography is, that uncharged analytes elute in a time interval which is given by the retention time of a solute that does not interact with the micelle, t_{eo}, and the retention time of a solute that is completely solubilized in the micelle, t_{mc}. This means that only the time interval t_{mc} - t_{eo} is available for the separation. A chromatogram of the separation of uncharged solutes by MEKC is shown in Fig. 5.12.

The capacity factor k' of the analyte can be calculated from the retention time as [230]

$$k' = \frac{t_i - t_{eo}}{t_{eo} \cdot \left(1 - \frac{t_i}{t_{mc}}\right)} \tag{5-7}$$

t_i retention time of analyte i

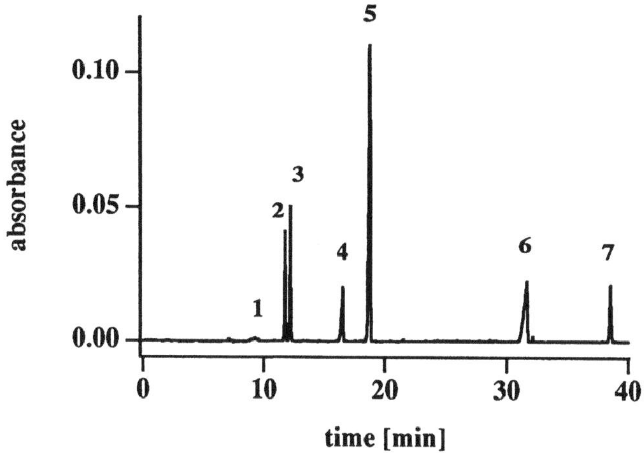

Fig. 5.12. Separation of uncharged solutes by MEKC: (**1**) methanol, (**2**) phenol, (**3**) benzyl alcohol, (**4**) benzene, (**5**) nitrobenzene, (**6**) toluene, (**7**) Sudan III. **1** represents the elution time t_{eo} of a solute with no interaction with the micelle and **7** the elution time t_{mc} of the micelle. Instrument: Beckman P/ACE 2000; experimental conditions: fused silica capillary, 57 cm x 75 μm i.d., hydrodynamic injection for 1 s, field strength 263 $V \cdot cm^{-1}$, temperature 25 °C, UV detection at 214 nm, electrolyte system 20 mM sodium phosphate buffer with 50 mM SDS, pH 8.0

For t_{mc} becomes infinite, Eq. 5-7 is equivalent to the well-known definition of k' in liquid chromatography. The time window of elution for uncharged solutes can be expressed by the ratio t_{eo}/t_{mc}. The smaller this value is, the larger is the time window available for the elution of the analytes. The capacity factor can also be calculated from the electrophoretic mobilities. For neutral solutes k' is given by:

$$k' = \frac{\mu_i}{\mu_{mc} - \mu_i} \tag{5-8}$$

μ_{mc} electrophoretic mobility of the micelle [$cm^2 \cdot V^{-1} \cdot s^{-1}$]

For charged analytes an electrophoretic separation mechanism is superimposed to chromatography. Thus, the net mobility μ_i' of an ion is the weighted average of the mobility of the micellar phase μ_{mc} and its own mobility in the aqueous phase μ_i [233].

$$\mu_i' = \left(\frac{k'}{k'+1}\right) \cdot \mu_{mc} + \left(\frac{k'}{k'+1}\right) \cdot \mu_i \qquad (5\text{-}9)$$

In this case k' is to be calculated as

$$k' = \frac{\mu_i' - \mu_i}{\mu_{mc} - \mu_i'} \qquad (5\text{-}10)$$

A number of assumptions are made when deriving this equation. For instance, it is assumed that the mobility of the micelle does not change with the solubilization of a solute. In addition, secondary chemical equilibria with buffer constituents are presumed not to occur. For low micelle concentrations the capacity factor is directly proportional to the micelle concentration as given in Eq. 5-11.

$$k' = P \cdot V\,([S] - CMC) \qquad (5\text{-}11)$$

P partition coefficient
V molar volume of the surfactant [$L \cdot mol^{-1}$]
[S] total surfactant concentration [M]
CMC critical micelle concentration [M]

Eq. 5-11 holds for both non-charged and ionizable solutes. It is obvious from this equation that the retention and the selectivity of solutes can be manipulated by the total concentration of the surfactant in the buffer system. By plotting the k' of a solute versus the total surfactant concentration, a linear relationship should result from which the partition coefficient and the critical micelle concentration can be calculated from the slope and the y-intercept, respectively (Fig. 5.13.). Intercepts of the plots at k' = 0 are found at an average concentration of 6 mM which can be interpreted as the CMC under these conditions. This value is lower than the reported value of 8.1 mM in water at 25 °C. The deviation may be due to the buffer components which increase the ionic strength and the polarity of the system. The partition coefficient can be calculated from the slope taking the molar volume of SDS, V = 0.2515 $L \cdot mol^{-1}$, into account [230]. Table 5.5. summarizes the partition coefficients at 25 °C.

As already mentioned above, MEKC exhibits a limited elution range defined by the ratio of t_{eo} and t_{mc}. In this respect MEKC most closely resembles size exclusion chromatography (SEC). The maximum number of peaks which can be resolved in MEKC is given by the peak capacity n. n is defined by the time window of elution and depends on the efficiency of the separation system according to:

$$n = 1 + \frac{\sqrt{N}}{4} \cdot \ln \frac{t_{mc}}{t_{eo}} \tag{5-12}$$

N can be calculated from a chromatogram using Eqs. 2-29 or 2-30.

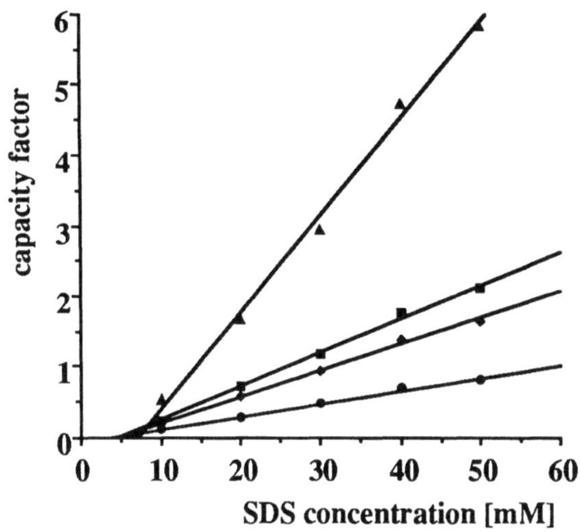

Fig. 5.13. Dependence of capacity factor k' on the concentration of SDS. Experimental conditions are from Fig. 5.12. except the concentration of SDS in the buffer system which is 10, 20, 30, 40 and 50 mM. *Circles* - benzyl alcohol, *diamonds* - benzene, *squares* - nitrobenzene and *triangles* - toluene

Solute	P
benzyl alcohol	72
benzene	148
nitrobenzene	189
toluene	545

Table 5.5. Partition coefficients P at 25 °C calculated from the slopes of k' versus the SDS concentration in Fig. 5.13.

The resolution of two analytes is derived by inserting Eq. 5-7 into the classic equation for calculating the resolution in elution chromatography [230]:

$$R = \frac{\sqrt{N}}{4} \cdot \frac{\alpha - 1}{\alpha} \cdot \frac{k_2'}{1 + k_2'} \cdot \frac{1 - \frac{t_{eo}}{t_{mc}}}{1 + \left(\frac{t_{eo}}{t_{mc}}\right) \cdot k_1'} \tag{5-13}$$

The resolution in MEKC is highly dependent on the term t_{eo}/t_{mc}. To improve the resolution it is advantageous to decrease t_{eo}/t_{mc} from values of 0.25 or so typically found in the literature to values of 0.03 or even 0.01. This can be accomplished by reducing or suppressing the EOF by adding organic modifiers such as methanol to the buffer system or to use coated capillaries. In practice, the resolution of two peaks in the chromatogram is obtained by applying Eq. 2-48.

Foley [234] presented a theory for the optimization of the capacity factor and the corresponding surfactant concentration with respect to the resolution of non-charged analytes. Accordingly, the capacity factor for optimum resolution can be calculated as

$$k'_{opt} = \sqrt{\frac{t_{mc}}{t_{eo}}} \qquad (5\text{-}14)$$

By dividing k'_{opt} by $t_{mc} - t_{eo}$ the capacity factor for the best resolution per time unit is obtained. It is independent of the specific time window of a given separation system and lies in the range of 1.2 to 2. As the author pointed out, separations obtained by using the best capacity factor for resolution per time unit instead of resolution might be slightly poorer with respect to the resolution but should be much faster.

The optimum surfactant concentration $[S]_{opt}$ depends on the analyte partition coefficients and on t_{mc}/t_{eo} as follows

$$[S]_{opt} = \frac{\sqrt{\frac{t_{mc}}{t_{eo}}}}{V \cdot P} + CMC \qquad (5\text{-}15)$$

Besides the surfactant concentration, a number of effects have an influence on the separation in MEKC, mainly the type of surfactant, the temperature, the buffer pH and buffer additives.

5.3.2 Effect of the Type of Surfactant

The choice of surfactant in MEKC defines the physico-chemical nature of the micelles, analogous to the choice of stationary phase in HPLC. However, in contrast to HPLC a change of the micellar phase in MEKC requires only the capillary to be rinsed with the new buffer system.

SDS micellar systems are reported to be similar to an octadecylsilane (ODS) stationary phase in HPLC for moderately water-soluble compounds [235]. This system is very stable and suitable for a broad range of solutes. Smaller anionic surfactants with shorter alkyl chains such as sodium decylsulphate (STS) are less efficient and show poor retention reproducibility. On the other hand, the use of surfactants with longer alkyl chains, e.g. C_{14}, is limited because of the low solubility in aqueous media.

Hint: *Surfactants of the SDS type form precipitates with potassium and alkaline earth metal ions. Therefore, these cations must be avoided as buffer constituent.*

Surfactants such as cetyltrimethylammonium chloride (CTAC) are useful for larger molecules where the solubilization in the buffer system is limited. Note, that the migration directions of both the electroosmotic flow and the electrophoretic migration of the cationic micelle are the opposite to those in the SDS system. Because of the adsorption of CTAC to the fused silica and the charge reversal, the electroosmotic flow is from the cathode to the anode and the electrophoretic migration of the micelle is from the anode to the cathode.

Swedberg [236] investigated the use of non-ionic and zwitterionic surfactants in MEKC. The addition of octylglucoside or CHAPS showed several advantages over ionic surfactants. Both compounds have less influence on the magnitude of EOF. Even at high concentrations they do not contribute to the conductivity of the buffer and thus, to the electric current and heat generation. Finally, these compounds are known to stabilize protein structures, particularly those of membrane proteins.

Bile salts are an interesting class of surfactants which have not attained much attention in CE so far. Table 5.6. shows the structures of the most important

Table 5.6. Structures of bile salts

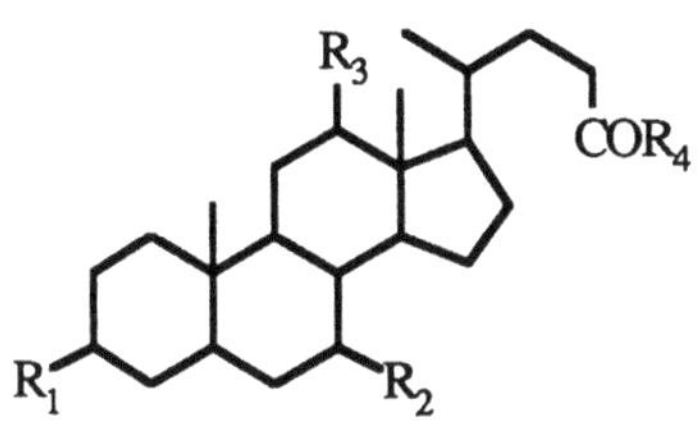

Bile salt	R_1	R_2	R_3	R_4
sodium cholate	OH	OH	OH	ONa
sodium taurocholate	OH	OH	OH	$NH(CH_2)_2SO_3Na$
sodium deoxycholate	OH	H	OH	ONa
sodium taurodeoxycholate	OH	H	OH	$NH(CH_2)_2SO_3Na$
sodium dehydrocholate	O	O	O	ONa

bile salts. Their unique structure and aggregation properties provide many advantages over alkyl surfactant type [237]. Moderately retained solutes are extremely well resolved. Improved stability of the micelles in the presence of organic solvents increases the application range to more hydrophobic solutes of low solubility in water. Owing to the fact that bile salts are optically active compounds, they are mainly used as chiral selector for enantio-separation (see Sect. 7.10). The separation power of bile salts is demonstrated in Fig. 5.14. for the resolution of fourteen active ingredients.

By using mixed micellar systems the scope and the application range of MEKC can be further enlarged.

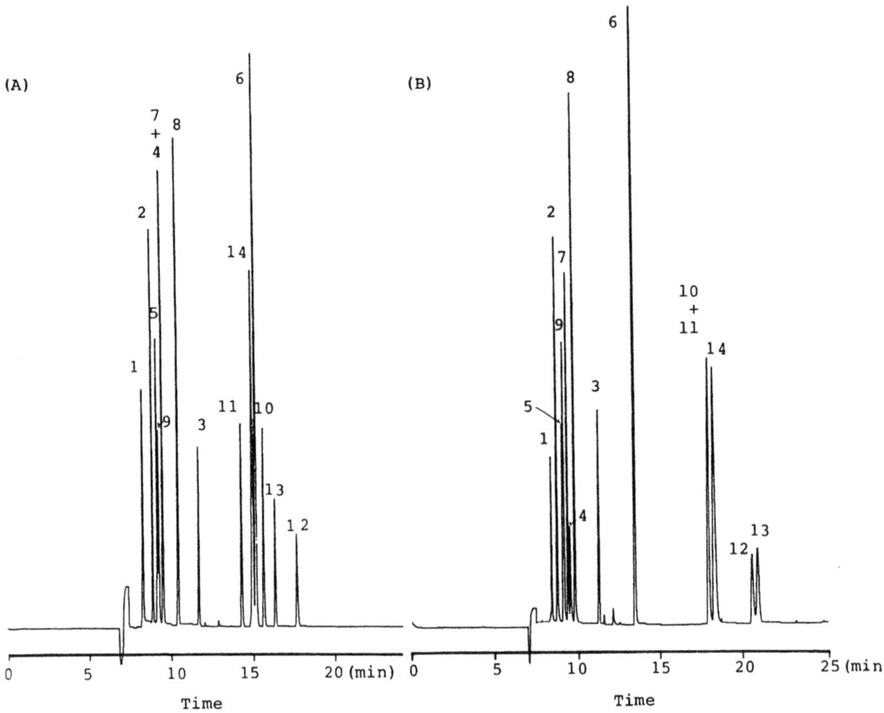

Fig. 5.14. Separation of 14 active ingredients by MEKC using bile salts. Solutes: (**1**) caffeine, (**2**) acetaminophen, (**3**) sulpyrin, (**4**) trimetoquinol, (**5**) guaifenesin, (**6**) naproxen, (**7**) ethenzamide, (**8**) phenacetin, (**9**) isopropylantipyrine, (**10**) noscapine, (**11**) chlorpheniramine, (**12**) tipepidine, (**13**) dibucaine and (**14**) triprolidine. Experimental conditions: fused silica capillary, 65 cm x 50 µm i.d., hydrodynamic injection, voltage 20 kV, UV detection at 210 nm, electrolyte system 0.02 M phosphate - borate, pH 9.0, containing (**a**) 0.1 M sodium cholate and (**b**) 0.05 M sodium deoxycholate. (Reprinted according to Ref. 238 with permission of Elsevier Science Publishers)

5.3.3 Effect of Temperature

The temperature has an influence on the separation in MEKC in several respects. First, micelles are only formed if a critical micelle temperature (so-called Kraft temperature) is exceeded. This temperature is dependent on type and concentration of the surfactant. In general, a temperature increase causes a reduction in the concentration of free micelles due to an increase in CMC. Secondly, the temperature influences the partition coefficient which can be described by

$$\ln P = -\frac{\Delta H}{RT} + \frac{\Delta S}{R} \tag{5-16}$$

ΔH enthalpy difference associated with the micellar solubilization [$J \cdot mol^{-1}$]
ΔS corresponding entropy change [$J \cdot K^{-1} \cdot mol^{-1}$]
R gas constant [$J \cdot K^{-1} \cdot mol^{-1}$]
T temperature [K]

As shown by Terabe and coworkers [230] plots of the logarithm of the partition coefficients versus the reciprocal temperature (Van't Hoff plots) give straight lines from which ΔH and ΔS are to be calculated from the slopes and the y-intercept, respectively. Values of ΔH, ΔS and the Gibbs free energy ΔG for some neutral compounds are listed in Table 5.7. One can readily see that an average value for the enthalpy of solubilization is approximately -13 $kJ \cdot mol^{-1}$. By inserting this value into Eq. 5-16, one can predict a factor of about 2 for the decrease in P when the temperature is raised from 25 °C to 65 °C.

Table 5.7. Enthalpy, entropy and Gibbs free energy changes, calculated from the partition coefficients at different temperatures at 0.1 M SDS, pH 7.0 (according to Ref. 230).

Solute	ΔH [$kJ \cdot mol^{-1}$]	ΔS [$J \cdot K^{-1} \cdot mol^{-1}$]	ΔG (25 °C) [$kJ \cdot mol^{-1}$]
phenol	-13	-7.8	15.3
nitrobenzene	-10	8.5	7.5
toluene	-17	-5.7	18.7

The influence of the temperature on the capacity factor and the efficiency was studied by Balchunas and Sepaniak [239] (Table 5.8.). The observed temperature-related changes of k' were more pronounced for low SDS concentrations. As the temperature increases, the k' value consistently decreases. The change of the theoretical plate number can be explained by contributions of mass transfer in the mobile phase owing to low solute diffusivity, dispersive temperature gradients within the capillary and the polydispersity of the micelles. The improved

Table 5.8. Influence of temperature on capacity factor k' and efficiency N of derivatized N-butylamine for two SDS concentrations (according to Ref. 239).

T [°C]	0.015 M SDS		0.1 M SDS	
	k'	N	k'	N
28	1.20	12 160	3.26	530 000
39	0.93	58 600	3.03	404 800
47	0.86	24 100	2.87	123 200
57	0.77	6 800	2.60	102 600

efficiencies at low SDS concentration, obtained if temperature increases from 28 °C to about 40 °C, can be attributed to enhanced solute diffusivity. However, if the temperature is further increased, contributions of Joule heating and temperature gradients become more pronounced leading to stronger band broadening. At high SDS concentrations mass transfer does not limit the efficiency because of the higher micelle concentration in the mobile phase, but dispersion effects dominate as temperature is raised. Consequently, higher temperatures decrease the efficiency. A dramatic drop in efficiency is observed for temperatures higher than 40 °C.

Hint: *A capillary temperature of 30 - 40 °C appears to be optimal if moderate SDS (10 - 100 mM) and buffer (10 - 20 mM) concentrations as well as capillaries of 75 µm i.d. or below are used. Temperatures higher than 40 °C should be avoided.*

The effect of temperature on the separation is shown in Fig. 5.15. The same electrophoretic system was used as described in Fig. 5.12. except the temperature which was 40 °C. As one can readily see, increasing the temperature from 25 °C (Fig. 5.12.) to 40 °C enhances the analysis time by a factor of about two at a simultaneously small decrease of resolution.

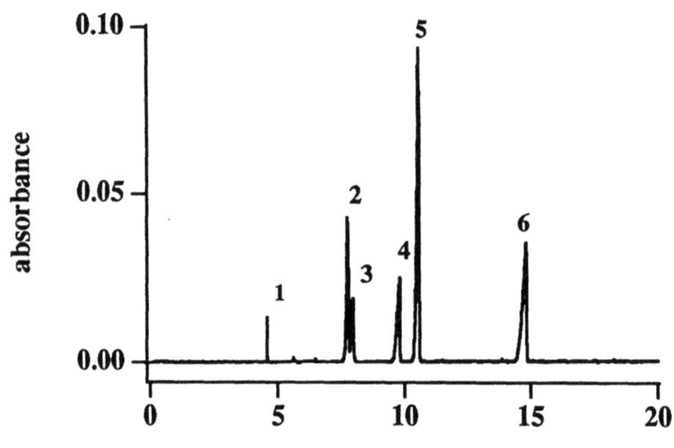

Fig. 5.15. Separation of uncharged analytes by MEKC using SDS. Experimental conditions as described in Fig. 5.12. but with a column temperature of 40 °C

5.3.4 Effect of Buffer pH

The pH influences the separation in MEKC by changing the electroosmotic flow and the electrophoretic mobility of weak acids or bases (see also Sects. 2.5 and 3.3.2). Otsuka et al. [231] investigated the influence of pH on the separation of chlorinated phenols. Under pH conditions, where solutes are estimated to be partially ionic, capacity factors decreased with increasing pH, whereas those of neutral solutes remained almost constant regardless of the pH. As discussed by the authors, electrostatic repulsion between the anionic solutes and the SDS micelles suppresses micellar solubilization and causes the decrease of k'.

5.3.5 Effect of Buffer Additives

The influence of organic modifiers on retention in MEKC is much more complicated to describe than in reversed phase HPLC because the partition coefficient P as well as the phase ratio between the micelles and the aqueous phase β is concerned. The selectivity in MEKC is altered by the addition of organic solvents such as methanol or acetonitrile [238, 240-242]. The modifier alters the retention mechanism by changing the polarity of the aqueous phase. Consequently, the partition coefficient of the solutes shifts. In addition, organic solvents such as methanol or isopropanol increase the time window by increasing the viscosity of the buffer system (see also Sect. 3.3.7) and, thus, slowing down the electroosmotic flow or by simply increasing t_{mc} as in the case of acetonitrile. By adding methanol an improvement of the resolution was found by Wu et al. [240], which is explained by an increase in the time window. Owing to a smaller value of t_{eo}/t_{mc}, obtained by the addition of methanol or acetonitrile, resolution increases in accordance to Eq. 5-13. Gorse et al. [241] found an extension of t_{eo}/t_{mc} from 0.30 at 0% organic modifier to 0.15 at 20% methanol and 0.18 at 20% acetonitrile. This is illustrated in Fig. 5.16.

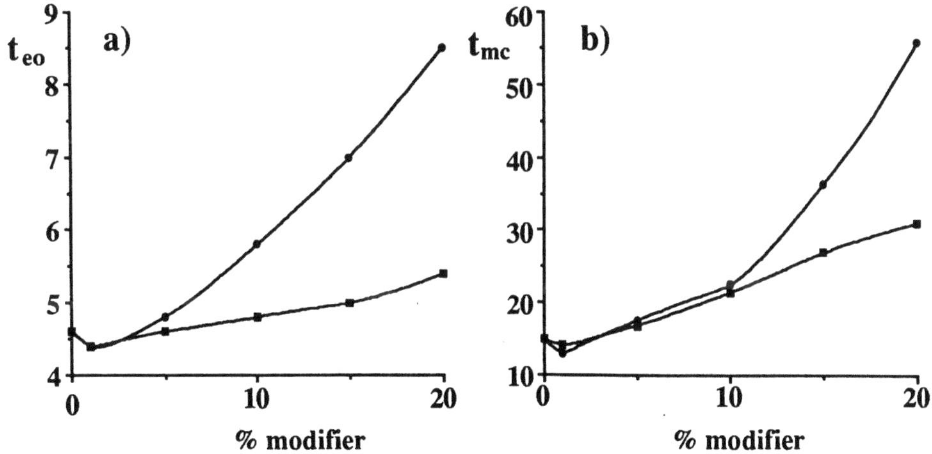

Fig. 5.16. t_{eo} (a) and t_{mc} (b) as a function of the proportion of methanol (circle) and acetonitrile (square)

Although MEKC exhibits a time window, it suffers from the same "general elution problem" as conventional elution chromatography. This means that peak broadening increases with increasing retention times. Thus, MEKC also profits from a gradient elution mode similar to LC. Gradient elution with a stepwise profile can easily be accomplished by pipetting the gradient solvent containing increasing amounts of e.g. isopropanol during electrophoresis into the buffer reservoir at the injection side [239].

Besides organic solvents a number of other additives are used to change the selectivity which can be classified in those that

> modify the water structure (e.g. chaotropic agents)
> form mixed micelles
> form complexes with the analytes.

Terabe et al. [243] added urea as chaotropic agent up to 8 M to the mobile phase in SDS-MEKC. They found that the capacity factor k' decreased logarithmically with increasing urea concentration. In addition, the elution time window was extended and resolution was enhanced by the addition of urea. Hydrophobic solutes which were strongly solubilized by the micelle could be resolved with this approach. The addition of glucose [244] has been reported to extend the elution range as well. As claimed by the authors, glucose decreases the partition coefficient of solutes with high capacity factors.

Mixed micelles composed of SDS and optically active sodium dodecylvalinate (SDVal) or sodium dodecylalanate (SDAla) were used to separate chiral compounds [245] (see Sect. 7.10). A number of catechols were well separated by a mixed micellar system consisting of SDS and sodium octylsulphate in the presence of borate buffers [246]. Nishi et al. [247] successfully separated water-soluble vitamins and antibiotics with mixed micelles of SDS and tetraalkylammonium (TAA) salts. The addition of TAA to an SDS micellar system shortens the migration times of cationic solutes and increases that of anionic solutes comparable to ion-pair chromatography.

Karger and coworkers [248] separated nucleosides and oligonucleotides by adding metal ions such as Mg(II) or Cu(II) to the running buffer. The metal ions are attracted to the micelle surface by Coulombic forces and differential metal complexation of the oligonucleotides with the micellar surface brings about the separation.

Complex formation with cyclodextrins (CD) is a very promising approach to alter electrophoretic migration of analytes (see also Sect. 3.3.6.3). CD in MEKC are extremely effective for the separation of chiral compounds, closely related aromatic hydrocarbons [249], peptides [250], drugs [251] and amino acids [252]. In a buffer solution containing micelles and CD, an analyte is distributed among three phases, namely the aqueous phase, the micellar phase and the CD cavity. Basically, CD are neutral and move with the velocity of the aqueous phase. However, owing to the hydrophobic nature of the CD cavity, free surfactants will form inclusion complexes and, thus, modify the migration velocity.

Based on the different stability constants of the inclusion complexes of lipophilic solutes with the CDs, migration times decrease and selectivity is manipulated.

5.4 Capillary Isotachophoresis

Isotachophoresis (ITP) is the electrophoretic counterpart of displacement chromatography and belongs to the basic electrophoretic principles. A fundamental attribute which distinguishes ITP from ZE is the fact that all sample zones migrate with the same electrophoretic velocity if equilibrium (steady state) is established. This is expressed by the name "iso-tachos". The basic principles of isotachophoresis are explained in Sect. 2.1.2.

Capillary isotachophoresis (CITP) features certain advantages over CZE. There are lower requirements in sharp injection zones of highly concentrated analyte solutions. Due to the zone sharpening and concentrating effects, determined by the concentration of the leading ion, diffusional phenomena play a negligible role in the steady state of ITP. As a consequence, wider bore capillaries, e.g. 200 μm i.d. can be used without significant loss in efficiency, albeit narrow bores are recommended from the theoretical point of view.

The main disadvantages of CITP originate from its separation principle. Only anions or cations can be separated in one run. The selection of the discontinuous buffer system, consisting of a leading and a terminating electrolyte, is done primarily on the basis of their mobility. At steady state conditions the sample zone consists of only the ionic analyte and the counterion. This desalting effect in combination with the concentrating effect may cause problems in practice, e.g. precipitation of proteins.

In principle CITP can be carried out using commercial CE equipment. Although recent studies have pointed out that CITP can be performed in the presence of electroosmotic flow, conventionally, the EOF has to be suppressed for CITP. This can be accomplished i) by increasing the viscosity, e.g. by the addition of hydroxymethyl cellulose, ii) by separating the capillary from the buffer reservoirs using semipermeable membranes, iii) by using tubes made from PTFE or similar materials, or iv) by using coated fused silica capillaries.

UV detection is hardly useful in CITP because the individual bands of the analytes migrate one behind the other like a moving train through the detector. Thus, zones with about the same UV response are not detected as individual bands. Using a universal conductivity detector, a response signal is obtained which is proportional to the conductivity of the solution. The time-related output, the isotachopherogram, depicts the bands as a stepwise decrease of the conductivity of the bands (see also Fig. 2.2.d). A step in the isotachopherogram is characterized by its height and its length. The step height is a qualitative property of the substance and can be taken for identification. Quantitative analysis is based on the step length. Comparing the step length of a compound with the calibration curve of standard solutions, the concentration can be calculated. This unusual form of a detector output may be one reason why CITP has not gained much attention as an analytical separation technique. However, with the growing interest in CZE, CITP is going through a renaissance, particularly in combination with CZE as a two-dimensional technique (see Sect. 5.7.3).

A more detailed description of CITP is beyond the scope of this book. The interested reader is referred to several excellent monographs which present comprehensive reviews about theory and applications of CITP [3, 253].

5.5 Capillary Isoelectric Focusing

Isoelectric focusing (IEF) represents a unique technique among the electrophoretic modes. In contrast to CZE or CITP, where separation is based on differences in electrophoretic mobility, in CIEF analytes are separated on the basis of their isoelectric point (pI). Thus, IEF is limited to amphoteric analytes and is almost exclusively used for the resolution of proteins or polypeptides. The chromatographic counterpart of IEF is chromatofocusing. The basic principle of IEF is described in Sect. 2.1.3.

Separation in IEF is based on the electrophoretic migration behavior of an ampholyte in a pH gradient. This pH gradient is established by carrier ampholytes under the influence of an electric field. The ampholytes form Gaussian concentration distribution curves extending over a finite pH region, with maxima at their isoelectric points. Adjacent isoelectric zones will overlap considerably, and if a sufficient number of components are available any required pH range can be spanned. It is obvious that a very large number of components are required if a pH region is to be covered completely without gaps of low conductance. Commercial carrier ampholyte mixtures (available from Bio-Rad, Pharmacia, Sigma, Serva, etc.) consist of several hundreds of constituents each with an individual pI value. More acidic ampholytes possess higher negative net charges and accumulate close to the anode, whereas basic ampholytes concentrate near the cathode. In principle, neither special anolyte solutions nor catholyte solutions are required to establish the pH gradient. In practice, however, the carrier ampholytes must be insulated from the electrode reservoirs, because oxidation and reduction processes at the electrodes can alter the composition of the buffer solution. The use of an acid in the anode compartment (anolyte) and a base in the cathode compartment (catholyte) insulate the anode and the cathode from the separation compartment. The carrier ampholyte has to meet a number of requirements:

- amphoteric nature with pI values in the pH range of interest
- high enough conductivity to carry the current

- high buffer capacity at the pI
- low UV absorbance
- no disturbing interactions with the analyte
- high solubility in water.

Commercial carrier ampholytes are polyamino polycarboxylic acids synthesized by random polymerization of acrylic acid and polyethylene polyamine in water [254]. Such ampholytes cover the pH range from about 3 to 10 with molecular weights ranging from 300 to 1000. Ampholytes with narrower pH ranges are obtained by electrophoretic fractionation of the mixture above. The main disadvantage of these carrier ampholytes is their relatively high UV absorbance at short wavelengths and their heterogen distribution of UV absorbance after establishment of the pH gradient which makes detection at 280 nm necessary. The mechanism of zone formation in IEF has been analyzed in detail by Svensson [255]. The process involves the establishment of an equilibrium between solute concentration by electrophoretic migration near the pI and solute dispersion by diffusion. At equilibrium

$$\frac{c \cdot \mu \cdot I}{\kappa \cdot A} = D \cdot \frac{dc}{dx} \tag{5-17}$$

A cross sectional area of the capillary [m^2]
x distance of the pH gradient along the capillary [m]

If κ and A are constant and with $I/\kappa \cdot A = E$ it follows

(5-18)

For μ being a function of x, integration can be carried out. If the origin for x is taken at the zone maximum, then $C = C_0$ for $x = 0$, where C_0 is the maximum concentration. Now it follows

$$C = C_0 \cdot \exp\left[\frac{D}{E} \cdot \frac{d(pH)}{d\mu} \cdot \frac{dx}{pH}\right] \tag{5-19}$$

Eq. 5-19 has the form of a Gaussian curve with a variance given by the term in brackets:

$$\sigma^2 = \frac{D}{E} \cdot \frac{d(pH)}{d\mu} \cdot \frac{dx}{pH} \tag{5-20}$$

A sharp zone, characterized by a small variance is thus favoured by a high field strength, by a low diffusion coefficient and by high values of $d\mu/d(pH)$, the rate of change of mobility with pH, and by $d(pH)/dx$, the slope of the pH gradient. Of these variables, the diffusion coefficient and $d\mu/d(pH)$ are intrinsic properties of the analyte, so that only the pH gradient and the field strength can be varied experimentally. Although an increase of $d(pH)/dx$ sharpens the focused zone, it also crowds adjacent zones together, as in all gradient methods. Therefore, resolution is not greatly affected. As in other electrophoretic techniques, zone sharpening is improved by a high value of the ratio of E/D. Thus, IEF is particularly favourable for macromolecules such as proteins with low diffusion coefficients. By inserting experimental values in Eq. 5-20, Vesterberg and Svensson [256] calculated that a pI difference of 0.05 units will be necessary for a complete separation of two ampholytes with baseline resolution (4σ).

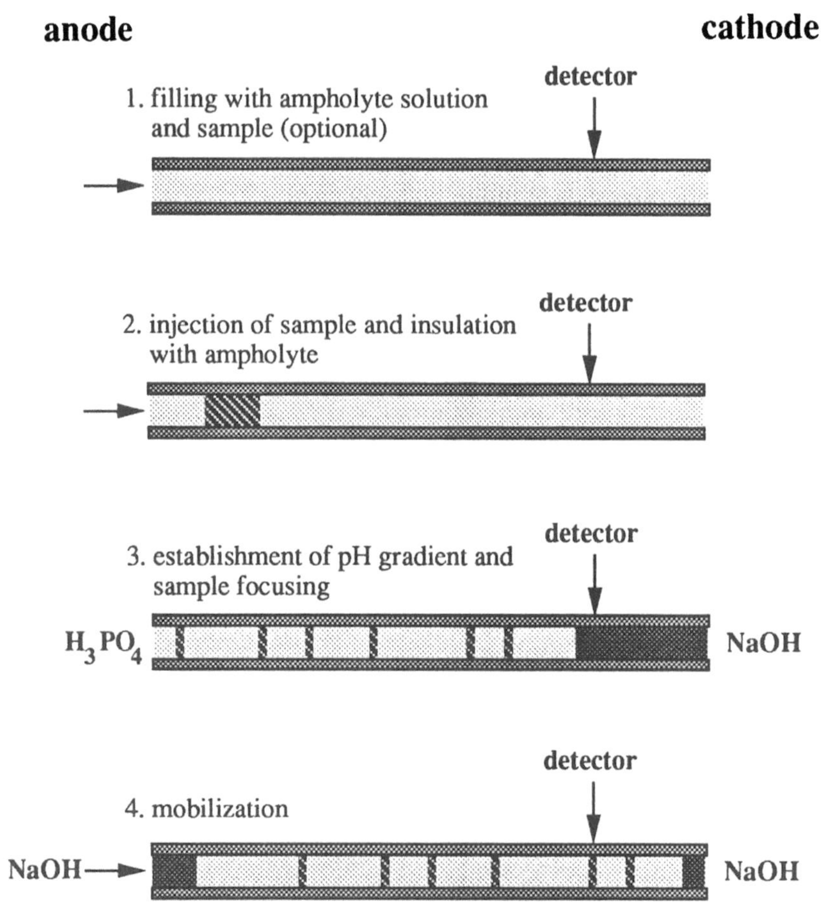

Fig. 5.17. Schematic representation of the procedure for CIEF

Isoelectric focusing in capillaries (CIEF) was first reported by Hjertén and Zhu [257]. In CIEF the key issue is the electroosmotic flow. Elimination or at least reduction is crucial in order to obtain stable focused zones. For this purpose, Hjertén [29] developed capillary coatings on the basis of methyl cellulose or polyacrylamide which efficiently eliminated EOF (for details see Sect. 5.1.2). The procedure to carry out CIEF can be divided in four steps (Fig. 5.17.):

1. Filling of the capillary and the electrode reservoirs:

> Fill the capillary completely with the carrier ampholyte solution of the desired pH range. Typical concentrations of ampholytes are 1 - 2%. Optionally, dissolve the sample in the carrier ampholyte. The sample solution should be desalted before mixing with the ampholyte solution.

> Fill the buffer reservoir at the anode with an acid such as phosphoric or aspartic acid (i.e. 0.05 M). Correspondingly, fill the buffer reservoir at the cathode with a base such as NaOH (0.02 M) or arginine (0.05 M).

> Because mobilization of the focused bands is in one direction only, analytes which are focused in the segment between the detection window and the end of the capillary (the "blind" end) remain undetected during mobilization. To circumvent this, add 0.5 - 1% of a basic compound such as tetramethylene ethylenediamine (TEMED) to the carrier ampholyte to extend the pH gradient in the basic range to pH 12. TEMED becomes concentrated at the cathodic end of the capillary during focusing, shifting the desired pH gradient from the capillary end to a point before the detection window. Thus, basic proteins were focused before passing the detector [258].

2. Sample injection (optional):

> If the sample has not already dissolved in the ampholyte solution as described under 1., it can be alternatively injected hydrodynamically as a solution of sample in carrier ampholyte into the tube. Subsequently, insulate the sample plug from the anode buffer reservoir by injecting a small volume of ampholyte solution.

3. Simultaneous establishment of the pH gradient and focusing of the analytes:

> Apply a voltage of 30 kV to establish the pH gradient and to focus the analytes. The accumulation of TEMED at the cathodic end of the capillary prevents focusing at the "blind" end of the capillary.

> Monitor the current as it decreases with the time. Completion of the focusing process (steady state) is indicated by a minimal current flow which does not change anymore. Since there is no evident factor which shows the end of focusing, a second, longer run which leads to the same peak pattern proves the steady state has been achieved.

4. Mobilization of the gradient through the detector:

Theoretically, there is no electrophoretic migration in the capillary after the steady state is reached. Therefore, the entire gradient has to be moved through the detector cell for detection of the analyte's bands. To accomplish mobilization several procedures are known [257-259]. Hjertén [257] described two methods:

- First, apply a pressure which pushes the whole solution through the capillary, with the voltage constantly applied to avoid band broadening during elution.
- Alternatively, replace the acid at the anode buffer reservoir by a base (see 4. Fig. 5.17.) or the base in the cathode vessel by an acid to elute the gradient electrophoretically. When an acid at the anodic buffer reservoir is replaced by a base, e.g. NaOH (20 mM), the sodium ions migrate toward the cathode causing an increase of the conductivity. The change of the pH leads to slow titration of both ampholytes and analytes which become negatively charged and begin to migrate (cathodic mobilization). A reversed process occurs for anodic mobilization.

Both mobilization techniques, pressure elution and electrophoretic elution, are comparable with respect to resolution [257]. Hydrodynamic elution, however, is not feasible in gel-filled capillaries.

Zhu et al. [258] accomplished mobilization of the separated bands after focusing by replacing the anolyte or catholyte solution against a neutral salt solution such as NaCl (100 mM). As a consequence of the migration of sodium or chloride ions into the capillary H^+ or OH^- ions are replaced and the pH changes beginning at the capillary end and gradually progresses deeper into the tube. An alternative mobilization using zwitterionic agents was developed which proved to be superior to the salt mobilization in terms of efficiency and useful pH range.

Mazzeo and Krull [259] discussed disadvantages of the electrophoretic mobilization described above. First, polyacrylamide coated capillaries via Si-O-Si bonds show poor stability especially at alkaline pH. Further, mobilization of the zones involves additional operation steps which may have an influence on the separation result. Because of these reasons, the authors have explored an alternative approach for performing CIEF in uncoated capillaries. Controlling of EOF rather than eliminating is claimed to bring about successful focusing and separation. By their experimental set-up a slow EOF which allows focusing of the bands before migrating through the detector cell is maintained by adding appropriate amounts of methyl cellulose to the carrier ampholyte. The separation of four model proteins is shown in Fig. 5.18. With the method described in Fig. 5.18. proteins up to a pI of 4.8 were successfully focused. However, for even more acidic proteins broad peaks and poor resolution was found. The reproducibility of the migration times was better than 3 %, that of peak area better than 8%. A plot of migration time versus pI for the proteins in Fig. 5.18. gives a straight line (Fig. 5.19.) which allow the pI value of an unknown compound to be determined.

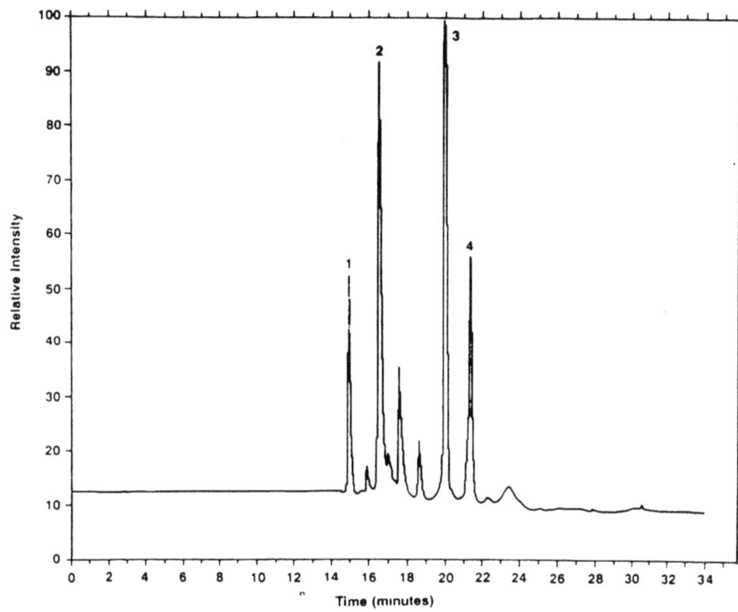

Fig. 5.18. Separation of model proteins by CIEF [259]. Capillary: uncoated fused silica 60 cm x 75 μm i.d. (L_D 40 cm). Anolyte 10 mM H_3PO_4, catholyte 20 mM NaOH, voltage 30 kV. UV detection at 280 nm. Carrier ampholyte 1 mg/mL of each protein, 5% ampholyte 3-10, 0.1% methyl cellulose, 1% TEMED. Identification: (**1**) cytochrome c, pI 9.6; (**2**) chymotrypsinogen, pI 9; (**3**) myoglobin, pI 7.2; (**4**) myoglobin, pI 6.8. (Reprinted with permission of the American Chemical Society)

Hint: *If the sample is dissolved in the entire carrier ampholyte transient multiple peaks of one single compound resulting from the concentration at the conductivity boundaries may be detected. To make sure that steady state is reached the experiment should be repeated with longer focusing times. If the same electropherograms are obtained the steady state has been achieved.*

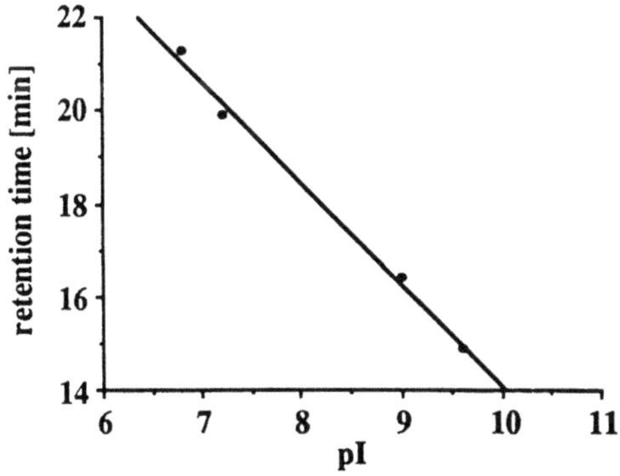

Fig. 5.19. Plot of migration time versus pI value for the proteins in Fig. 5.18.

Protein precipitation is a well known problem in polyacrylamide gel isoelectric focusing. It results for two reasons. Commonly proteins are focused with very high concentration factors at a form of zero net charge where solubility is usually low. In addition, stabilizing counterions are removed from the proteins in the focused zones. In CIEF precipitation of proteins is reflected by extremely sharp peaks or spikes in the electropherogram. This effect can be suppressed by adding nonionic detergents such as Triton X-100 [258], urea or ethylene glycol to the carrier ampholyte.

Although CIEF is a young technique and the potential has not yet been fully explored, it is already a high-resolution separation technique.

5.6 Electrochromatography

Tsuda defined electrochromatography (EC) as follows: "Electrochromatography should remain restricted to electrophoretic procedures where sorptive interactions with the support constitutes a major factor" [260]. Accordingly, separation in electrochromatography is based on electrophoretic processes and interactions with a stationary phase such as adsorption, partition and gel permeation in a packed narrow bore capillary. According to this definition, conventional HPLC packing material as well as polymer networks can be regarded as stationary phase. Because capillary gel electrophoresis (CGE) is treated in a separate section (see Sect. 5.2) this chapter is devoted only to electrochromatography with HPLC packing materials using silica gel particles.

The principle of EC is in close analogy to conventional HPLC with packed columns except the mobile phase is driven by electroosmosis rather than by pressure. A schematic representation of the separation principle is shown in Fig. 5.20. Partitioning or adsorption of the analytes occurs in the same way as in HPLC. Thus, selectivity in EC and HPLC is identical for neutral analytes but differs for ionic analytes due to additional electrophoretic migration. A comparison between HPLC and EC with respect to some characteristic attributes is given in Table 5.9. In contrast to pressure-driven LC the inner diameter of the capillary in EC is limited to not more than 200 μm because of heat generation. On the other hand, the particle diameter of the silica gel in HPLC is limited by the resulting pressure increase to about 3 μm, whereas EC is feasible with particles down to at least 1.5 μm diameter as shown in an excellent paper by Knox and Grant [261]. Actually, it was found that the EOF is unaffected by the particle size. It is important to note that also surfaces of ODS bonded silica exhibit sufficient silanol groups to generate an electroosmotic flow.

There is an important difference in the flow profile of pressure-driven and electrically driven chromatography. While a parabolic flow profile results in pressure-driven systems with the flow velocity being zero at the wall and twice the mean flow rate in the center, a flat flow profile arises in EC (see also Sect.

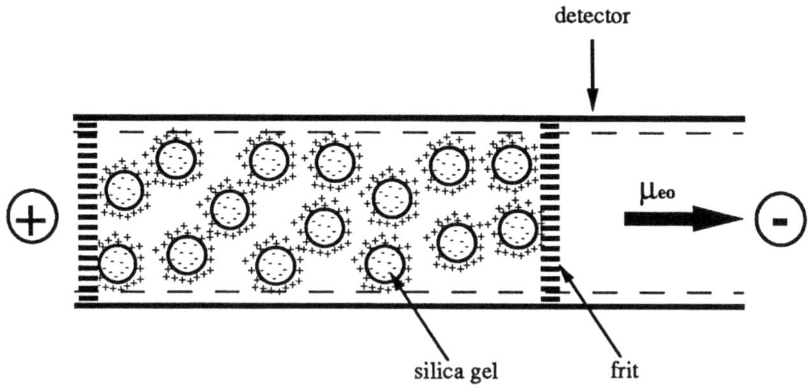

Fig. 5.20. Schematic representation of electrochromatography in capillaries packed with silica gel particles

Table 5.9. Characteristic attributes of HPLC and EC

	HPLC	EC
tube diameter	optional	< 200 μm
particle size	> 3 μm	< 1 μm possible
flow profile	laminar	plug
motion of liquid	balance between pressure and viscous resistance	balance between electric force and viscous drag
characterization of solutes	rel. retention time (k')	rel. retention time (k')
linear velocity	≈ 1 mm·s^{-1}	≈ 1 mm·s^{-1}

2.5). For packed capillaries the same principles of EOF generation apply as for open tubes. However, there exist numerous flow channels of disparate size, shape and direction. Although the zeta potential of silica will be the same regardless whether it is an open tube or a packed capillary, the electroosmotic velocity will be lower in packed capillaries than in open tubes because of two reasons [262]. Firstly, alignments of the channels in a packed capillary is usually not axial, so that the effective field strength will be E·cos Θ, where Θ is the angle between the axis of the channel and the axis of the capillary. Secondly, silica gel particles are, in general, porous. But the electroosmotic flow will take place outside of the pores since they are so small that the surface double layer overlap. As a consequence of the overlap, EOF is strongly reduced in the pores. According to Knox and Grant, this effect decreases the EOF by a factor of 0.5 to 0.7. The velocity of the electroosmotic flow v_{eo} in a packed capillary can be described by

$$v_{eo} = \frac{V_0 \cdot \omega}{V_m} \cdot \frac{\zeta \cdot \varepsilon_0 \cdot \varepsilon_r}{\eta} \cdot E \qquad (5\text{-}21)$$

V_0 volume of the mobile zone (outside particles) [cm^3]
V_m volume of the mobile phase [cm^3]
ω tortuosity factor

Obviously, the electroosmotic flow velocity will be a maximum in non-porous media where $V_0/V_m = 1$. With increasing porosity of the stationary phase material the first term in Eq. 5-21 becomes smaller with the consequence that v_{eo} decreases. Taking both effects into account, electroosmotic flow velocities for packed capillaries are about 40 - 60% of those for open tubes.

Until now packed capillaries for use in EC are not commercially available. In the following, we describe a technique for the preparation of appropriate columns which was first reported by Knox and Grant [262] and later slightly modified by Yamamoto et al. [263].

The capillary, i.e. 50 - 100 µm i.d., is packed using the slurry packing method. The procedure encompasses 5 steps (Fig. 5.21.):

1. Formation of a porous frit at one end of the tube to retain the silica gel during the slurry packing:

> Moisten a small amount of spherical silica gel, e.g. Merck Superspher Si 60 (4 µm), with a dilute solution of sodium silicate. Only so much of liquid should be added until a paste is formed which is just noticeably moist.
> Introduce the silica into the capillary by repeatedly pushing one end of the tube into the paste. A length of ca. 0.5 mm of the capillary end should be filled.
> Sinter the packing by carefully heating with a small microtorch flame for ca. 15 s.

2. Slurry packing of the capillary with silica particles:

> For the slurry suspend about 200 mg of 4 µm silica gel (e.g. Merck) in 2 mL acetonitrile under ultrasonication to get a homogeneous dispersion.
> Pump the slurry into the capillary by using a liquid chromatographic pump (flow rate: 0.1 $mL \cdot min^{-1}$, pressure: ca. 400 bar).
> After the tube is completely filled, release the pressure slowly to avoid sudden change in pressure across the packed bed.

3. Preparation of a second frit some distance from the end of the capillary to leave an empty part of the capillary for UV detection:

> Dry the packing with a helium stream and incinerate the polyimide coating in a cold part of the butane flame.

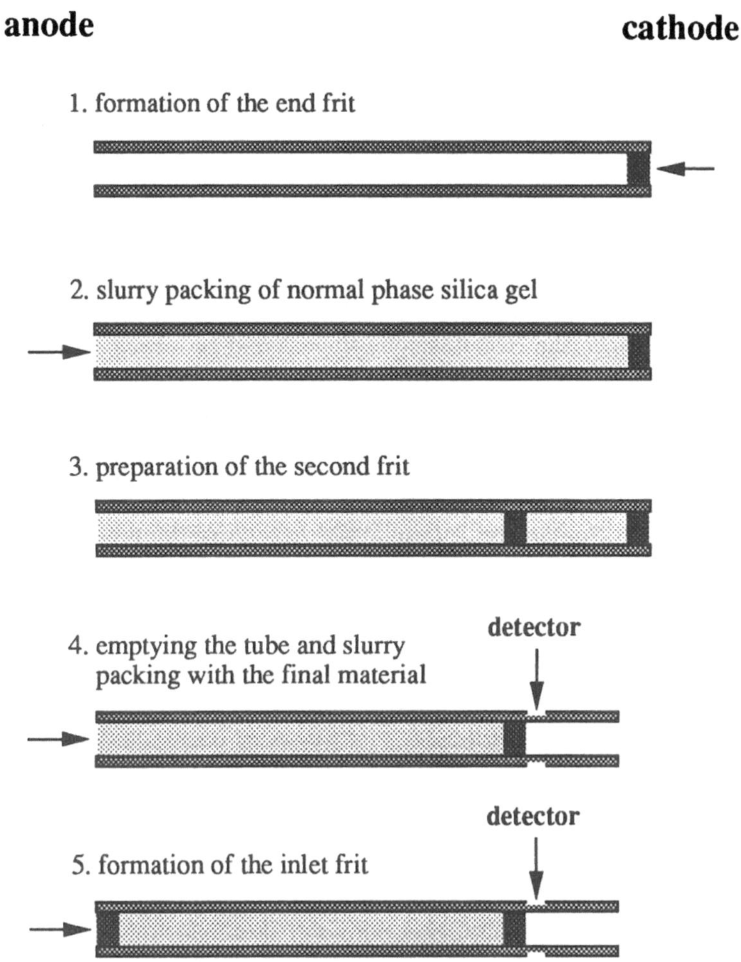

Fig. 5.21. Schematic representation of the procedure for preparing packed capillaries suitable for electrokinetic chromatography. For details see text

> Sinter the frit by heating the packing in the middle of the butane flame (about 20 s) until the particles just begin to glow red. Slowly rotate the tube during this procedure to allow a uniform fusing.

4. Emptying of the capillary and slurry packing with the final packing material:

> Cut off the first frit at the end and reconnect the capillary to the pump. Empty the tube from each side by pumping distilled water through the tube. Dislodging of the packing is eased by applying ultrasound.

> Burn the polyimide coating away to make the detection window.

> Pack the capillary with the desired packing material as described in 2.

5. Closing the inlet of the capillary with a frit:

> Prepare a frit at the inlet of the capillary by dipping the end of the tube into a sodium silicate solution and push it into the sintering mixture as described in 1.

> Sinter the frit by carefully heating with a microtorch.

A number of factors influence the electroosmotic flow velocity. As in open-tubular CE the field strength, pH, ionic strength and organic modifiers of the mobile phase have an effect on EOF and the selectivity of the separation system. Theoretically, the electroosmotic flow should be related linearly to the applied field strength. However, owing to temperature effects at higher field strengths a similar concave curve is found as for open tubes (see also Fig. 3.9.a). In addition, a sigmoidal curve as in open tubes is obtained for the influence of the EOF on the pH (see Fig. 2.12.). The influence of the ionic strength is critical in EC. On the one hand the electrolyte concentration has only little influence on EOF in the range from $4 \cdot 10^{-5}$ M to $2 \cdot 10^{-1}$ M but decreases for

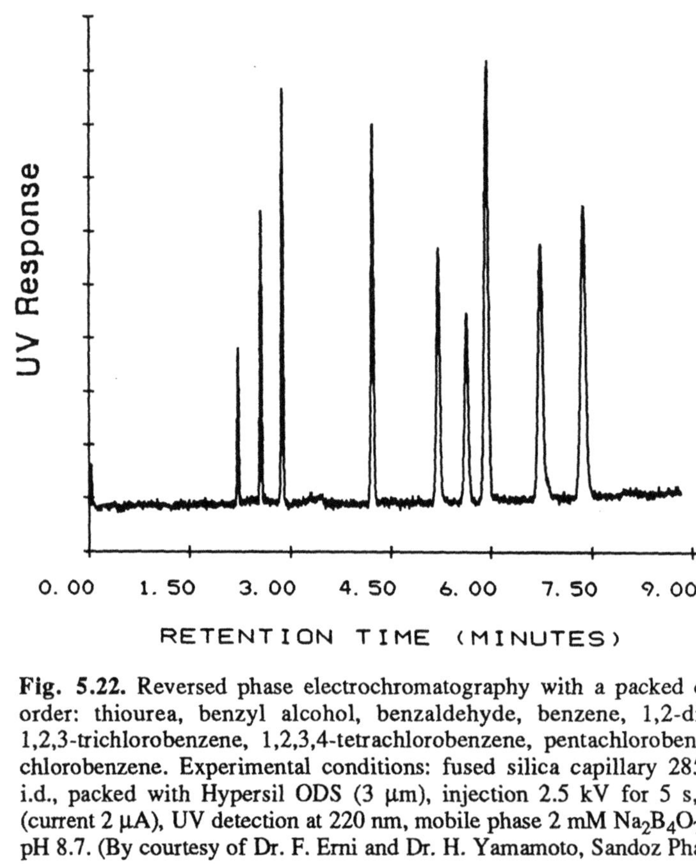

Fig. 5.22. Reversed phase electrochromatography with a packed column. Elution order: thiourea, benzyl alcohol, benzaldehyde, benzene, 1,2-dichlorobenzene, 1,2,3-trichlorobenzene, 1,2,3,4-tetrachlorobenzene, pentachlorobenzene and hexachlorobenzene. Experimental conditions: fused silica capillary 285 mm x 50 µm i.d., packed with Hypersil ODS (3 µm), injection 2.5 kV for 5 s, voltage 45 kV (current 2 µA), UV detection at 220 nm, mobile phase 2 mM $Na_2B_4O_7$ - 80% CH_3CN, pH 8.7. (By courtesy of Dr. F. Erni and Dr. H. Yamamoto, Sandoz Pharma Ltd. Basel)

tigher concentrations, on the other hand significant enhancement of the efficiency occurs with increasing buffer concentration. Additionally, if the buffer concentration is too high, gas bubbles occur because of overheating. For sodium tetraborate a concentration of 2 - 4 mM reveals to be optimal. Fig. 5.22. demonstrates the potential of EC.

A more detailed discussion of the theory of band broadening [262] and the influence of thermal effects [21] and of particle diameter [261] on the separation efficiency in EC is presented elsewhere. Tsuda [264] introduced the concept of pressurized flow electrochromatography using an HPLC pump in order to solve the problem of gas bubbles generated at the electrodes.

Hyphenated Techniques

In general, hyphenated (coupled) techniques in separation science have evolved as combinations of two or more unrelated methods being interfaced to provide informations about the same sample. Appropriate coupling of the various techniques available depends on the separation problem and the informations required. Different purposes for hyphenation of CZE with suitable techniques are discussed in the following.

For samples of minor complexity where structural informations about the components are requested, the first method of the coupled system could represent a separation step whereas the second technique provides spectral informations. Techniques of this category are capillary electrophoresis-diode array UV detection, capillary electrophoresis-mass spectrometry (CE-MS) or capillary electrophoresis-Fourier transform infrared spectrometry (CE-FTIR). By coupled operation of these techniques one obtains two sets of data, separation and spectral data, of the same sample.

Very complex samples need to undergo more than one separation technique to minimize peak overlap. However, several criteria need to be observed to obtain maximal informations from coupled systems. First, both techniques should have independent separation mechanisms (orthogonality of separation) based on as different sample properties as possible. However, the aspect of orthogonality causes an interesting problem for two dimensional (2-D) systems. The more dissimilar both separation mechanisms are, the more dissimilar their operation is. Consequently, the more difficulties will arise to couple both systems. Reversed phase HPLC and CZE are highly orthogonal separation techniques. While the first separates solutes based on their hydrophobicity, the latter separates solutes on the basis of charge and size.

A third need for hyphenation results from the limited concentration sensitivity of CZE. Therefore, CZE can hardly be used for trace analysis of biological samples without sample pretreatment. Capillary isotachophoresis is also based on electromigration of the analytes, but this technique is advantageous because of the higher sample loadability and the concentration effect. These features and the similarity of both separation systems makes CITP an ideal technique for coupling with CZE.

5.7.1 Capillary Electrophoresis - Mass Spectrometry (CE - MS)

Mass spectrometry (MS) plays an important role in the analytical and structural characterization of biological substances. A large number of separation techniques has been combined with MS including gas chromatography, liquid chromatography, supercritical fluid chromatography and, recently, capillary electrophoresis. The biggest advantage of using a mass spectrometer as the detector for CE is not only its high sensitivity, but also its high selectivity: both molecular weights and structural informations can be provided together with the migration times. In the following we will shortly present the precautions that have to be taken when coupling a CE device to a mass spectrometer.

The two most common ionization techniques which have been used so far in combination with CE are continuous-flow fast atom bombardment (CF-FAB) and atmospheric pressure electrospray ionization (ESI). These techniques do not expose the analyte to excessive heat and provide very mild ionization conditions that ensure molecular weight determination. A special form of the electrospray ionization interface is the pneumatically-assisted electrospray or ion spray interface. Whereas a CF-FAB interface often provides some additional fragmentation information, an ESI interface typically produces only protonated or deprotonated molecule ions with little or no fragmentation. The most significant advantage of the ESI interface is the applicability to higher molecular weight compounds which are impractical by CF-FAB methods. Additional structural information can easily be obtained by tandem mass spectrometry (MS-MS).

A special junction is needed to facilitate the coupling of the low CE buffer flow to the different MS interfaces. CZE flow rates are commonly too low for reproducible operation of most ions sources which require liquid flows of 2 - 5 µL/min. The first CE-MS device was based on an ESI interface developed by Smith and coworkers [266]. In this system, no cathodic buffer reservoir is used. Instead, electrical contact was made directly to the solvent in the column through an electrospray needle at the column outlet. Following this initial work they reported an improved ESI interface which incorporates a sheath-flow liquid electrode [267, 268] which is shown in Fig. 5.23. The sheath-flow electrode of liquid allows the composition and flow rate of the electrosprayed solution to be different than that of the electrophoresis system, which is desirable when working with aqueous solutions. The electric contact for the buffer at the low voltage end of the capillary is made by a coaxial sheath-flow of organic solvent, such as acetonitrile, methanol, 2-propanol or acetone with the addition of acetic acid and in some instances a small percentage of water. The sheath-flow liquid constitutes the large majority of the electrosprayed liquid. The electric contact serves to define both the separation voltage as well as the ESI voltage, which is

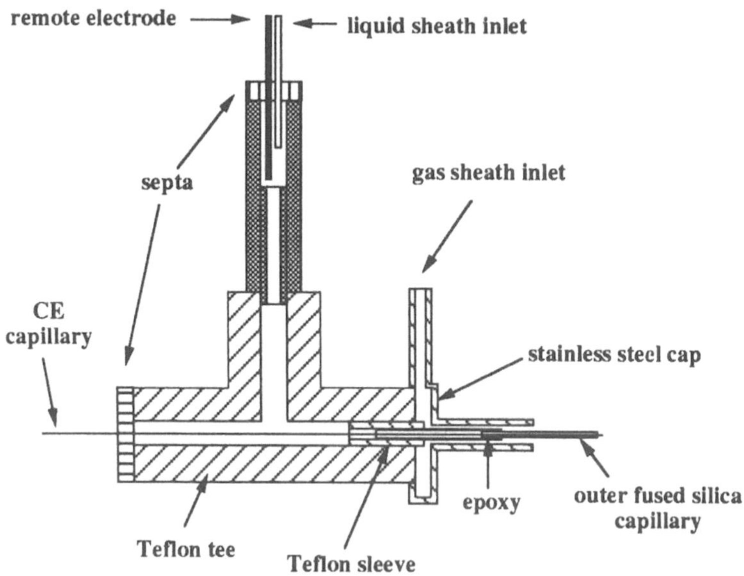

Fig. 5.23. Schematic illustration of a sheath-flow electrode as a junction for CE-MS. (Reprinted from Ref. 268 with permission of Elsevier Science Publishers)

typically in the range of 4 - 6 kV for positive ions and - 5 kV for negative ions. The potential across the capillary is the difference between that applied at the high voltage end and that of the sheath-flow electrode. The contact is made remotely to eliminate metal parts near the end of the capillary. The sheath liquid and the stainless steel electrospray electrode are introduced through the same arm of a Teflon tee. A precise pulse-free liquid flow of 2 - 30 $\mu L \cdot min^{-1}$ is provided by a syringe pump. An additional flow of gas is used on occasion as a sheath around the capillary terminus. The purpose of this gas flow is to add oxygen or another gas to suppress discharges in the negative ion mode of operation and to provide a cooling for the sheath liquid flow at CE high currents. The ESI source consists of a 50-μm i.d. uncoated fused silica capillary used for CE that protrudes 0.2 - 0.4 mm from a concentric fused silica capillary of 200 μm i.d. Both capillaries are fixed to the Teflon tee with epoxy. The ESI source tip is mounted ca. 1.5 cm from the ion sampling nozzle of the sampling orifice inlet to the quadrupole mass spectrometer (not shown).

Besides this sheath-flow arrangement, the ion spray interface can be coupled to the CE capillary via a liquid junction [269-270]. The buffer in the liquid junction facilitates the electric contact between a remote electrode and the separation capillary and the transfer capillary to the ion-spray interface, respectively. Most often, it is the same as the separation buffer for CE. The end of the separation capillary is placed in the liquid junction device and aligned to the 75 μm i.d., ca. 6 - 7 cm long transfer capillary under a microscope (Fig. 5.24.).

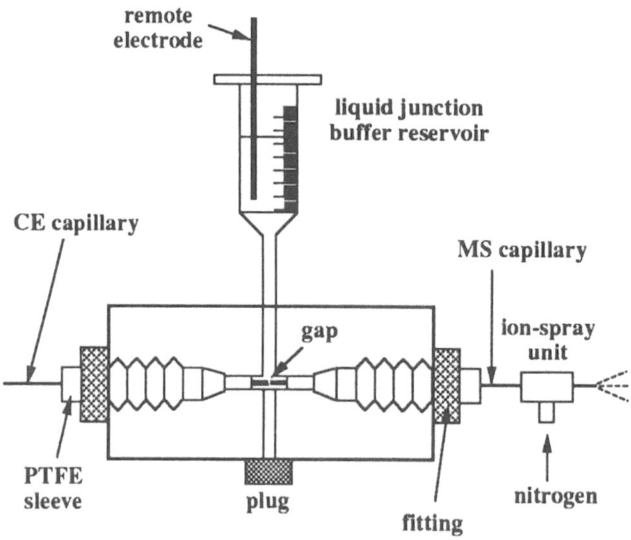

Fig. 5.24. Schematic presentation of the liquid junction ion-spray CE-MS interface. (Reprinted from Ref. 270 with permission of Elsevier Science Publishers)

Alignment of the gap (typically 10 μm) between the exit of the separation capillary and the transfer capillary is the most critical aspect of the coupling device. On the one hand, the gap must be wide enough to allow entry of the liquid junction buffer, while, on the other hand, it must be small enough to minimize extra-column band broadening. The buffer flow is provided by a syringe serving as a buffer reservoir which is placed on top of the junction gap. The remote electrode is placed in the buffer reservoir.

Alternative CE-MS designs are based on CF-FAB mass spectrometry. The two most commonly CF-FAB interfaces are based again on a liquid junction [271] and a coaxial sheath flow [272]. Figure 5.25. shows a schematic diagram of a Plexiglas liquid junction as it was used by Nichols et al. [270]. It consists of a 0.062" vertical static reservoir intersected by a horizontal 0.062" cylindrical hole with a 0.062" o.d. · 0.031" i.d. PTFE insert at one end. The 45 cm x 75 μm i.d. fused silica transfer capillary self-aligns with the low voltage end of the CE capillary in the PTFE sleeve within the liquid junction. The gap between the two capillaries is about 50 μm. The junction is grounded through an electrode in the bottom of the vertical reservoir which is connected to the frame of both the CE system and the mass spectrometer. The liquid junction is filled with matrix-buffer solution which contains the running buffer and as much as 20% glycerol for the FAB matrix. The transport of the analytes to the ionization source through 45 cm of capillary tubing causes extracolumn band broadening.

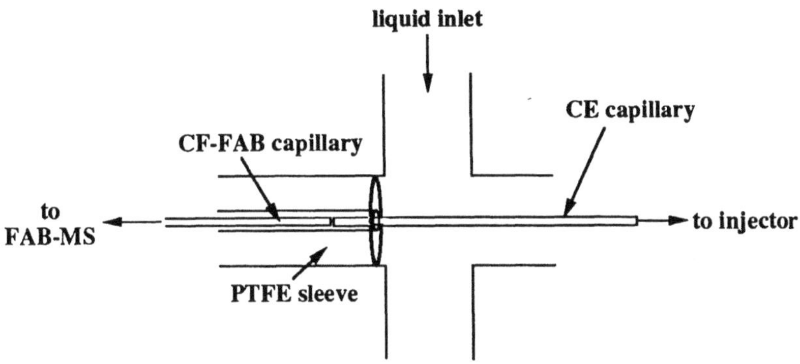

Fig. 5.25. Liquid junction for a CE-CF-FAB interface according to Ref. 270

In the coaxial sheath flow interface described by Moseley et al.[272] a sheath capillary surrounds the CE capillary. Both terminate at the tip of the FAB probe. The coaxial sheath flow interface offers the advantage of electrophoretic separation up to the site of ionization inside the mass spectrometer's vacuum chamber, thus not decreasing the separation efficiency. Furthermore, the composition and the flow rates of the two liquid streams can be independently optimized. Nevertheless it involves careful control of the very small dimensions and balanced flow of the FAB matrix with the running buffer through the very small annular space between the coaxial capillaries. Liquid junction coupling offers a simpler experimental setup because the matrix is introduced externally to the mass spectrometer vacuum system. The only critical point is the liquid junction gap.

Compatibility problems between CZE and MS may arise from the buffer system used in CZE. Non-volatile buffers such as sodium phosphate or borate widely used in CE are less suitable for CZE-MS coupling. The generation of gas-phase ions is less effective under these conditions. Moseley et al. [272] studied the effect of potassium phosphate over a concentration range of 0.01 to 0.05 M on the intensity of the protonated molecular ion (MH^+) and the competing potassium adduct (MK^+). According to their results, the intensity of (MH^+) ion signal decreases with increasing potassium concentration in the CZE buffer, commensurate with an increasing (MK^+) signal.

Volatile CZE buffers such as ammonium acetate, triethylamine or trifluoroacetic acid do not interfere with the ion generation in the gas-phase. Thus, most buffer systems described in the literature are based on ammonium acetate with varying proportions of e.g. trifluoroacetic acid or triethylamine to adjust the pH. The buffer concentration has an enormous influence on the detection signal. For instance, a 15 mM ammonium acetate buffer yielded in a two-fold increase in response of the API-MS system compared to a 20 mM buffer [270]. Moreover, ammonium acetate was found to be superior over ammonium formate, although the reasons for this finding are unknown. High contents of organic solvents, typically methanol or acetonitrile, in the buffer system are recommended to facilitate the ion-spray process.

5.7.2 Liquid Chromatography - Capillary Electrophoresis (LC - CE)

Hyphenated systems composed of two separation techniques are called 2-D systems because they provide information about the sample composition in two dimensions. One has to differ between two principle kinds of 2-D systems (Fig. 5.26.). First, the truly 2-D system as it is given in 2-D slab gel electrophoresis, where the first dimension represents isoelectric focusing and the second step SDS-gel electrophoresis. 2-D slab gel electrophoresis provides the highest peak capacity (some thousands) of any separation system known. The overall peak

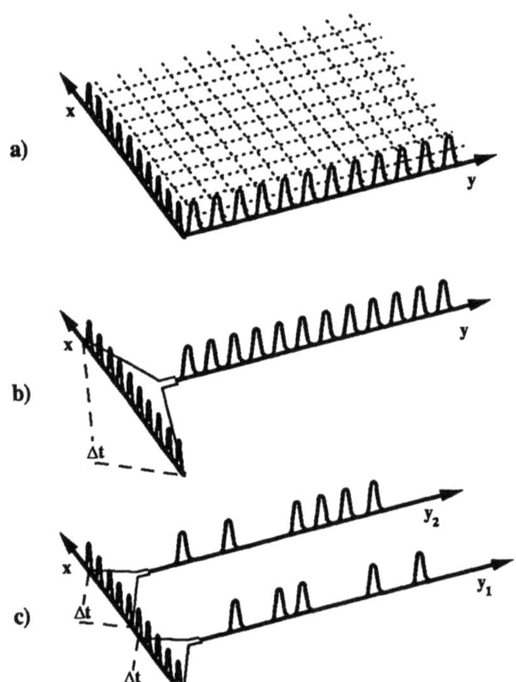

Fig. 5.26. Schematic representation of (**a**) a truly two-dimensional system, (**b**) a purely tandem system composed of two coupled separation systems and (**c**) coupled system with small Δt interval and twice injections in the second dimension. x represents the first dimension, y the second dimension. The number of boxes in (**a**) is approximately equal to the product of the peak capacities $n_x \cdot n_y$

capacity of a 2-D system is about equal to the product $n_x \cdot n_y$ where n is the peak capacity of the individual dimension (Fig. 5.26.a). The second kind represent arrangements, where two separation columns are coupled such that the eluent stream from the first column is fed directly into the second column (purely tandem system, shown in Fig. 5.26.b). In such arrangements resolution obtained in the first system can partially or completely nullified by the second step because of another retention mechanism. Thus, for complex sample mixtures a significant improvement in resolution is hardly obtained by tandem systems [273].

Coupled systems such as HPLC-CZE lie somewhere between 2-D and tandem separation systems, depending on the time interval Δt which is cut from

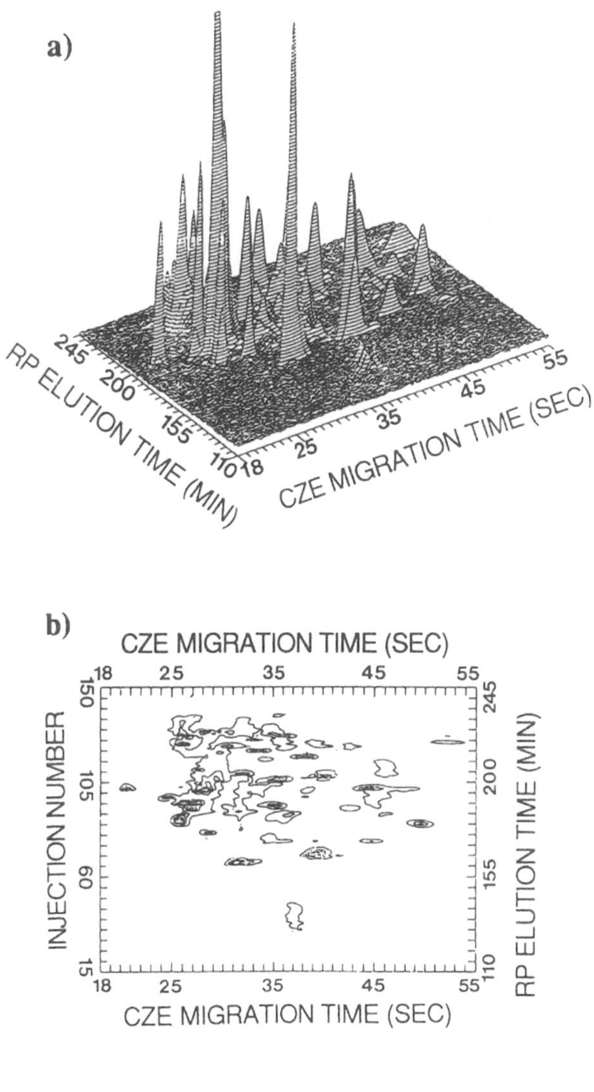

Fig. 5.27. 2-D separation of fluorescamine labeled tryptic digest of ovalbumin as surfer chromato-electropherogram (**a**) and contour plot (**b**). First dimension RP-HPLC, second dimension CZE. Tic marks on the injection number axis represent five injections and 5 min each. (According to Ref. 274 with permission of the American Chemical Society)

the first column to feed the second column (Fig. 5.26.c). If Δt is small, e.g. few peak widths, then the coupled system resembles that of a truly 2-D system. If, however, Δt is large, e.g. many peak widths, even well resolved peaks from the first step may become worse resolved in the second step. Such coupled systems approach purely tandem operation if Δt extends to cover more and more the whole elution spectrum. To approach the peak capacity of a truly 2-D system many columns in the second dimension are required to cover the elution range of the first step or, alternatively, the total separation time of the second system has to approximate the sampling time interval Δt. Although 2-D systems are unexcelled in resolution power, coupled systems are more flexible. "The capacity (of 2-D systems) can be compared to that of mapping the sky with a wide angle low resolution telescope" whereas coupled systems "can be utilized like a high resolution telescope with a zoom lens for seeing, as desired, the utmost detail in certain regions of the sky" [273].

As mentioned above the peak capacity substantially improves with the speed of the second separation step. Bushey and Jorgenson [274] set up a coupled system consisting of a reversed phase HPLC as the first step followed by fast CZE. They calculated a peak capacity of approximately 35 for HPLC with an elution time window from 120 min to 260 min at a peak width of 4 min. Taking the second CZE-step into account with a migration time ranging from 18 s to 55 s and a peak width of 3 s, they estimated an overall peak capacity for their coupled system of at least 420. Their separation of a tryptic digest is shown in Fig. 5.27.

The interface of the LC column with the CZE capillary of 50 μm i.d. represents a six-port valve made of Hastelloy C (Valco Instruments). The entire valve is electrically grounded and serves as the anode of the CZE system. The two valve positions are schematically shown in Fig. 5.28. In the "inject" position (Fig. 5.28.a), the effluent from the LC column goes directly to waste while

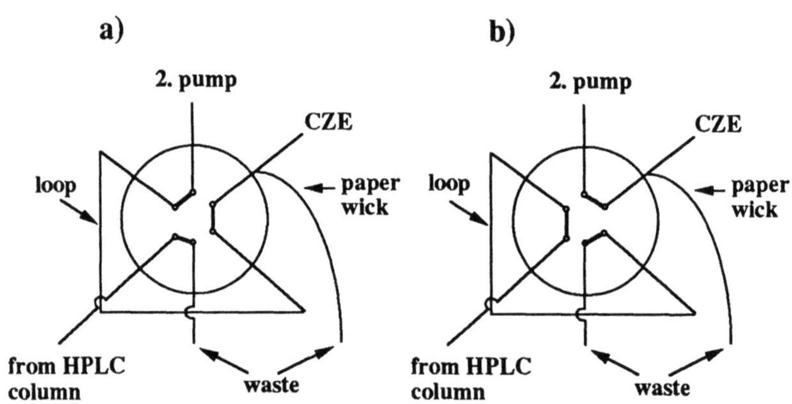

Fig. 5.28. Six-port valve as interface for HPLC-CZE coupling with (**a**) "inject" and (**b**) "separate" position

the second pump flushes the contents of the loop past the end of the capillary for electrokinetic injection. A paper wick transports excess solution to waste. In the "separate" position (Fig. 5.28.b), the effluent from the LC column fills the loop while the second pump flushes fresh buffer past the anodic end of the capillary. A ± 30 kV power supply is used in the negative voltage mode.

5.7.3 Capillary Isotachophoresis - Capillary Electrophoresis (CITP - CE)

One of the major drawbacks of capillary zone electrophoresis is its limited sample loadability and low concentration sensitivity of the UV detector. Commonly, sample concentrations in the range of 10^{-3} to 10^{-5} M are required in CZE to provide a reliable detection signal. Thus, trace analysis is hardly feasible in CZE without sample pretreatment. In addition, complexity of biological matrices and artifact formation are other important reasons to perform a sample clean-up prior to electrophoresis.

Capillary isotachophoresis is closely related to CZE because separation in both techniques is based on differences in electrophoretic mobility. However, CITP has the advantage of high loadability (e.g. some µL) and is an efficient method for sample concentration. Hence, CITP should be an ideal technique for sample pretreatment for CZE.

The first coupling of CITP-CZE was reported by Kaniansky and Marak [275]. Commercial CITP equipment was modified for this purpose. The interface of CITP and CZE was a T-piece allowing a heart-cut injection of the sample zones after isotachophoresis. Although both systems were equipped with capillaries of 0.3 mm i.d., impressive separations were demonstrated. Dolnik et al.[276] described a tandem CITP-CZE system. Through a judicious choice of leading, terminating and background electrolytes isotachophoretic preconcentration was performed on-line with CZE. With this approach a 120-nL sample could be injected and analyzed subsequently by CZE. The change from CITP to CZE was accomplished by exchanging the terminating buffer by an appropriate background buffer. Foret et al. [277] reported on-line isotachophoretic sample preconcentration in CZE by inserting a CZE capillary into a column with a large inner diameter. An increase in detectability of more than 200-fold was obtained by this set-up.

Probably the most sophisticated CITP-CZE system was developed by Stegehuis et al. [278, 279]. Their system featured in terms of detectability, reproducibility and ease of operation. A schematic diagram of the device is shown in Fig. 5.29. The fused silica capillary (50 µm i.d. and 220 µm o.d.) is inserted into the ITP capillary as close as possible to the detector cell without interfering the optical pathway. The ITP capillary consists of two parts, a separation part made of PTFE with an inner diameter of 450 µm, an outer diameter of 700 µm and 250 mm length. The detection part is a fused silica capillary with an i.d. of 320 µm, an o.d. of 450 µm and 50 mm length. Injection into the ITP system is

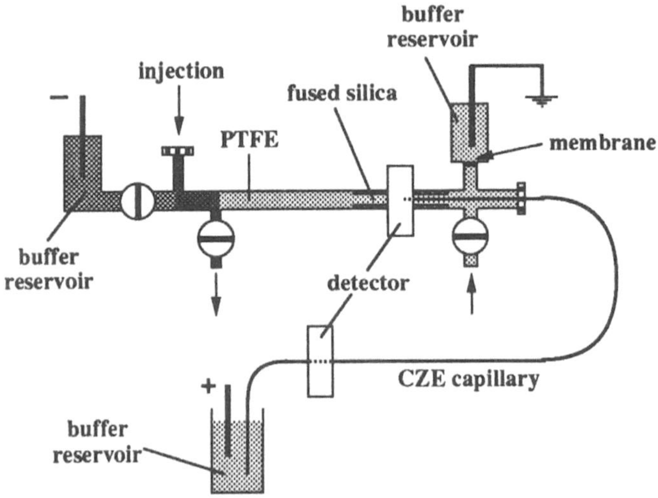

Fig. 5.29. Schematic representation of the CITP-CZE system. (According to Ref. 279 and with permission from Elsevier Science Publishers)

done by a syringe, whereas injection into the CZE system is performed electrokinetically or hydrodynamically. By positioning the CZE capillary exactly behind the detector of the CITP system, accurate timing of the injection is feasible. While the use of PTFE as separation capillary strongly reduces the EOF, the fused silica detection capillary accelerates the EOF such that hydroxypropylmethyl cellulose (HPMC) has to be added to the electrolyte system. An HPMC concentration of 0.05% (w/w) was found to be optimum with respect to reduction of EOF and unacceptable increase of viscosity.

Table 5.10. Optimized experimental conditions for the determination of fluorescein labeled angiotensin III (according to Ref. 279).

	CITP	CZE
	terminator: 0.01 M β-alanine/ $Ba(OH)_2$, pH 10.4, 0.05% HPMC leader: 0.01 M Cl^-/ TRIS, pH 9.2, 0.05% HPMC	background buffer: 0.01 M Cl^-/ TRIS, pH 9.2, 0.05% HPMC
voltage	5 kV	25 kV
capillary	L_T: 25 cm L_D: 22.5 cm	L_T: 50 cm L_D: 30 cm
injection	10 μL (syringe)	10% of the injected amount by ITP

Optimized experimental conditions for the separation of fluorescein labeled angiotensin III (FITC-A III) from deproteinized plasma sample are given in Table 5.10. The separation of FITC-A III by CZE and CITP-CZE coupling is shown in Fig. 5.30. As one can readily see, the detection limit is improved by a factor of 1000 by the preconcentration with CITP.

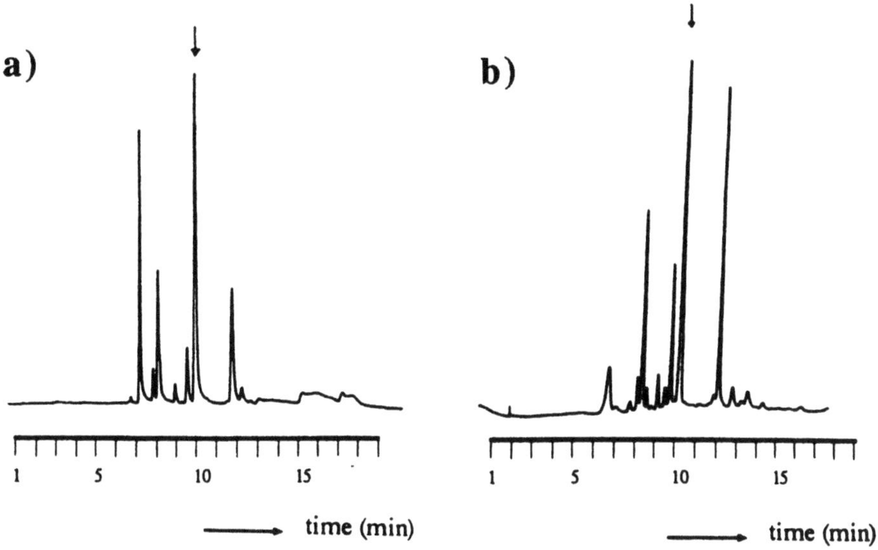

Fig. 5.30. Electrophoretic separation of fluorescein labeled angiotensin III sample (**a**) with a concentration of 5 μg/mL in off-line CZE and (**b**) with 5 ng/mL by CITP-CZE coupling. The *arrow* indicates the angiotensin III. Experimental conditions are given in Table 5.10. and in the text. (With permission from Ref. 279)

5.8 Special Techniques

5.8.1 Capillary Affinity Electrophoresis

"Affinity electrophoresis in a broad sense denotes all techniques in which some kind of biospecific interaction between an electrophoresed component (receptor) and another component present in the medium (ligand) occurs" [280]. As a consequence of this interaction the electrophoretic mobility of the substance will change. Affinity electrophoresis is mainly used for analytical purposes (i) to detect those components of a sample which undergo ligand complexation, (ii) to determine influences such as pH, temperature or complex inhibiting factors and (iii) to study quantitatively the complex formation.

The magnitude of the mobility change of the complexed component caused by the complex formation depends on various properties of the ligand. If the mass of the ligand is small compared to the receptor, i.e. a small substrate of an enzyme, the change in mobility of the enzyme will be small or even not measurable. This holds especially if the ligand is neutral or exhibits a small net charge. Pronounced effects on the electrophoretic mobility can be measured if the ligand and the receptor possess comparable size, if the ligand is highly charged, or if the receptor enables simultaneous complex formation with more than one ligand and vice versa.

While affinity electrophoresis performed in slab or rod gels has been used for a long time [280, 281] capillary affinity electrophoresis (CAE) is a relatively new technique. Compared to other techniques for the determination of association constants such as UV absorbance, nuclear magnetic resonance, differential scanning calorimetry, etc., CAE has a number of advantages [282]:

- Only small quantities in the ng range of receptor and ligand are required.
- The technique does not require high purity or a known concentration of the receptor, because the association constant is based on electrophoretic migration rather than on peak area.
- Association constants of more than one receptor can be measured in the same run.
- The technique is capable of distinguishing between those subunits of a protein that bind to the ligand and those that do not bind or are denatured.
- Measurements are performed in free solution where parameters like pH, ionic strength and temperature are well controlled.
- Because no stationary phases or gels interfere with the receptor, CAE is suitable for even labile proteins.
- Commercial automated equipment is available allowing overnight series analysis.

In general, the association reaction of let us say a protein with a ligand is an equilibrium reaction. Provided a monovalent reaction mode the association constant can be determined experimentally based on the following theoretical derivation [284]. According to the law of mass action the association constant K_a of the protein-ligand interaction can be written as

$$K_a = \frac{[PL]}{[P]\cdot[L]} \tag{5-22}$$

[P] concentration of the protein [M]
[L] concentration of the ligand [M]
[PL] concentration of the protein-ligand complex [M]

The molar fraction or dissociation degree α of the complex is then given by

$$\alpha = \frac{[PL]}{[P]+[L]} \qquad (5\text{-}23)$$

Combining Eqs. 5-22 and 5-23 one can derive

$$\frac{1}{\alpha} = \frac{1}{K_a \cdot [L]} + 1 \qquad (5\text{-}24)$$

The observed migration velocity v_{eff} of the protein in the fused silica capillary is given by the electroosmotic velocity and the electrophoretic velocity v_P according to

$$v_{eff} = v_{eo} - v_P \qquad (5\text{-}25)$$

If a ligand is added to the buffer system the observed migration velocity v'_{eff} will be changed to

$$v'_{eff} = v_{eo} - v'_P \qquad (5\text{-}26)$$

v'_P electrophoretic velocity of the protein at a given ligand concentration

The term v'_P can be expressed as the sum of the velocity of the protein v_P and the complex v_{PL} (where v does not change anymore with increasing ligand concentration), multiplied by their molar fractions as follows:

$$v'_P = (1 - \alpha) \cdot v_P + v_{PL} \qquad (5\text{-}27)$$

From the derived equations above, the molar fraction can be written as

$$\alpha = \frac{v_{eff} - v'_{eff}}{v_{PL} - v_P} \qquad (5\text{-}28)$$

By substituting α in Eq. 5-24 by that in Eq. 5-28 and $\Delta v = v_{PL} - v_P$ one obtains

$$\frac{1}{K_a \cdot [L]} + 1 = \frac{\Delta v}{v_{eff} - v'_{eff}} \qquad (5\text{-}29)$$

Since $v'_{eff} = L_D/t'_P$ and $v_{eff} = L_D/t_P$, where t'_P and t_P are the migration times of the protein in the presence and absence of the ligand, Eq. 5-29 can be written as

$$\frac{1}{t'_P - t_P} = \frac{L_D}{\Delta v \cdot t_P^2 \cdot K_a} \cdot \frac{1}{[L]} + \frac{L_D - \Delta v \cdot t_P}{\Delta v \cdot t_P^2} \tag{5-30}$$

According to Eq. 5-30 a plot of $1/(t'_P - t_P)$ versus $1/[L]$ should be linear with a slope of $\Delta v \cdot t_P^2 \cdot K_a$. Since

$$\Delta v = v_{PL} - v_P = L_D \cdot \left(\frac{1}{t_{eo}} - \frac{1}{t_{PL}}\right) \cdot \left(\frac{1}{t_{eo}} - \frac{1}{t_P}\right) = L_D \cdot \frac{t_{PL} - t_P}{t_P \cdot t_{PL}} \tag{5-31}$$

t_{eo} time required by the EOF to migrate to the detector [s]

t_{PL} migration time of the complex where t does not change anymore with increasing ligand concentration [s]

From Eqs. 5-30 and 5-31 the association constant can be calculated as

$$K_a = \frac{t_{PL}}{t_P} \cdot \frac{1}{t_{PL} - t_P} \cdot \frac{1}{A} \tag{5-32}$$

Thus, the association constant of a complex formation can be calculated by using migration times t_P and t_{PL} and the slope of the straight line of a plot as given by Eq. 5-30. The value of t_{PL} can be obtained from an electrophoretic run where the migration time does not change anymore with increasing concentration of ligand (see Fig. 5.31.a).

It has to be noted that the theoretical treatment given above is based on the premise that the ligand is bound to a single site on the protein molecule (mono-

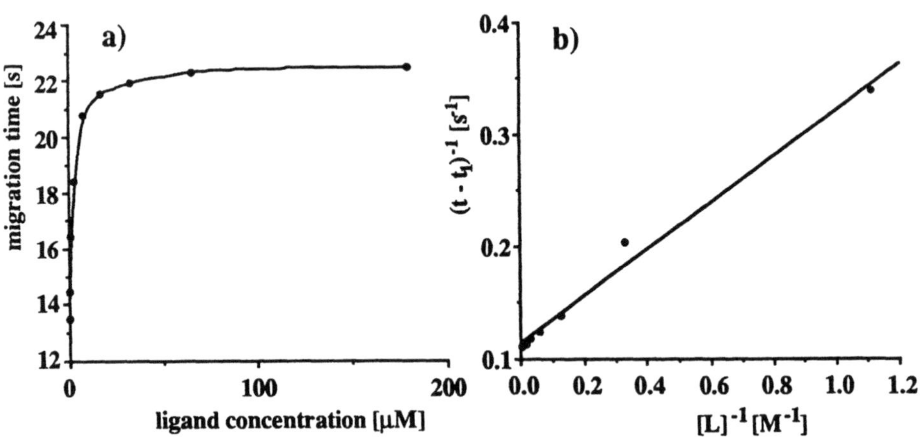

Fig. 5.31. (**a**) Plot of the migration time of bovine carbonic anydrase as a function of the ligand concentration in the mobile phase. Data are taken from Ref. 282; (**b**) plot of $(t - t_1)^{-1}$ versus $[L]^{-1}$ for the data of (**a**)

valent mode interaction). From this derivation the association constant is obtained from a series of experiments, where a receptor and an electroosmotic flow marker is subjected to electrophoresis. Maintaining all other experimental conditions constant, only the concentration of the ligand in the buffer system is varied in a given concentration range. Eq. 5-32 does not contain a concentration term of the sample. Hence, association constants can be principally measured at any receptor concentration which is still detectable.

For the choice of the buffer system some aspects should be taken into account. First, protein-ligand complexation occurs only at non-denaturing conditions. Depending on the protein of interest, the buffer system must not contain any additives such as urea, EDTA, SDS or heavy metals which may denature the protein or interfere with the complexation. Therefore, one should ensure in advance that buffer components do not act as inhibitors as it is often the case with phosphate or TRIS. Second, the complexation of a protein is strongly dependent on the pH of the medium. As a rule of thumb, protein-ligand complexation constants are highest under biological conditions. Third, mobility changes of the protein by the complexation with the ligand become most prominent if the mobility of the non-complexed protein is small or even zero. This holds usually around the pI of the protein.

Whitesides and coworkers [282] investigated the complexation of bovine carbonic anhydrase B with alkylbenzenesulfonamides as ligand by CAE. By using the data from their electropherograms the following relationship, given in Fig. 5.31.a, between the ligand concentration and the change in migration time was found. From these data a plot according to Eq. 5-30 was drawn (Fig. 5.31.b). From the slope of the straight line and the migration times at zero concentration (t_P) and 180 μM (t_{PL}) of ligand an association constant of $0.71 \cdot 10^6$ M^{-1} was calculated which is in good agreement to the value calculated by the authors using Scatchard analysis ($0.48 \cdot 10^6$ M^{-1}).

Honda and coworkers [283] investigated the interaction of β-galactose-specific lectin with lactobionic acid. They found that the reproducibility of the determination of association constants is fairly high with a relative standard deviation of 3.6% compared to other techniques such as equilibrium dialysis (more than 50%).

Chu et al. [284] and Carpenter et al. [285] studied independently the binding constants of vancomycin of di- and multiple peptides. Their values are in good agreement with values reported in the literature using different techniques.

Guttman and Cooke [286] incorporated ethidium bromide as soluble ligand into a polyacrylamide gel to manipulate the selectivity of DNA restriction fragments in CGE. Ethidium bromide is used as a selective intercalating agent for double-stranded DNA molecules. Since ethidium bromide is positively charged it causes a reduction of the electrophoretic mobility of DNA. In addition to the change of the DNA net charge, the molecular mass of DNA is increased up to 12% due to complex formation. Since the complexation of ethidium bromide is not a proper receptor-ligand interaction with a defined reaction scheme, this approach is not suitable for any further studies on the recognition mechanism.

Although CAE is still a new technique for studying recognition mechanisms the first results indicate that it has a tremendous potential for this purpose. Adsorption of proteins to the fused silica wall, which might be a potential limitation, can be overcome by using appropriately coated capillaries.

5.8.2 Sample Stacking

The technique of sample stacking for on-column sample concentration in a single, continuous support buffer has been well known in electrophoresis for a long time. After it was introduced in CE by Mikkers et al. [41] in 1979, it was 10 years before it was reinvestigated by the Lauer's group [287] and by Chien and Burgi [288]. It offers an alternative to pre-column concentration procedures and CITP. In its simpliest form, sample stacking or field-amplified CZE is carried out by hydrodynamically injecting a long plug of sample dissolved in water or a lower concentration buffer 1 into a capillary filled with higher concentration buffer 2. When a voltage is applied across the capillary a higher electrical field strength is set up in the sampling compartment than in the separation compartment due to its low concentration of ions. Assuming that a length $x \cdot L_T$ of the capillary is filled with buffer 1 (sampling compartment) and a length $(1-x) \cdot L_T$ with buffer 2 (separation compartment) the local electric field strengths E_1 and E_2 in the two regions are given by:

$$E_1 = \frac{\gamma \cdot E_0}{\gamma \cdot x + (1-x)} \quad \text{and} \tag{5-33}$$

$$E_2 = \frac{E_0}{\gamma \cdot x + (1-x)}\,, \tag{5-34}$$

where $\gamma = \kappa_2/\kappa_1$, the ratio of the specific conductances of buffer 1 and buffer 2, respectively, and E_0 is the original uniform field strength. While the field strength E_2 in the separation compartment is lower than the original field strength E_0, the field strength E_1 in the sampling compartment will be increased by a field enhancement factor f:

$$f = \frac{\gamma}{\gamma \cdot x + (1-x)} \tag{5-35}$$

Furthermore, it can be seen from Eqs. 5-33 and 5-34 that the ratio between the field strengths in the two compartments is equal to γ and will remain constant independently from the capillary length. If the sample is prepared in water, γ can easily reach several hundred. As a consequence of the high electric field strength in the sampling compartment the sample ions move very rapidly toward the boundaries between sampling and separation compartment (see also Sect. 3.1.4). Once the ions pass through the boundary they experience a lower electric field strength and immediately slow down. Now, electrophoretic separation takes place according to the different electrophoretic mobilities of the analytes. Because the flux of the ions across the boundary must be conserved and the electrophoretic velocities are proportional to the electric field strength, the concentration of the sample ions will be enhanced by the factor γ:

$$c_2 = \gamma \cdot c_1 \tag{5-36}$$

Index 1 refers to the sampling compartment and 2 to the separation compartment. Because the total number of sample ions must be conserved, the length of the sample zone must be reduced by the same factor γ. Neglecting diffusion the effective plug length l_{eff} after stacking is:

$$l_{eff} = \frac{1}{\gamma} \tag{5–37}$$

This narrow zone of ions moves through the separation compartment separating into individual zones by conventional ZE. The stacking mechanism occurs for both positively and negatively charged analytes. In conventional CZE with the injection end at the anode and a cathodic EOF toward the outlet of the column, the cations stack up in front of the sample plug, the anions in the rear, and the neutral species are left in the sampling compartment and coelute with it (Fig. 5.32.). To perform sample stacking of both positively and negatively charged analytes in the presence of cathodic EOF, follow this procedure:

> Prepare the sample solution by dissolving the sample in the separation buffer which has been diluted ca. 10 times. Alternatively, use your low concentrated sample solution. Take care that the conductivity is about 10 times lower than that of the separation buffer.

> Place the anodic end of the capillary in the sample vial and the cathodic end in the separation buffer reservoir. Inject hydrodynamically so much sample that the sample plug width is ca. 10 times the diffusion limited peak width (see Eq. 3-23). See Fig. 5.32.a.

> Replace the sample vial by the separation buffer reservoir and apply high voltage. See Fig. 5.32.b.

Cations stack up in front of the sampling compartment, anions in the rear (Fig. 5.32.c). If the analytes pass the boundaries to the separation compartment, their electrophoretic separation begins. Since the electroosmotic velocity in the separation compartment is higher than the electrophoretic velocities of the analytes, both the anions and the cations as well as the sampling compartment are pushed to the cathode (Fig. 5.32.d). Figure 5.33. shows the stacking effect for a

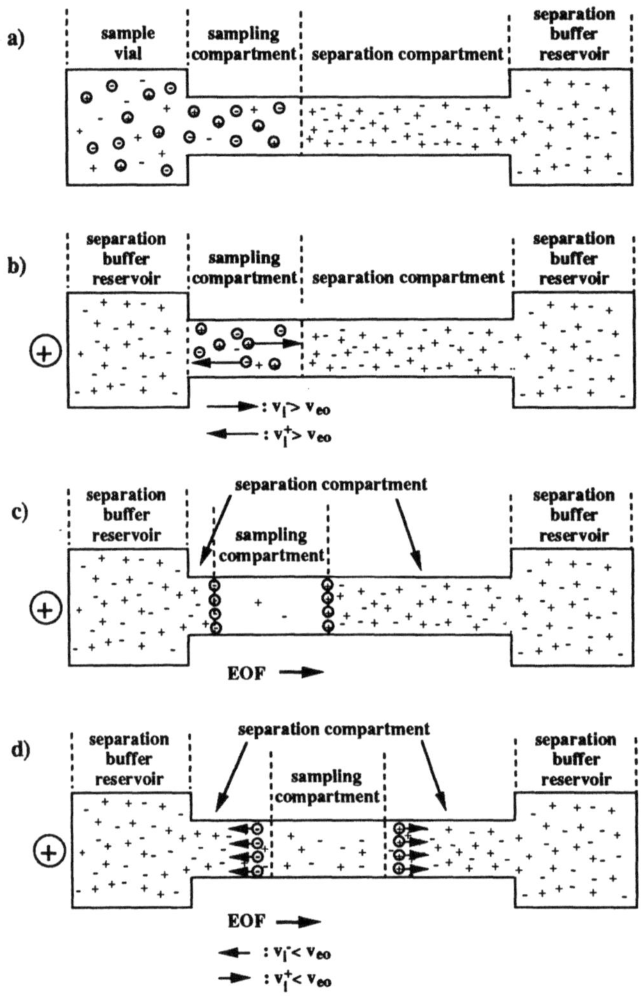

Fig. 5.32. Sample stacking after hydrodynamic injection at the anode. For details see text

positively charged peptide. In this example the injection time is kept constant and the concentration of the sample buffer solution is stepwise increased. By keeping the injected amount constant in all runs it can be seen that efficiency is improved significantly by amplifying the electric field in the sampling compartment.

Theoretically, the amount of stacking is proportional to the field enhancement factor: the larger the difference in concentrations, the narrower the sample zone and the higher the sample concentration. Consequently, a rather long

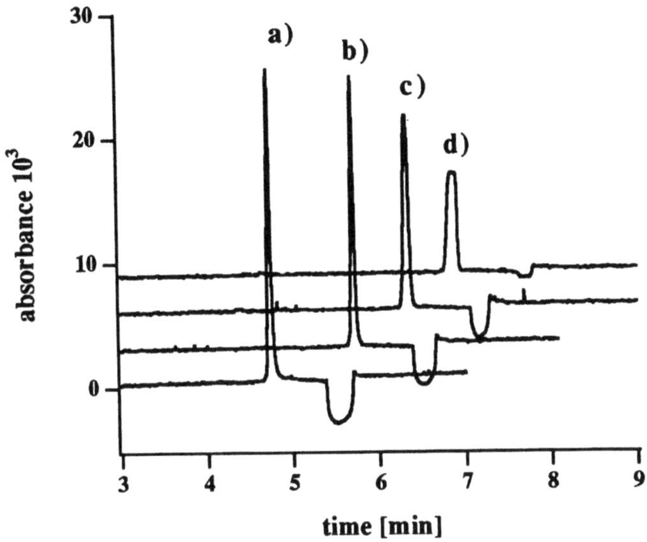

Fig. 5.33. Sample stacking of a positively charged peptide. Instrument: Beckman P/ACE 2000. Experimental conditions: fused silica capillary 100 cm x 75 μm i.d., hydrodynamic injection for 20 s, voltage 25 kV, UV detection at 200 nm, temperature 25 °C, electrolyte system 30 mM sodium phosphate, pH 7.0. The sample is dissolved in (**a**) 5 mM, (**b**) 10 mM, (**c**) 15 mM and (**d**) 25 mM sodium phosphate buffer, pH 7.0. For clarity reasons overlayed plots are shifted along the time axis

sample plug prepared in water should be stacked into a very thin zone. In the presence of EOF, however, a hydrodynamic backflow with a laminar flow profile is generated because the local electroosmotic velocity in the sample plug is greater than the bulk EOF velocity in the separation compartment [288, 289]. The laminar flow will broaden the sharp analyte zone generated by the stacking process. The differences between bulk EOF and the local electroosmotic velocity in the sample plug is increased by increasing field strength and decreasing conductivity of the sample buffer. Hence, stacking and broadening counteract to yield an optimum related to the concentrations of the sample buffer and the separation buffer and the sample plug length. Another problem occuring with long sample plugs is the redistribution of the field strength. Since almost all the voltage is dropped across the sampling compartment, the field strength in the separation compartment will approach zero. Consequently, the electrophoretic velocities of the stacked analytes will also approach zero. Therefore, the analytes will remain at the concentration boundary between the sampling and the separation compartment and will only move through the capillary with the EOF. No separation will occur. Figure 5.34. shows the electropherograms for the separation of PTH-Asp and PTH-Glu at four different injection lengths. As indicated by the migration time of the water plug, the average EOF velocity increases as the sample plug length increases. At the same time, the electrophoretic velocities of the analytes in the separation compartment decrease resulting in the peaks getting closer to the water plug and to each other. In Fig. 5.34.a the sample plug is so long that no more separation takes place; the analytes move with the EOF past the detection window. The optimal stacking conditions for this example are those in Fig. 5.34.d. Under these conditions the analytes migrate at the correct time in comparison with a run without stacking conditions. Burgi and Chien [289] have suggested that the optimal condition for sample stacking with hydrodynamic injection at the anode in the presence of EOF would be to prepare the sample in a buffer concentration 10 times less than that used for the separation and to choose a sample plug width up to 10 times the diffusion limited peak width (see Eq. 3-23). Using these parameters, a 10-fold improvement in detection sensitivity can be achieved without any loss in resolution [289, 290].

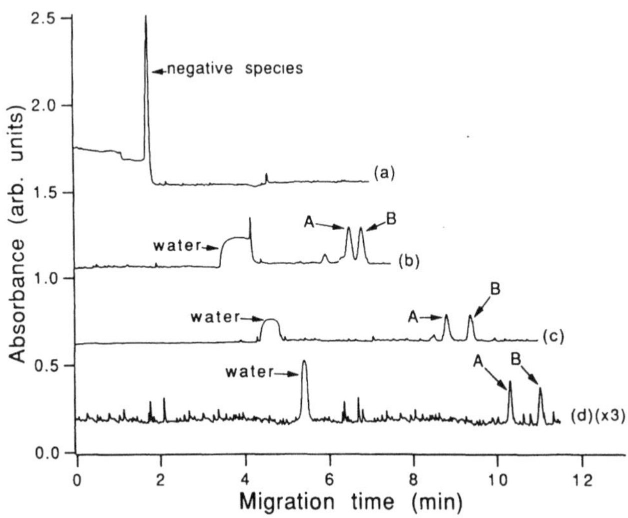

Fig. 5.34. Electropherograms for the separation of PTH-Asp and PTH-Glu at four different injection lengths. Experimental conditions: fused silica capillary 100 cm x 50 μm i.d. (L_D = 35 cm), voltage 30 kV (current 8 μA), hydrodynamic injection, UV-detection at 265 nm, electrolyte system 100 mM MES - 100 mM His, pH 6.1. Peak A is $4.6 \cdot 10^{-5}$ M PTH-Asp, peak B is $3.4 \cdot 10^{-5}$ M PTH-Glu. The sample is dissolved in water. Sample injection lengths (**a**) 35, (**b**) 7, (**c**) 3.5 and (**d**) 0.7 cm. Panel (**d**) is expanded vertically by a factor of three. (With permission of Ref. 44)

To increase the amount of sample injected above a factor of 10 in the presence of EOF, the non-uniform distributions of both the EOF velocity and the field strength have to be eliminated. This can be accomplished by pumping the sample buffer out of the column by means of the electroosmotic flow while the stacking process is in progress [44, 291]. To realize that the polarity of the electrodes is reversed after injection, until the sample buffer is pushed almost completely out of the column. This method is useful for either positively or negatively charged ions but cannot be used on both species simultaneously. For concentration and subsequent separation of negatively charged ions (Fig. 5.35.), the EOF has to be directed toward the cathode (as it is the case in an uncoated open-tube fused silica capillary). Assuming hydrodynamic injection at the anodic end (Fig. 5.35.a), cations migrate to the front of the sample plug and anions to the rear (Fig. 5.35.b). If reversed polarity is applied after injection, the direction of EOF is toward the anode. First, the positive ions are pushed out of the column, followed by the sample buffer. Since the negative ions which usually migrate with the EOF obtain a much higher velocity inside the sampling compartment because of the field enhancement, they overcome the EOF and stack inside the column at the front boundary of the sampling compartment during the sample buffer removal. Therefore one can apply reversed polarity immediately after injection and have not to wait until the stacking process is completed (Fig. 5.35.c). When the sample buffer is almost completely removed from the column, the polarity of the electrodes is switched again. Now, the negatively charged analytes migrate very fast over the remaining sampling compartment to its rear boundary and stack there again before their separation begins (Fig. 5.35.d). To concentrate and separate positively charged analytes, the charge of the silica wall has to be made positive by dynamic coating with cationic surfactants like CTAB or TTAB (≈ 1 mM), thus changing the direction of the EOF. In the latter case, negatively charged species are stacked at the front boundary of the sampling compartment and will be pushed out of the column followed by the sample buffer.

Hint: *To indicate the moment when the sample buffer is almost completely removed from the capillary the current is monitored and compared with the current of a column completely filled with the separation buffer. First, the current decreases during injection. As the sample buffer is then pushed out of the column, the current increases again. If it reaches about 95% of the current of the pure separation buffer, the polarity is switched again, and separation can occur.*

With this technique a several hundred fold increase in the amount of sample injected has been achieved by injecting a sample plug as long as two-thirds of the total column length. Chien and Burgi [44] have calculated the maximum filled length l_{max} possible without any loss of analyte ions as

$$l_{max} = -\frac{\mu_i}{\mu_{eo}} \qquad (5\text{-}38)$$

If, for instance, the electrophoretic mobility of the analyte is half the electroosmotic mobility, up to 50% of the column can be filled with the sample solution without any loss of sample. In practice, however, it is sometimes easier to fill the whole capillary with the sample solution (e.g. with a syringe or a wa-

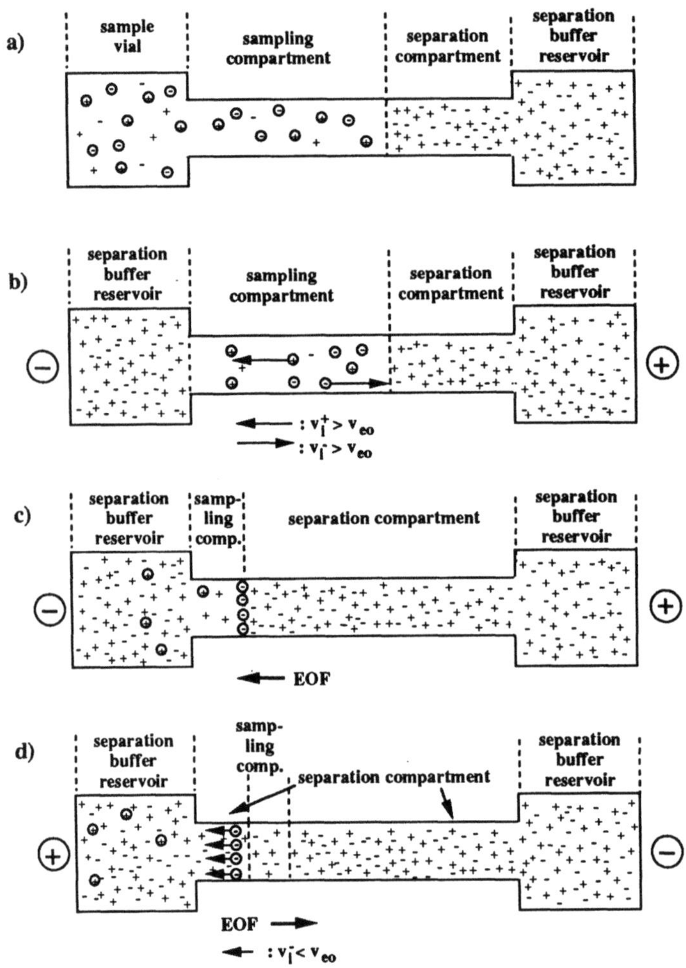

Fig. 5.35. Sample stacking of anions with subsequent removal of the sample buffer using EOF. A long plug of sample dissolved in water or the low concentrated separation buffer is injected hydrodynamically into the capillary (**a**). High voltage with reversed polarity is applied, the sample buffer and positively charged sample compounds are removed. Anions migrate against the EOF to the stationary boundary which remains behind in the capillary (**b+c**). The polarity is switched again to normal configuration, when the sample buffer is almost completely removed (**d**).

shing step) and perform sample stacking and removing of the sample buffer simultaneously by inserting the capillary into the separation buffer reservoirs. At the beginning of the run there is no stationary boundary at which analyte ions can be concentrated. Consequently, some of the analytes will be carried out of the column by the EOF. As the separation buffer slowly replaces the sample buffer, the electric field in the remaining sample buffer zone will begin to increase rapidly. If the length of the sample buffer zone reaches l_{max}, the analytes begin to migrate in opposite direction to the EOF and stack at the end of the sample buffer zone which remains behind in the column.

Hint: *The method of sample stacking of extremely large injection volumes should only be used in the case of very diluted samples. If the sample concentration gets too high, overloading occurs which manifests itself in electrophoretic dispersion .*

Note, that this technique can only be used for either cations or anions, but not for the simultaneous analysis of both.

Vinther and Søeberg [47] have investigated dispersion processes under stacking conditions. They found that with increased stacking power, also dispersion is increased. Therefore they recommend to use only moderate stacking conditions which means an only slightly lower conductivity of the sample solution in comparison to the separation buffer. Furthermore, they suggest to apply a low voltage during the stacking period and to increase the voltage afterwards in order to speed up analysis.

Sample stacking can also be applied to electrokinetic injection [292, 293]. If the sample is dissolved in a buffer of lower concentration than the separation buffer or in water, the electric field at the injection point is much higher than in the capillary according to Eqs. 5-33 and 5-34. For a short injection time, where $x \ll 1$ and $\gamma \cdot x \ll 1$, E_2 changes very little, and E_1 is enhanced by γ. If the conductivitites of the sample and the separation buffer differ significantly (γ is very large) such that $x \ll 1$ but $\gamma \cdot x \gg 1$, the field enhancement is inversely proportional to the plug length.

In the absence of EOF, sample stacking for either positive or negative ions can be performed by choosing the proper polarity of the electrodes. In practice, a much smaller than predicted signal enhancement is found when the column is switched directly between the separation buffer and the sample solution, probably due to distortions of the electric field at the injection point. Moreover, only ions migrating in the same direction as the EOF can be concentrated in the presence of EOF. Under normal polarity with the EOF toward the outlet of the column, the high field strength at the injection point pushes away negatively charged ions. With reversed polarity the negative ions would be pulled into the capillary, but they would be carried out immediately after reaching the boundary between sample and separation compartment by the reversed EOF. The same will happen for positively charged ions in a coated capillary with an anodic EOF toward the capillary outlet. These problems can be circumvented by hydrodynamically injecting a short plug of water or low-conductivity buffer prior to sample introduction (Fig. 5.36.). By selecting the proper polarity of the electrodes during injection, either positive (Fig. 5.36.b-d) or negative ions (Fig. 5.36.e-g) can now be concentrated. If the polarity is switched after the injection of positive ions both positive and negative ions can be enhanced at the corresponding boundary (performing all steps of Fig. 5.36. from a-h). Using electrokinetic injection with a water plug, the LOD for PTH-arginine could be improved from 10^{-5} M to 10^{-7} M [292].

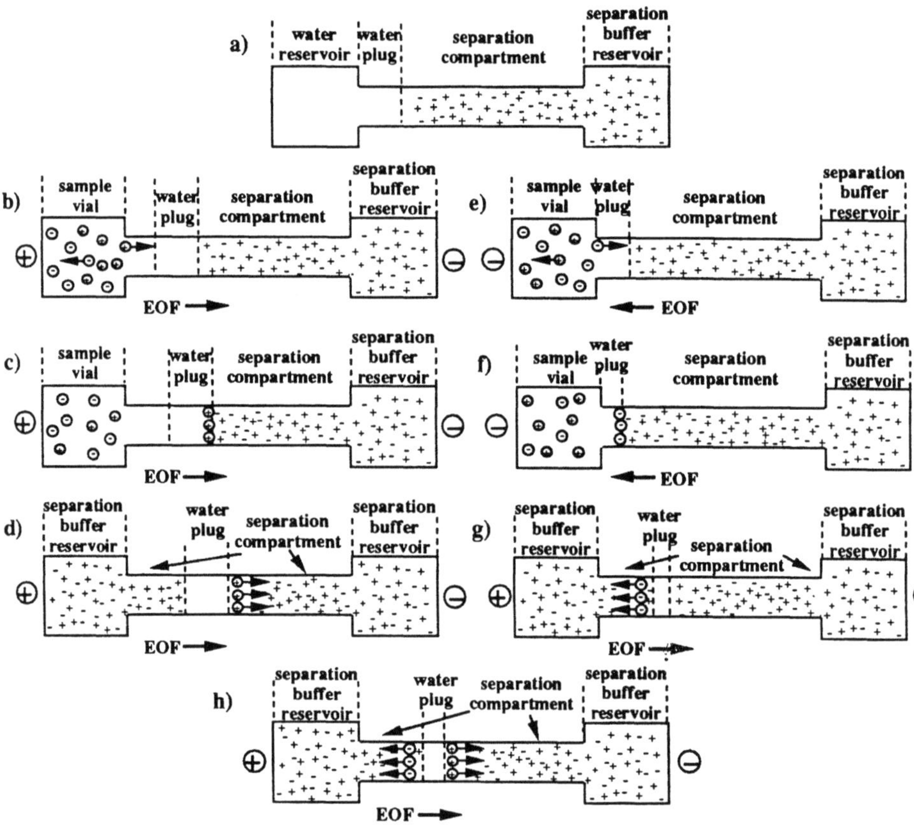

Fig. 5.36. Sample stacking after electrokinetic injection of positively (**b** - **d**) and negatively (**e** - **g**) charged analytes, respectively. For the sake of simplicity the counter charges are not shown in the water plug region. (**a**) Hydrodynamic injection of a short plug of water or low-conductivity buffer. (**b, e**) Electrokinetic injection of the sample by choosing the proper polarity. (**c, f**) Stacking of the analytes at the front boundary of the water plug. (**d, g**) Migration of the analytes out of the water plug and start of separation. Before leaving the water plug region negatively charged analytes migrate to the rear boundary of the water plug when the polarity of the arrangement is switched (not shown). (**h**) Start of the separation after polarity-switching electrokinetically injection for the simultaneous analysis of cations and anions

Hint: Since the water plug moves out of the capillary from the injection end during injection of negative ions (see Fig. 5.36.e,f), its length should be long enough so that a part of it remains inside the column after injection.

Neglecting diffusion, the effective length of the analyte zone after stacking is

$$l_{eff} = \left(\frac{\mu_{eo}}{\gamma} + \mu_i \right) \cdot E_0 \cdot t \qquad (5\text{-}39)$$

For large γ the contribution from EOF can be neglected, and l_{eff} is simply proportional to μ_i. On the other hand, in conventional electrokinetic injection the zone length is dominated by the EOF (see Eq. 4-3). Thus, the zone length becomes narrower than in conventional injection if using the stacking effect. Consequently, the amount of sample can be enhanced further by longer injection times or at higher injection voltages. In practice, injection times of 30 s (for U = 5 kV) and voltages of 30 kV are recommended [292].

Another interesting feature of sample stacking after electrokinetic injection is the very effective charge discrimination. Although there is a large bias against negative ions in conventional electrokinetic injection, they still can migrate into the column as long as their mobility is smaller than the electroosmotic mobility. By using sample stacking only those ions migrate into the capillary that have positive mobilities with respect to the EOF.

Jandik and Jones [154] reported another electrokinetic injection procedure which can be used to improve sensitivity in the analysis of inorganic ions. If the separation buffer has a higher mobility than any analyte ion, it acts as a leading electrolyte during electrokinetic injection, which means that an isotachophoretic distribution of the analytes is created after electromigration into the capillary. Especially in situations where the sample concentration is in the nanomolar range and the concentration of the leading electrolyte is adjusted to 5 - 20 mM, a high enrichment factor is achieved. In solutions containing total ionic concentrations in the nanomolar range, the sample conductivity becomes too low to allow sufficient charge throughput for ionic transfer from the sample solution into the capillary. To increase the ionic strength of the sample by simultaneously allowing isotachophoretic preconcentration, a suitable ion is added to the sample matrix which has a lower mobility than any analyte ion of the same charge, thus functioning as the terminating electrolyte.

6 Qualitative and Quantitative Analysis

6.1 General Aspects

It is the objective of any analytical separation process to provide appropriate answers to the following two questions:

1. What is the composition of the sample?
2. What is the concentration of the components in the sample?

While the first question refers to qualitative analysis the second one is designated as quantitative analysis.

Qualitative analysis in CE provides information about the identity of a peak in the electropherogram. This purpose can be accomplished by comparing the retention time or electrophoretic mobility of the particular peak with the experimental data of the known compound. If the same retention times or mobilities are obtained, both compounds may be identical. To identify a peak with certainty in an electropherogram the following procedure has to be performed:

- Spike the sample solution with the pure compound prior to injection. The resulting peak in the electropherogram should be higher than the original peak without any indications of a shoulder.
- Perform the same procedure using another separation mechanism, e.g. MEKC or reversed phase HPLC. If no indication of an additional peak is found, peak identity is rather certain.
- Complete certainty is obtained if the peak of interest is analyzed by additional techniques such as CZE-MS coupling which provide information about the chemical structure.

Comparison of peaks by their retention times requires highly constant experimental conditions. Small changes in operational parameters such as temperature, buffer pH, buffer ionic strength, capillary pretreatment, etc., alter the mobility of the analytes which may lead to misidentification of peaks. Although reproducibility of migration times is usually high in CE with less than 1% relative standard deviation (RSD), improved results are obtained by performing the following suggested procedures.

First, frequent buffer regeneration is essential to avoid depletion of the ionic strength during sequence analysis. Second, temperature influences by Joule heating are minimized by working at constant current mode instead of constant voltage as discussed in Sect. 3.2.1. Highly reproducible retention times necessitate effective temperature control to eliminate or minimize temperature dependent variables (e.g. viscosity, pH, etc.). Third, retention times should be corrected by eliminating the proportion of EOF. Due to the impact of EOF, electrophoretic mobility is a much more rugged migration parameter. The RSD of mobility is much less than the error in retention time (1% versus 2 - 2.5%) as examined by Smith et al. [294]. However, in MEKC retention time appears to be more reproducible than mobility. Fourth, preconditioning of the fused silica capillary has a great effect on migration reproducibility. The influence of capillary rinsing and frequency of rinsing was studied in Ref. 294. The best reproducibility was found when the capillary was rinsed between each run. The nature of the rinsing medium appeared to be of little importance. Sodium hydroxide, buffer or a methanolic solution gave similar results, but the combination of NaOH with methanol seemed to destabilize the migration behavior. NaOH is recommended for conditioning new capillaries to remove contamination from the tube surface.

Quantitative analysis provides information about the amount or concentration of a substance in a given sample. This information is obtained from the height or, preferably, area of a peak in the electropherogram. While the peak height can be read directly from the electropherogram, state-of-the-art determination of the peak area requires an integration system. The output of an integrator is the mathematical product of the detection signal (usually in mV) with the corresponding time interval and represents the peak area. The unknown amount or concentration c_x of a substance can be calculated by correlating the peak area a_x with the peak areas $a_{1...n}$ of reference samples with known contents. This procedure can be done by two methods: (i) external standard or (ii) internal standard.

For the external standard method a reference solution (external standard) with known concentrations of all compounds of interest is injected and the peak areas are determined with an integration system. Ideally several concentrations of the reference solution should be measured to determine the linearity of the detector for a given concentration range. Subsequently, the sample is injected and the peak area is determined in the same way as for the external standard. By plotting the concentration versus the peak area (Fig. 6.1.a) a straight line should be obtained from which the concentration of the sample can be calculated. Since peak areas of consecutive runs are compared in the external standard method, highly reproducible and accurate injection volumes are required.

For the internal standard method a known substance (internal standard) is added to both the reference solution and the sample solution. The concentration of the analyte in the sample is determined by comparing the peak ratios of the internal standard and the analyte with the corresponding peak ratios of the reference solutions. An example is given below:

The concentration of phenylalanine in a sample solution shall be determined by using tyrosine as an internal standard (not present in the sample). Use the following procedure:

> Prepare reference solutions of phenylalanine with defined concentrations of e.g. 2.5 mM, 5.0 mM, 7.5 mM and 10.0 mM in water.
> Add a known amount of tyrosine (e.g. to a final concentration of 5 mM) to each reference solution.
> Analyze each reference solution and determine the peak areas of both compounds. Calculate the concentration ratios and the corresponding ratios of the peak areas and draw a calibration curve as shown in Fig. 6.1.b.
> Add a known amount of tyrosine (to a final concentration of 5 mM) to the sample solution and analyze it.
> Calculate the ratio of the peak areas of phenylalanine and tyrosine.
> Determine the concentration ratio of the sample solution by using Fig. 6.1.b and calculate the concentration of phenylalanine.

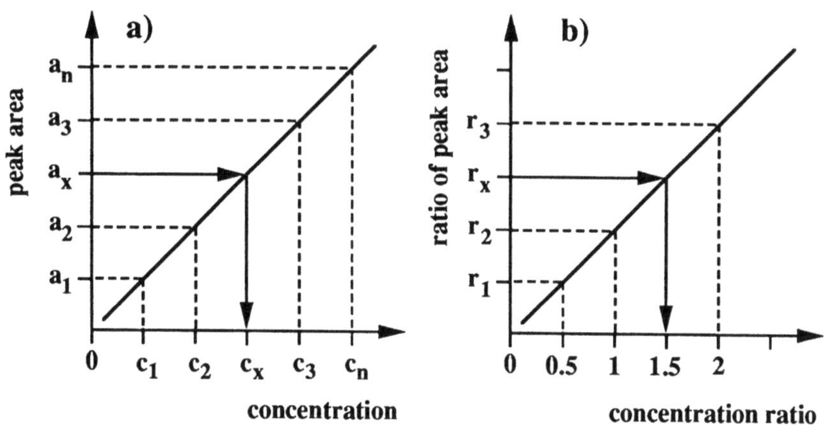

Fig. 6.1. Schematic calibration curve for quantitative analysis with an external standard (**a**) and an internal standard (**b**)

The internal standard has to meet a number of requirements. Firstly, the substance must absent in the original sample. Secondly, it has to a be pure, well defined substance with an electrophoretic and detection behavior similar to the analyte. Thirdly, it has to be separated in the electropherogram from all other components of the sample. Because the internal standard is added to the sample solution which is actually injected, variations of the injection volume do not influence the reproducibility of the method as far as detection is performed in the linear range of the detector. Therefore, the internal standard is the method of choice for quantitative analysis in CZE.

Another important point needs to be discussed [295]. In chromatographic separation techniques all components of a sample pass the detector cell at a velocity equal to the flow rate of the mobile phase. Consequently, the time related peaks that are recorded correspond to the actual width of the components after they elute out of the column. In CE the situation is markedly different. A sample zone migrates at a velocity that is given by the electrophoretic velocity of the component and the superimposed electroosmotic velocity of the background electrolyte. Analytes which migrate in the same direction as the electroosmotic flow move faster than neutral species. In contrast, compounds migrating in the opposite direction to the electroosmotic flow move slower than neutral analytes. Because each zone migrates at an individual velocity, the time needed for them to pass the detection cell differs. Thus, different time-related peak widths in the electropherogram may not necessarily denote different zone volumes of the analytes. Fig. 6.2. depicts two schematic electropherograms of three peaks at which the second represents an electroosmotic flow marker. Thus, zone 1 migrates faster and zone 3 slower than the EOF. Figure 6.2.a represents the time-related electropherogram as might be recorded by the detector. Figure 6.2.b shows the spatial peak width which is obtained from Fig. 6.2.a by a mathematical correction according to Eq. 4-8 (Sect. 4.2.2). Note that the spatial width of the first peak is larger than the actual detector response. In turn the spatial width of the third peak is smaller than the detector response.

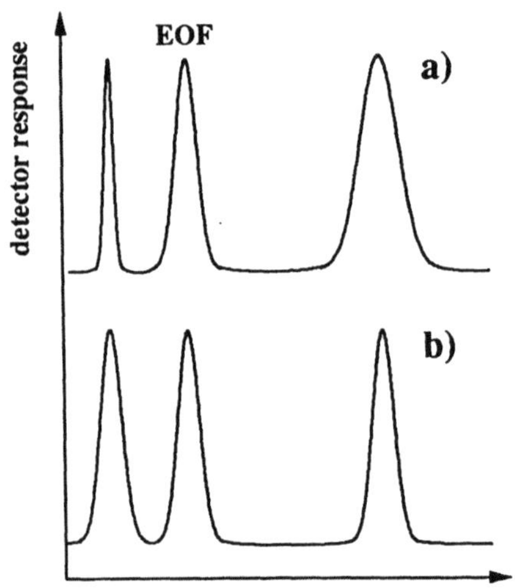

Fig. 6.2. Schematic electropherogram of three components as a function of time (**a**) and length (**b**). EOF represents an electroosmotic flow marker

6.2 Influence of Injection

The most critical point in quantitative analysis is the injection of the sample. As already discussed in Sect. 4.1 two injections modes are available in CE: (i) electrokinetic and (ii) hydrodynamic injection.

Two problems arise in electrokinetic injection. Firstly, due to mobility differences of the analytes a discrimination of the injected sample components occurs. Ions with high electrophoretic mobilities migrate faster into the capillary during injection than ions with low mobilities. Oppositely charged ions are even repelled from migrating into the tube. These ions only enter the capillary if the electroosmotic flow is sufficiently high. As a consequence of the discrimination, a bias results in the peak area that can principally be corrected for if the ion mobilities are known [296]. This phenomenon is shown in Fig. 6.3. A sample consisting of benzylamine (BNH_2), benzyltrimethylammonium chloride (BTA), benzyl alcohol (BOH) as electroosmotic flow marker, acetylsalicylic acid (ASA) and benzoic acid (BzA) is injected electrokinetically (Fig. 6.3.a) and hy-

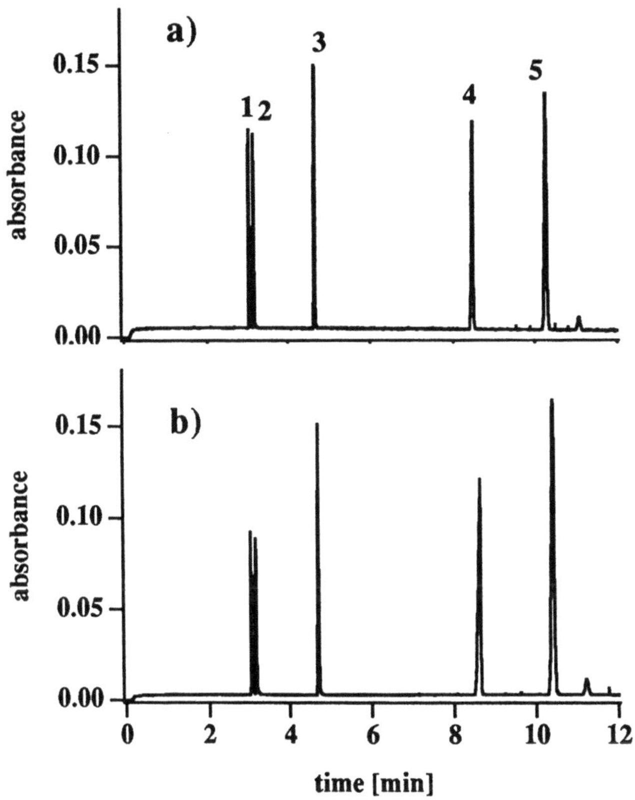

Fig. 6.3. Relative detector responses of electrokinetic injection (**a**) and hydrodynamic injection (**b**). Elution order of the analytes: (**1**) benzylamine, (**2**) benzyltrimethylammonium, (**3**) benzyl alcohol, (**4**) acetylsalicylic acid and (**5**) benzoic acid. Instrument: Beckman P/ACE 2000; experimental conditions: fused silica capillary 57 cm x 75 μm i.d., electrokinetic injection for 2 s at 5 kV (**a**) and hydrodynamic injection for 1 s (**b**), temperature 30 °C, field strength 300 $V \cdot cm^{-1}$, UV detection at 200 nm, electrolyte system 50 mM sodium phosphate buffer, pH 7.0. Benzyl alcohol is used as EOF marker

drodynamically (Fig. 6.3.b). As one can readily see, peaks 4 and 5 (ASA and BzA) are smaller in Fig. 6.3.a than in 6.3.b due to the discrimination of the anionic species by the electrokinetic injection.

The second problem using electrokinetic injection is the run-to-run variations based on changes of the injection voltage, injection time and sample composition. RSD's of the peak area by electrokinetic injection are unacceptably high as shown in Table 6.1. For three commercial instruments RSD's of the peak areas are calculated for BTA and BzA after hydrodynamic and electrokinetic injection. Throughout, electrokinetic injection gave dramatically worse RSD's than hydrodynamic injection which makes this injection mode unsuitable for quantitative work. Somewhat better results were obtained at longer injection times with all instruments. However, even then precision is too low for quantitation. It is interesting to notice that significantly better RSD's were calculated for the cationic species than for anions, although the data were calculated from the same experiments. One reason for this fact might be that migration velocity variations were more pronounced in the case of BzA than in the case of BTA.

Table 6.1. Relative standard deviations of the peak areas in % of (a) benzytrimethylammonium (BTA) and (b) benzoic acis (BzA) after hydrodynamic and electrokinetic injection using three commercial instruments. The RSD was calculated from 7 injections. For experimental conditions see Fig. 6.3. except the capillary dimension which varied

a) BTA

Injection mode	Condition	Beckman	Spectra-Physics	ABI
hydrodynamic	ca. 5 nL	1.8	3.2	2.2
hydrodynamic	ca. 20 nL	0.8	0.9	0.3
electrokinetic	1 s at 5 kV	6.8	3.3	4.1
electrokinetic	10 s at 5 kV	2.8	2.0	3.7

b) BzA

Injection mode	Condition	Beckman	Spectra-Physics	ABI
hydrodynamic	ca. 5 nL	2.8	3.6	2.3
hydrodynamic	ca. 20 nL	3.6	1.1	1.2
electrokinetic	1 s at 5 kV	10.2	7.4	18.4
electrokinetic	10 s at 5 kV	6.9	3.2	10.3

Dose and Guiochon [296] described a procedure to correct electrophoretic responses for the run-to-run variations by using two internal standards. Applying their concept, electrokinetic injection is claimed to provide high accuracy and a precision of < 1% RSD. As the authors pointed out, their procedure appears to maintain the high degree of accuracy and precision even when significant errors in injection voltage, time, etc., occur.

Another correction procedure is described by Lee and Yeung [297]. By monitoring the electrophoretic current during injection and separation quantitative precision is improved. Effects of sample conductivities on the amounts of analytes injected electrokinetically are nullified by proper corrections with the measured conductivities of sample and buffer solutions together with the migration times of the analytes and an electroosmotic flow marker

Moring et al. [298] investigated the linearity of peak area versus injection time for both electrokinetic and hydrodynamic injection mode. Good linear correlation was found for the latter mode up to an injection time of 20 seconds. However, under electrokinetic injection conditions a non-linear relation was found. With increasing injection time the curve showed a convex shape which was attributed to a continuous change of the sample composition and ionic strength during injection.

In hydrodynamic injection a defined volume is introduced into the capillary by means of a pressure difference (see Sect. 4.1.1 for details). Although strong efforts are made by the manufacturers to properly control this process, variations of the injected volumes at short injection times are too high to allow quantitation (Table 6.1.) especially for pharmaceutical purposes where precision in the range of less than 1% are required. Even if higher sample volumes (e.g. 20 nL) are injected, RSD's higher than 1% are obtained (Table 6.1.b). Again, as in electrokinetic injection, analysis of the anion (BzA) give worse results than of the cation (BTA). For the time being, the only way out of this problem is to work with an internal standard where we usually calculate RSD's of smaller than 1%.

6.3 Method Validation

As any other analytical method, qualitative and quantitative analysis in CE requires proper validation of the method. This validation provides information on performance characteristics such as precision, accuracy, linearity, limit of quantitation and selectivity. For routine work additional data about day-to-day reproducibility, ruggedness and short-term sample stability for series analysis are required.

Data on precision of a method are commonly required for the reproducibility of migration times (see Table 3.6.) and for peak area (see Table 6.1.). These data are obtained by multiple injection of the same sample at experimental conditions as close as possible. At least triplicate injections have to be performed to calculate the mean value and the relative standard deviation.

Accuracy can be proved by two ways. First, a placebo solution of the sample is spiked with a known amount of the reference substance and the concentration is calculated from the peak area. The value obtained should lie within an interval defined as the mean ± 3 x RSD of a value found by multiple injection of the reference standard. Second, the accuracy is obtained from the y-intercept of a linearity plot as shown in Fig. 6.1. If the origin of the axis is found within a 95% confidence interval, accuracy of the method should be satisfactory.

Linearity and limit of quantitation are strongly dependent on the performance of the detector used. For routine work, a UV detector should be linear over a range of at least two orders of magnitude. In contrast to the detection limit, limit of quantitation is not directly related to the detection noise. It is moreover a value close to the detection limit which should be chosen such that it allows quantitation with sufficient precision. This holds usually for a concentration of three to five times the detection limit.

7 Applications

Although capillary electrophoresis is a relatively new separation technique an immense number of practical applications have already appeared in various journals. Giving a comprehensive overview on this topic resembles the labor of Sisyphus in Greek mythology. Since new applications are published daily, it seems to be impossible to cite all the applications in this chapter. For the sake of lucidity and brevity citations are presented in tables. Thus, information is intended to be hints for introducing a specific area of application rather than to provide a detailed survey of the current literature. The classification into distinct categories is made arbitrarily in the hope that this form of presentation gives the best overview. Keep in mind, that the best separation conditions can only result in highly efficient runs, if the capillaries are conditioned before use and if they are rinsed after each run with appropriate solutions (see Sect. 4.3).

7.1 Small Ions

Although the main interest in CE research is located in bioanalytical and pharmaceutical applications, several interesting developments have also been made for the separation of inorganic ions and other charged low molecular weight compounds. Especially capillary ion analysis (CIA, Waters' trade name for capillary electrophoresis of small ions in combination with indirect UV-VIS absorbance) which has been developed and systematically improved by Jandik and coworkers [155] has proved to be very powerful. This technique is based both on the proper modulation of the EOF to speed up analysis and on the proper choice of the background electrolyte to allow indirect detection of low absorbing ions with high efficiency. Whereas relatively slow migrating negatively charged analytes like peptides, proteins and other large biomolecules are conventionally separated in the presence of a cathodic EOF, separations of highly mobile anions would either last an extremely long time or could not be performed at all (in the cases where $-\mu_i > \mu_{eo}$) under these conditions. Therefore, one key in the analysis of highly to moderately mobile anions is the elimination or reversion of the EOF by dynamically coating the capillary wall with a cationic surfactant (see Sect. 2.5). The analysis is then performed under reversed polarity of the electrodes with the anode as the detection side.

In the case of cationic separations, no modulation of the EOF is needed. To reduce electrophoretic dispersion, special electrolyte systems matching the mobilities of the ions to be analyzed were developed. Since indirect UV absorbance and fluorescence detection have turned out to be the most versatile detection modes for ion analysis by means of CE, the electrolyte system has to meet also the requirements needed for indirect detection systems (see Sect. 4.2.6 and Table 4.3.). The separation of cations can furthermore be improved by adding adequate weakly complexing agents to the buffer solution. Electrolyte systems for special applications of CIA are commercially available from Waters under the trade name NICE-Pak. The very high efficiency of CIA for the analysis of highly mobile ions is illustrated in Fig. 7.1. showing the separation of 30 anions in only 3 minutes with plate numbers ranging from $5 \cdot 10^5$ - 10^6.

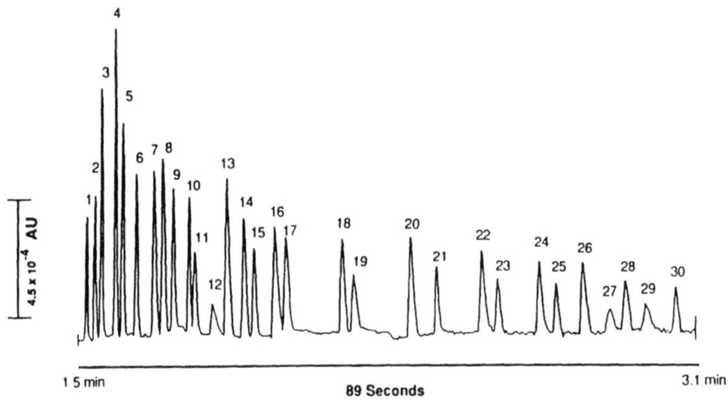

Fig. 7.1. CIA of 30 anions: (**1**) thiosulphate, (**2**) bromide, (**3**) chloride, (**4**) sulphate, (**5**) nitrite, (**6**) nitrate, (**7**) molybdate, (**8**) azide, (**9**) tungstate, (**10**) monofluorophosphate, (**11**) chlorate, (**12**) citrate, (**13**) fluoride, (**14**) formate, (**15**) phosphate, (**16**) phosphite, (**17**) chlorite, (**18**) galactarate, (**19**) carbonate, (**20**) acetate, (**21**) ethanesulphonate, (**22**) propionate, (**23**) propanesulphonate, (**24**) butyrate, (**25**) butanesulphonate, (**26**) valerate, (**27**) benzoate, (**28**) D-glutamate, (**29**) pentanesulphonate and (**30**) D-gluconate. Experimental conditions: fused silica capillary, 60 cm (L_D 52 cm) x 50 µm i.d., voltage 30 kV, indirect UV detection at 254 nm, electrolyte system 5 mM chromate, 0.5 mM NICE-Pak OFM Anion-BT, adjusted to pH 8.0 with 100 mM NaOH. (Reprinted with permission of Ref. 156)

Besides inorganic metal cations, CIA in combination with indirect UV absorbance or fluorescence detection can also be used for the analysis of low molecular weight aliphatic amines. If the compounds to be analyzed show sufficient UV absorptivities, other methods than CIA. In the following, some applications of CIA, conventional CZE and MEKC for the analysis of metal cations (Table 7.1.), inorganic anions, aliphatic and aromatic carboxylic acids (Table 7.2.) and low molecular weight amines and aminoalcohols (Table 7.3.) are summarized in tabular form. For biogenic amines such as catecholamines see Table 7.10.

Table 7.1. Metal cations. The experiments are all performed with normal polarity with the detection at the cathodic end

Appl. No.	Compounds	Conditions	Remarks	Ref.
1	14 lanthanides, Li^+, Mg^{2+}, Na^+ and K^+	PAA coated capillary, 20 cm x 25 µm 30 mM creatinine-30 mM CH_3CO_2H, pH 4.8, 4 mM HIBA electrokin. inj. 8 s, 1.5 kV, U = 12 kV $UV_{ind.}$ at 220 nm; analysis time 5 min	c = 10^{-3} M of each dissolved in water, in the presence of α-hydroxyisobutyric acid (HIBA) as com-plexing agent	155 157
2	Nd, Pr, Ce, La, Mg and K in flint alloy	electrokin. inj. 8 s, 700 V, other conditions as in Appl. No. 1 analysis time 2 min	as nitrates, 145 mg of the alloy dissolved in 200 mL of 15 mM creatinine - 15 mM acetic acid	157
3	14 lanthanides, Rb^+, Ca^{2+}, Li^+, Mg^{2+}, Na^+ and K^+	capillary, 36.5 cm x 75 µm 10 mM creatinine, adj. to pH 4.4 with CH_3CO_2H, 4 mM HIBA hydrodyn. inj., U = 30 kV, $UV_{ind.}$ at 214 nm; analysis time 1.8 min	1-5 ppm of each cation dissolved in water, in the presence of HIBA as complexing agent	300
4	K^+, Ba^{2+}, Sr^{2+}, Na^+, Ca^{2+}, Mg^{2+} and Li^+	capillary, 60 cm x 75 µm 5 mM creatinine, adj. to pH 5.5 with CH_3CO_2H, 0.021 mM citrate hydrodyn. inj., U = 25 kV $UV_{ind.}$ at 214 nm; analysis time 2.5 min	in the presence of citric acid as complexing agent to improve separation (with the additional benefit to decrease μ_i so that it better matches µ of the coion)	300
5	K^+, Ba^{2+}, Sr^{2+}, Ca^{2+}, Na^+, Mg^{2+}, Mn^{2+}, Cd^{2+}, $Fe2^{2+}$, Co^{2+}, Pb^{2+}, Ni^{2+}, Li^+, $Zn^{2+}$and Cu^{2+}	capillary, 60 cm x 75 µm 5 mM creatinine, adj. to pH 4.4 with CH_3CO_2H, 6.5 mM HIBA hydrodyn. inj., U = 25 kV $UV_{ind.}$ at 214 (185) nm; analysis time 8 min	in the presence of HIBA (by using 185 nm [301] instead of 214 nm [300], sensitivity is improved ca. 1.7-fold)	300 301

Table 7.1. (continued)

Appl. No.	Compounds	Conditions	Remarks	Ref.
6	K^+, Na^+ and Li^+ in the presence of NH_4^+	capillary, 52 cm (L_D) x 75 µm 5 mM MES, adjusted to pH 8.15 hydrodyn. inj., U = 25 kV $UV_{ind.}$ at 214 nm; analysis time 1.8 min	at pH 8.15 where K^+ and NH_4^+ have different mobilities	300
7	K^+/NH_4^+ (not resolved), Na^+, Ca^{2+} and Mg^{2+} in apple vinegar	capillary, 63 (L_D = 55) cm x 75 µm 5 mM imidazole, adj. to pH 4.2 with H_2SO_4 hydrodyn. inj., U = 20 kV $UV_{ind.}$ at 214 nm; analysis time 3.5 min	$c(K^+/NH_4^+)$ = 1125, $c(Na^+)$ = 74, $c(Ca^{2+})$ = 114, $c(Mg^{2+})$ = 64 ppm	158
8	K^+, Na^+, Ca^{2+} and Mg^{2+} in mineral waters	capillary, 23 cm x 75 µm 3 mM imidazole, adj. to pH 6.0 with H_2SO_4 hydrodyn. inj., voltage 30 kV $UV_{ind.}$ at 214 nm; analysis time 25 s	$c(K^+)$ = 0.5-5 ppm, $c(Na^+)$ = 3-230 ppm, $c(Ca^{2+})$ = 4-140 ppm, $c(Mg^{2+})$ = 1-70 ppm	158
9	Cs^+, K^+, Ca^{2+}, Na^+, Mg^{2+}, Sr^{2+}, Ba^{2+} and Li^{2+} in the presence of NH_4^+	capillary, 80 cm (L_D = 55 cm) x 75 µm, coated with 1 mM $C_{12}H_{25}NH_2$ in MeOH/H_2O (10:90), adj. to pH 4.0 with H_3PO_4; 0.5 mM Ce(III)sulphate, 2.5 mM 18C6 hydrodyn. inj., U = 30 kV ind. fluor. 251/345nm; analysis time 4.5 min	c = 50 µM of each; 18-crown-6 (18C6) added to selectively interact with K^+, Ba^{2+} and Sr^{2+} allowing separation of K^+/NH_4^+ and Ba^{2+}/Sr^{2+}	302
10	NH_4^+, K^+, Ca^{2+},and Na^+ in a cola beverage	conditions as in Appl. No. 9, unless electrokin. inj.	see Appl. No. 9	303

Table 7.1. (continued)

Appl. No.	Compounds	Conditions	Remarks	Ref.
11	trace analysis of Mg^{2+}, Mn^{2+}, and Zn^{2+} in the presence of K^+, Na^+ as main cationic compounds in a fermentation broth	capillary, 60 cm x 75 µm 5 mM creatinine - 5 mM CH_3CO_2H, pH 4.2, 6.5 mM HIBA hydrodyn. inj., U = 20 kV $UV_{ind.}$ at 214 nm; analysis time 5.5 min	c(Mg^{2+}, Mn^{2+}, Zn^{2+}) = 10-100 ppb, c(Na^+) = 100-1000 ppm	155 300
12	K^+, Ca^{2+}, Na^+ and Mg^{2+} in industrial waste-water	conditions as in Appl. No. 11 analysis time 4.5 min	c(K^+) = 260 ppb, c(Ca^{2+}) = 58 ppb, c(Na^+) = 50 ppb, c(Mg^{2+}) = 8 ppb	155
13	Li^+ in the presence of K^+ and excess of Na^+ in human serum of a patient on lithium therapy	capillary, 70 cm x 75 µm 20 mM MES - 20 mM His, pH 6.1 hydrodyn. inj., U = 25 kV conduct. detection; analysis time 6 min	plasma diluted 1:19 with separation buffer	140
14	Ca^{2+}, Na^+, Mg^{2+}, Ni^{2+} and Cd^{2+}	capillary, 70 cm x 75 µm 5 mM KCH_3CO_2, pH 5.0; U = 14 kV conduct. detection; analysis time 6 min	c ≈ $5 \cdot 10^{-5}$ M of each	142

Table 7.2. Inorganic anions and carboxylic acids. The experiments were performed with reversed polarity (detection at the anodic end) if not otherwise stated.

Appl. No.	Compounds	Conditions	Remarks	Ref.
15	$S_2O_3^{2-}$, Cl^-, SO_4^{2-}, $C_2O_4^{2-}$, SO_3^{2-}, CO_3^{2-}, $CH_3CO_2^-$, HCO_2^-, propionate and butyrate in kraft black liquor	capillary, 60 cm x 75 μm 5 mM Na_2CrO_4, 0.5 mM NICE-Pak OFM Anion-BT, adj. to pH 10 with 100 mM NaOH; hydrodyn. inj., U = 20 kV $UV_{ind.}$ at 254 nm; analysis time 5 min	sample 1:1000 diluted and passed through Waters Sep-pak C18 solid-phase extraction cartridge before analysis	155 304
16	formate, succinate, acetate, lactate, phosphate and propionate in a dental plaque sample	capillary, 60 cm x 75 μm 5 mM potassium phthalate, 0.5 mM NICE-Pak OFM Anion-BT, pH 5.6, hydrodyn. inj., U = 20 kV $UV_{ind.}$ at 254 nm; analysis time 4.5 min		155 304
17	Cl^-, SO_4^{2-} and NO_3^-	capillary as used in Appl. No. 9 1 mM 2,5-dihydroxybenzoic acid, 0.5 mM $Pb(CH_3CO_2)_2$, pH 4.3, electrokin. inj. 8s, 1.5 kV, U = 30 kV; ind. fluor. (314/389 nm); analysis time 3 min	Pb^{2+} forms neutral species with SO_4^{2-} thus shifting the migration time so that complete resolution of the 3 species is achieved	300 305
18	citric acid, tartaric acid, malic acid, succinic acid, acetic acid and lactic acid in Chablis wine	capillary, 100 cm x 75 μm 5 mM potassium phthalate, 0.5 mM NICE-Pak OFM Anion-BT, pH 7.0, hydrodyn. inj., U = 20 kV $UV_{ind.}$ at 254 nm; analysis time 15 min	c(citric) 127, c(tartaric) 2645, c(malic) 3291, c(succinic) 300, c(acetic) 260, c(lactic) 296 $\mu g \cdot mL^{-1}$, wine diluted 1:5000 in buffer	306
19	citric acid, malic acid, acetic acid and lactic acid in filtered tomato juice	conditions as in Appl. No. 18	c(citric acid) 3579, c(malic acid) 404, c(acetic acid) 99, c(lactic acid) 69 $\mu g \cdot mL^{-1}$	306

Table 7.2. (continued)

Appl. No.	Compounds	Conditions	Remarks	Ref.
21	Br^-, SO_4^{2-}, BF_4^-, F^- and $H_2PO_4^-$	capillary, 100 cm x 75 µm 10 mM Na_2CrO_4, 0.05 mM NICE-Pak OFM Anion-BT, pH 7.9, hydrodyn. inj., U = 20 kV $UV_{ind.}$ at 254 nm; analysis time 20 min	in 0.1% (v/v) HF solution containing a dissolved borophosphosilicate thin film (BPSG)	307
22	formic, acetic, propionic, butanoic, pentanoic, hexanoic and heptanoic acid	capillary, 35 cm x 75 µm 20 mM MES - 20 mM His, pH 6, hydrodyn. inj., U = 25 kV conduct. detection; analysis time 4 min	c = 1 mM of each	138
23	Br^-, NO_3^-, BrO_3^-, I^- and IO_3^-	capillary, 35 cm x 75 µm 20 mM KH_2PO_4 - 20 mM Tris, pH 7.0, 25 mM CTAC, hydrodyn. inj., U = 15 kV UV at 210 nm; analysis time 6 min	MEKC with cetyltrimethylammonium chloride (CTAC)	308
24	caprylic, sorbic, benzoic and propionic acid as food additives in an oyster sauce	capillary, 48 cm (L_D) x 50 µm 50 mM phosphate - 50 mM borate, pH 7.0, 10 mM TBA bromide, hydrodyn. inj., U = 15 kV UV at 190 nm; analysis time 16 min	tetrabutylammonium (TBA) serves as ion pairing agent; detection at the cathode	309
25	$S_2O_3^{2-}$, Br^-, Cl^-, I^-, SO_4^{2-}, NO_2^-, NO_3^-, ClO_3^-, SCN^- and F^-	capillary, 70 cm (L_D = 62) x 75 µm 5 mM chromate, 0.5 mM TTAB, hydrodyn. inj., U = 20 kV $UV_{ind.}$ at 254 nm; analysis time 4 min	tetradecyltrimethylammonium bromide (TTAB) reverses EOF; addition of ethylene glycol or MeOH improves selectivity but increases analysis time	310

Table 7.2. (continued)

Appl. No.	Compounds	Conditions	Remarks	Ref.
27	4-hydroxybenzoic, isovanillic, sinapic, ferulic, coumaric, iso-ferulic and salicylic acid	capillary, 46.7 cm x 75 µm 18 mM borate - 30 mM phosphate, 50 mM CTAB, pH 7.0, hydrodyn. inj., U = 20 kV, T = 40°C UV at 280 nm; analysis time 20 min	MEKC with cetyltrimethyl-ammonium bromide (CTAB) analyis of phenolic carboxylic acids in rye and whole rapeseed	312
28	4-hydroxy-3-methoxy-phenyl-glycol, 5-methoxy-indole-3-acetic acid, 5-hydroxyindole-3-acetic acid, 4-hydroxy-3-methoxy benzoic acid, vanillyl-mandelic acid, homogentisic acid and homovanillic acid	capillary, 50 cm (L_D) x 75 µm 200 mM acetate, pH 4.1, hydrodyn. inj., U = 25 kV UV at 214 nm; analysis time 12 min	for determination of these cate-cholamine metabolites in urine	313
29	hippuric, sorbic and orotic acid in rennet whey	capillary, 50 cm (L_D) x 75 µm 40 mM AMPD, adj. to pH 8.8 with 1 M BICINE, hydrodyn. inj., U = 25 kV UV at 254 or 280 nm; time 12 min	sample: 25 g of whey diluted with 20 mM AMPD-BICINE, pH 8.8, to 50 mL; 10 mL of this diluted with running buffer	314

Table 7.3. Low molecular weight amines and aminoalcohols (detection at the cathodic end)

Appl. No.	Compounds	Conditions	Remarks	Ref.
30	ammonium, dimethylamine, trimethylamine, diethylamine, triethylamine, diethanolamine and triethanolamine	capillary, 63 (L_D = 55) cm x 75 µm 5 mM imidazole, adj. to pH 4.5 with H_2SO_4, hydrodyn. inj., U = 25 kV $UV_{ind.}$ at 214 nm; analysis time 4 min	c = 1-3 ppm of each, simultaneous determination of K^+, Na^+, Li^+ and Ca^{2+}	158
31	NH_4^+, dimethyl-, tetramethyl-, propyl-, diethyl- and diethanolamine	capillary, 82.3 (L_D = 70.7) cm x 18 µm 0.38 mM quinine sulphate, 0.58 mN H_2SO_4, pH 3.7 electrokin. inj. 1s, 10 kV, U = 40 kV ind. fluorescence; analysis time 6 min	c = 25 µM of each	165
32	9 pyridinium salts	pyrex silica cap., 105 (L_D = 90) cm x 85 µm 50 mM Na_2HPO_4, I = 80 µA (ca. 14.2 kV) UV detection; analysis time 25 min		315
33	C_{12},C_{14}, C_{16} and C_{18} alkyltrimethylammonium	capillary, 50 µm 8 mM NaH_2PO_4, 3 mM SDS and 3 mM BDDAB in THF - H_2O (57.4:42.5); hydrodyn. inj., E = 18 kV·m^{-1}; $UV_{ind.}$ at 210 nm; time 20 min	benzyldimethyldodecylammonium bromide (BDDAB) serves as visualizing agent	316
34	C_{12},C_{14}, C_{16} and C_{18} alkylbenzyldimethylammonium	capillary, 50 µm 44 mM NaH_2PO_4 in THF - H_2O (57.4:42.5) hydrodyn. inj., E = 25 kV·m^{-1} UV at 210 nm; analysis time 18 min		316

7.2 Sulphonates and Alkylsulphates

The analysis of short chain alkylsulphonates can be performed again by CIA with indirect UV detection and reversed polarity using a low mobility electrolyte such as benzoate as visualizing agent and an EOF modifier. Long chain alkylsulphonates exceeding C_7, however, are not eluted by this approach. They can be analyzed under normal polarity using a very low mobility, UV absorbing electrolyte anion such as naphthalene sulphonate. Low absorbing alkylsulphates can also be determined by indirect UV detection using veronal as visualizing agent. Aryl- and alkylbenzenesulphonates can be analyzed by conventional CZE with UV detection due to their UV absorbing aromatic systems. High molecular weight polystyrene sulphonates (PSS) are separated in coated capillaries by adding hydroxyethyl cellulose to the buffer solution under reversed polarity. Some applications are shown in Table 7.4.

Table 7.4. Sulphonates and alkylsulphates

Appl. No.	Compounds	Conditions	Ref.
35	linear C_1 - C_7 alkylsulphonates, 10 ppm of each	capillary, 60 cm x 75 µm 10 mM benzoate, 0.5 mM NICE-Pak OFM Anion-BT, pH 6.0; hydrodyn. inj., U = 20 kV; $UV_{ind.}$ at 254 nm; detection at anode, time 5.5 min	304
36	linear C_4 - C_{12} alkylsulphonates, 25 ppm of each	capillary, 60 cm x 75 µm 10 mM naphthalenesulphonate, 30% CH_3CN, pH 10.0, hydrodyn. inj., U = 20 kV; $UV_{ind.}$ at 254 nm; detection at cathode, time 12 min	304
37	C_6, C_7, C_9- and C_{10} alkylsulphonates in alkylamido glycinate shampoo base, diluted 1:200 with water	conditions as in Appl. No. 36	304
38	C_9, C_{10}, C_{11}- C_{12} and C_{13} alkylsulphates in Teepol HB7, diluted 1:5000 with separation buffer	capillary, 70 cm x 50 µm 6 mM veronal, pH 8.6 hydrodyn. inj., U = 25 kV $UV_{ind.}$ at 240 nm; detection at the cathode, analysis time 12 min	161
39	4-amino-1-, 2-amino-1-, 5-amino-2-, 8-amino-2-naphthalenesulphonic acid and 1-naphthol-4-sulphonic acid	reversed phase capillary, 50 cm x 10 µm, coated with PS-264; 10 mM phosphate, pH 7.0, 1.25 mM tributylammonium chloride; hydrodyn. inj., U = 21 kV; LIF; detection at the anode, analysis time 6 min	317

Table 7.4. (continued)

Appl. No.	Compounds	Conditions	Ref.
40	linear C_2 - C_{12} alkylbenzene sulphonates	capillary, 57 cm x 50 µm 12.5 mM $Na_2B_4O_7$ - boric acid, pH 9.0, 30% CH_3CN hydrodyn. inj., U = 30 kV UV at 214 nm; detection at the cathode, analysis time 8 min	318
41	tert.-butyl, sec.-butyl, isobutyl and n-butylbenzene sulphonate	capillary, 57 cm x 50 µm 6.25 mM $Na_2B_4O_7$ - boric acid, pH 9.0, 50 mM SDS hydrodyn. inj., U = 25 kV UV at 214 nm; detection at the cathode, analysis time 8 min	318
42	sodium dodecylbenzenesulphonates with different sizes of alkyl side chains in a cleaning material	capillary, 57 cm x 75 µm 50 mM Tris - 50 mM GlyGly, pH 8.25, in 50% CH_3CN hydrodyn. inj., U = 30 kV UV at 206 nm; detection at the cathode, analysis time 13 min	319
43	8 PSS standards of MW 1800, 8000, 18 000, 46 000, 100 000, 400 000, 780 000 and 1 200 000 d, 375 $\mu g \cdot mL^{-1}$ of each in buffer	capillary, 50 cm x 50 µm, coated with an anorganic phase; 25 mM KH_2PO_4, pH 5.0, 10 $mg \cdot mL^{-1}$ HEC, electrokin. inj., I = 30 µA UV at 225 nm; detection at the anode, analysis time 10 min	320

7.3 Drugs and Natural Products

Drugs and natural products represent an extremely heterogeneous class of compounds covering acids, bases and neutral substances with molecular weights usually below 1000 Da. Although amino acids and many peptides belong to this class, they are separately discussed in Sect. 7.6.

Conventionally, pharmaceutical drug substances (active ingredients) and drug products (dosage forms) are analyzed for identity, content of active ingredient and purity. In addition the formation of degradation products after storage has to be determined to fix its shelf-life or expiration date. Conformation of identity is performed to verify that the sample on hand is identical to a reference standard. This is routinely done by comparing the migration time of the main peak with the migration time of the reference standard. The content of active ingredient is determined by relating the peak area of the sample peak to the peak area of the reference standard with known concentration (external standard) or with the peak area of an internal standard (see chapter 6). To fulfill this purpose, a method needs high specifity for the drug substance but relatively low sensitivity. Determination of impurities requires the quantitation of by-products from synthesis and degradation products in the presence of large amounts of the main components. In general, sensitivity of the detection system ought to be high enough to detect contaminants at levels of 0.01% and to allow quantification down to 0.05% of the major peak. Thus, purity analysis requires (i) high selectivity and efficiency to separate closely related compounds, (ii) a universal detection system and (iii) a broad dynamic range of detection covering at least three orders of magnitude.

Methods used in drug analysis have to be carefully validated concerning performance characteristics such as accuracy, precision, linearity, selectivity, limit of quantitation and ruggedness. Unfortunately, most applications given in literature lack this method validation. The following tables give an overview about the analysis of some antibiotics (Table 7.5.), analgesics (Table 7.6.), steroids (Table 7.7.) and selected drug substances (Table 7.8.). Natural products such as flavonoids (Table 7.9.), biogenic amines (Table 7.10.) and vitamins (Table 7.11.) are summarized subsequently.

Table 7.5. Antibiotics. Experiments are performed with normal polarity (detection at the cathodic end) if not otherwise stated.

Appl. No.	Compounds	Conditions	Remarks	Ref.
44	8 penicillins (amoxicillin, ampicillin, 6-aminopenicillanic acid, oxacillin, cloxicillin, ticarcillin, nafcillin and dicloxicillin)	capillary, 60 cm x 75 µm 20 mM NaH_2PO_4, adj. to pH 9.0 with $Na_2B_4O_7$, 50 mM SDS; hydrodyn. inj., U = 18 kV; UV at 214 nm; time 14 min	MEKC with SDS	321
45	7 penicillins (benzylpenicillin, ampicillin, carbenicillin, sulbenicillin, piperacillin, aspoxicillin and amoxicillin)	capillary, 65 (L_D = 50) cm x 50 µm 20 mM NaH_2PO_4, adj. to pH 9.0 with $Na_2B_4O_7$, 150 mM SDS hydrodyn. inj., U = 20 kV UV at 210 nm; analysis time 14 min	MEKC with SDS	322
46	aspoxicillin in plasma	conditions as in Appl. No. 45, but buffer: 20 mM NaH_2PO_4, adj. to pH 9.0 with $Na_2B_4O_7$, 150 mM SDS	MEKC with SDS; acetaminophen serves as internal standard	322
47	7 penicillins (as in Appl. No. 45)	capillary, 57 (L_D = 50) cm x 75 µm 100 mM borate, pH 8.3, 150 mM SDS hydrodyn. inj., U = 12 kV UV at 200 nm; analysis time 15 min	MEKC with SDS; c = 0.5 - 2 mg·mL^{-1} of each	323
48	9 cephalosporins (ceftazidime, cefotaxime, cefoperazone, cefmenoxime, cefpiramide, ceftriaxone, cefpimizole, cefminox and C-TA)	conditions as in Appl. No. 45, but buffer: 20 mM NaH_2PO_4, adj. to pH 9.0 with $Na_2B_4O_7$, 150 mM N-lauroyl-N-methyltaurate; analysis time 12 min	MEKC with N-lauroyl-*N*-methyltaurate; (improved reso-lution by adding tetraalkyl-ammonium salts as ion pai-ring agents [247])	322
49	9 cephalosporins (cephaloridine, D-(-)-hydroxyphenylglycine, cephadrine, cephadroxil, cephalexin, cephalosporin C, cefaloxin, cephalotin and 7-aminocephalosporic acid)	capillary, 60 (L_D) cm x 75 µm 30 mM NaH_2PO_4, adj. to pH 7.0 with $Na_2B_4O_7$ hydrodyn. inj., E = 275 V·cm^{-1} UV at 215 nm; analysis time 9 min		324

Table 7.5. (continued)

Appl. No.	Compounds	Conditions	Remarks	Ref.
50	9 sulphonamides (trimethoprim, sulphanilamide, sulphamethazine, sulphathiazole, sulphamerazine, sulphadiazine, sulphamethoxazole and sulphamethizole)	capillary, 60 (L_D) cm x 75 µm 30 mM NaH_2PO_4, adj. to pH 7.0 with $Na_2B_4O_7$ hydrodyn. inj., E = 240 $V \cdot cm^{-1}$ UV at 215 nm; analysis time 20 min		324
51	8 sulphonamides (sulphamethoxypyridazine, sulphachloropyridazine, sulphasalazine, sulphamerazine, sulphaguanidine, sulphadiazine, sulphaquinoxaline and sulphamethazine)	capillary, 50 cm x 50 µm 50 mM NaH_2PO_4 - 50 mM $Na_2B_4O_7$, pH 6.5, hydrodyn. inj., U = 15 kV UV at 210 nm; analysis time 20 min		325
52	10 sulphonamides in pork meat	capillary, 57 (L_D = 50) cm x 75 µm 20 mM NaH_2PO_4 - 20 mM $Na_2B_4O_7$, pH 7.0, hydrodyn. inj., U = 10 kV UV at 254 nm; analysis time 24 min		326
53	6 sulphonamides (sulphanilamide, sulphamethazine, sulphathiazole, sulphamerazine, sulphadimethoxine and sulphamethoxazole)	capillary, 100 cm x 75 µm 20 mM $NH_4CH_3\ CO_2$, pH 6.8, 20% (v/v) MEOH, hydrodyn. inj., U = 26 kV; UV at 254 nm, API-MS; time 6 resp. 26 min,	coupled CE-MS technique; $c = 4 \cdot 10^{-4}$ M of each	327
54	aminoglycosides (dihydrostreptomycin, lividomycin, amikacin, kanamycin, tobramycin and sisomycin)	capillary, 67 (L_D = 60) cm x 50 µm 10 mM imidazole, adj. to pH 5.0 with CH_3CO_2H, 50 $\mu L \cdot mL^{-1}$ FC 135 hydrodyn. inj., U = 12.5 kV $UV_{ind.}$ at 214 nm; analysis time 13 min	a fluorochem. surfactant (FC 135) serves as EOF modifier; reversed polarity with detection at the anodic end	328
55	tetracycline and its degradation products (4-epitetracycline, anhydrotetracycline and epianhydrotetracycline)	capillary, 20 cm x 25 µm 20 mM phosphate buffer, pH 3.9, 5 mM EDTA, electrokin. inj. 5 s, 10 kV, U = 10 kV; $UV_{ind.}$ at 265 nm; time 13 min	EDTA serves as ion pairing agent	329

Table 7.6. Analgesics, cold medicine formulas and opiates. Experiments are performed with normal polarity (detection at the cathodic end) if not otherwise stated.

Appl. No.	Compounds	Conditions	Remarks	Ref.
56	7 analgesics (acetaminophen, caffeine, benzamide, acetanilide, salicylamide, acetylsalicylate and salicylate)	capillary, 20 (L_D = 27) cm x 75 µm 60 mM $Na_2B_4O_7$,, adj. to pH 8.4, 60 mM SDS, 15% (v/v) MeOH hydrodyn. inj., U = 7.55 kV UV at 214 nm; analysis time 7 min	MEKC with SDS	75
57	codeine, caffeine, butalbital, aspirin and salicylate	capillary, 85 (L_D = 45) cm x 50 µm 50 mM phosphate buffer, pH 7.0 electrokin. inj., 10 s, 3.5 kV, U = 20 kV UV at 210 nm; analysis time 16 min	test for codeine in the presence of the other drugs	330
58	5 analgesics (caffeine, acetaminophen, acetylsalicylate, salicylamide and salicylate)	capillary, 60 cm x 50 µm 15 mM NaH_2PO_4,adj. to pH 11.0 with NaOH, 25 mM SDS hydrodyn. inj., U = 30 kV UV at 214 nm; analysis time 4 min	MEKC with SDS; c = 0.2 $mg{\cdot}mL^{-1}$	321
59	purity control of salicylamide	capillary, 60 cm x 50 µm 20 mM NaH_2PO_4, adj. to pH 11.0 with NaOH, 75 mM SDS hydrodyn. inj., U = 20 kV UV at 214 nm; analysis time 5.6 min	MEKC with SDS; c = 0.1 $mg{\cdot}mL^{-1}$	321

Table 7.6. (continued)

Appl. No.	Compounds	Conditions	Remarks	Ref.
60	12 cold medicines (caffeine, acetaminophen, sulpyrin, trimetoquinol·HCl, guaifenesin, naproxen, ethenzamide, phenacetin, isopropylantipyrine, noscapine, chlorpheniramine maleate and tipepidine hibenzate)	capillary, 65 (L_D = 50) cm x 50 µm 20 mM NaH_2PO_4, adj. to pH 9.0 with $Na_2B_4O_7$, 100 mM N-lauroyl-N-methyltaurate; hydrodyn. inj., U = 20 kV UV at 210 nm; analysis time 30 min	MEKC with N-lauroyl-N-methyltaurate	322
61	14 cold medicines (dibucaine·HCl, triprolidine·HCl, other solutes as in Appl. No. 59)	conditions as in Appl. No. 60, but buffer: 20 mM NaH_2PO_4, adj. to pH 9.0 with $Na_2B_4O_7$, 100 mM sodium cholate analysis time 18 min	MEKC with a bile salt; analysis of active ingredients in a dosage form (novapron granules)	238 322
62	10 substituted purines (theobromine, caffeine, paraxanthine, theophylline, 7-methylxanthine, 3-methylxanthine, 3-methyl uric acid, 1-methyl uric acid, 7-methyl uric acid and uric acid)	capillary, 90 (L_D = 70) cm x 75 µm 10 mM Na_2HPO_4 - 6 mM $Na_2B_4O_7$, pH 9, 75 mM SDS; hydrodyn. inj., U = 20 kV UV at 200, 240 and 280 nm simultaneously; analysis time 12 min	MEKC with SDS, analysis in serum, saliva and urine by direct injection without sample pretreatment	331
63	6 substituted purines (theobromine, theophylline, paraxanthine, xanthine, caffeine and uric acid)	capillary, 72 cm x 50 µm 25 mM phosphate, pH 8.0, 80 mM SDS hydrodyn. inj., U = 21 kV UV at 274 nm; analysis time 12 min	MEKC with SDS; analysis of theophylline in pretreated plasma	332
64	4 opiates (pholcodine, codeine, morphine and dextromethorphan)	capillary, 57 (L_D = 50) cm x 75 µm 50 mM triethylamine,50 mM CTAB, 20% CH_3CN; hydrodyn. inj., U = 15 kV UV at 215 nm; analysis time 20 min	MEKC with CTAB	333

Table 7.7. Steroids. Experiments are performed with normal polarity (detection at the cathodic end) if not otherwise stated.

Appl. No.	Compounds	Conditions	Remarks	Ref.
65	8 corticosteroids (hydrocortisone, triamcinolone, betamethasone, hydrocortisone acetate, dexamethazone acetate, triamcinolone acetonide, fluocinolone acetonide and fluocinonide)	capillary, 65 (L_D = 50) cm x 50 µm 20 mM NaH_2PO_4, adj. to pH 9.0 with $Na_2B_4O_7$, 100 mM sodium cholate hydrodyn. inj., U = 20 kV UV at 210 nm; analysis time 16 min	MEKC with bile salts	322
66	8 corticosteroids as in Appl. No. 65	capillary, 57 (L_D = 50) cm x 75 µm 100 mM borate buffer, pH 8.45, 100 mM sodium cholate; hydrodyn. inj., U = 12.5 kV UV at 210 nm; analysis time 23 min	MEKC with bile salts	323
67	8 corticosteroids as in Appl. No. 65, but cortisone acetate instead of triamcinolone	capillary, 65 (L_D = 50) cm x 50 µm 20 mM NaH_2PO_4 - 20 mM $Na_2B_4O_7$, pH 9.0, 50 mM SDS, 6 M urea hydrodyn. inj., U = 20 kV UV at 220 nm; analysis time 30 min	MEKC with SDS in the presence of urea	243
68	8 corticosteroids as in Appl. No. 65, but cortisone acetate instead of triamcinolone	capillary, 65 (L_D = 50) cm x 50 µm 20 mM NaH_2PO_4, adj. to pH 9.0 with $Na_2B_4O_7$, 50 mM SDS, 4 M urea, 15 mM γ–CD; hydrodyn. inj., U = 20 kV UV at 220 nm; analysis time 18 min	MEKC with SDS, urea and γ–cyclodextrin (shorter analysis time as in Appl. No. 67)	249
69	4 testosterone esters (testosterone propionate, phenylpropionate, isocaproate and decanoate)	capillary, 60 cm x 50 µm 50 mM $Na_2B_4O_7$, adj. to pH 9.0 with boric acid, 30 mM SDS, 50% CH_3CN hydrodyn. inj., U = 25 kV UV at 254 nm; analysis time 19 min	MEKC with SDS	334

Table 7.8. Some selected drugs. Experiments are performed with normal polarity (detection at the cathodic end) if not otherwise stated.

Appl. No.	Compounds	Conditions	Remarks	Ref.
70	alendronate in dosage forms	capillary, 70 (L_D = 64) cm x 75 µm 1.6 mM HNO_3, 2 mM $CuSO_4$;hydrodyn. inj., U = 25 kV; UV at 210 nm; time 10 min	$CuSO_4$ serves as complexing agent of the bisphosphonate drug	335
71	anthracyclines in human plasma (daunorubicin, doxorubicin and epirubicin)	capillary, 70 (L_D = 65) cm x 75 µm 100 mM sodium phosphate buffer, pH 4.2, 70% (v/v) CH_3CN; electrokin. inj. 5 s, 12 kV, U = 20 kV; LIF 476.5/595 nm; time 9 min	c ≈ 0.1 µg·mL of each	335
72	antiepileptic drugs in human plasma (ethosuccimide, primidone, valproic acid, phenobarbital, phenytoin and carbamazepine)	capillary, 72 (L_D = 50) cm x 50 µm 25 mM sodium phosphate buffer, pH 9.0, 50 mM SDS; hydrodyn. inj., U = 30 kV, T = 30 °C; UV at 210 nm; analysis time 13 min	MEKC with SDS; hexobarbital serves as internal standard	337
73	9 antihistamines in pharmaceuticals (pheniramine, doxylamine, methapyrilene, thonylamine, triprolidine·HCl, dimenhydrinate, cyclizine, promethazine and (±)-chloropheniramine)	capillary, 47 (L_D = 40) cm x 50 µm 50 mM NaH_2PO_4 - 50 mM $Na_2B_4O_7$, 100 mM TBA bromide, 10 mM ß-CD, pH 7.5 hydrodyn. inj., U = 30 kV, T = 25 °C UV at 214 nm; analysis time 7 min	MEKC with SDS, tetrabutylammonium (TBA) and ß-cyclodextrin; linear conc. range 100 - 500 ppm	338
74	3 anti-inflammatories (naproxen, ibuprofen and tolmetin)	capillary, 47 (L_D = 40) cm x 50 µm 30 mM NaH_2PO_4 - 9 mM $Na_2B_4O_7$, pH 7.0 electrokin. inj. 2 s, E = 275 $V \cdot cm^{-1}$ UV at 215 nm; analysis time 16 min		324
75	3 anti-inflammatories as in Appl. No. 74	capillary, 47 (L_D = 40) cm x 50 µm, coated with linear PAA 80 mM MES - 30 mM Tris, pH 6.1 electrokin. inj. 2 s, E = 275 $V \cdot cm^{-1}$ UV at 215 nm; analysis time 15 min	detection at the anode; reversed migration order compared to Appl. No. 74	324
76	assay and purity control of diltiazem	conditions as in Appl. No. 61 (60), but pH 8.0; analysis time 12 min		322

Appl. No.	Compounds	Conditions	Remarks	Ref.
77	8 barbiturates (barbital, aprobarbital, butabarbital, amobarbital, hexobarbital, pentobarbital, secobarbital and methohexital)	capillary, 47 (L_D = 40) cm x 50 µm 10 mM $Na_2B_4O_7$, adj. to pH 8.5 with boric acid, 50 mM SDS; electrokin. inj. 2 s, E = 250 $V \cdot cm^{-1}$ UV at 215 nm; analysis time 15 min	MEKC with SDS	324
78	7 barbiturates in human serum and urine (barbital, allobarbital, phenobarbital, butalbital, (+)-thiopental, (-)-thiopental, amobarbital and pentobarbital)	capillary, 90 cm x 75 µm 9 mM $Na_2B_4O_7$ - 15mM NaH_2PO_4 , pH 7.8, 50 mM SDS; hydrodyn. inj., U = 30 kV, T = 40 °C; UV-multiwavelength 195-320 nm; analysis time 17 min	MEKC with SDS; c = 100 $\mu g \cdot mL^{-1}$ of each	339
79	4 benzodiazepines (chlordiazepoxide, flurazepam, diazepam and prazepam)	capillary, 100 cm x 75 µm 15 mM $NH_4CH_3CO_2$, adj. to pH 2.5 with CF_3CO_2H, 15% (v/v) MEOH hydrodyn. inj., U = 26 kV, UV at 254 nm, API-MS; analysis time 7 and 31 min, resp.	coupled CE-MS technique	327
80	cimetidine and ranitidine (as internal standard) in serum after solid phase extraction	capillary, 60 cm x 50 µm 9.4 mM NaH_2PO_4 , 3.3 mM tris(hydroxymethyl)aminomethan, pH 6.4, 9.8 mM HTAB hydrodyn. inj., U = 20 kV UV at 228 nm; analysis time 15 min	MEKC with hexadecyltrimethylammonium bromide (HTAB); detection at the anode	340
81	enalapril maleate in its *cis-/trans-* isomers	capillary, 57 (L_D = 50) cm x 75 µm 80 mM $Na_2B_4O_7$, adj. to pH 8.5, 100 mM SDS; hydrodyn. inj., U = 16 kV, UV at 254 nm; analysis time 40 min	MEKC with SDS	341
82	8 bis(amidinohydrazones)	capillary, 68 cm x 75 µm 50 mM phosphate, pH 7, 1 mM CTAB hydrodyn. inj., U = 22 kV, UV at 280 nm; analysis time 15 min	MEKC with CTAB; reversed polarity with detection at the anode	342 243

Table 7.9. Flavonoids and flavonoid-O-glycosides. Experiments are performed with normal polarity (detection at the cathodic end) if not otherwise stated.

Appl. No.	Compounds	Conditions	Remarks	Ref.
83	hesperidin, naringin, rutin, isoquercitrin, hyperosid, quercitrin and luteolin in the presence of ferulic and caffeic acid	capillary, 52 (L_D = 30) cm x 50 µm 150 mM $Na_2B_4O_7$, adj. to pH 10 with NaOH hydrodyn. inj., U = 18 kV, T = 32 °C UV at 254 nm; analysis time 16 min	borate complexation	344
84	rutin in *Sambuci flos*	conditions as in Appl. No. 83; time 8 min	$c \approx 10^{-7}$- 10^{-8} M	344
85	quercetin, morin, chrysin, hesperetin, naringenin, rutin and naringin	capillary, 57 (L_D = 50) cm x 75 µm 50 mM NaH_2PO_4 - 9 mM $Na_2B_4O_7$, pH 7.5, 10 mM SDS; hydrodyn. inj., T = 30 °C; UV at 210 nm; analysis time 20 min	MEKC with SDS; voltage not given in text	345
86	naringin, rutin, naringenin, hesperetin, morin and chrysin	capillary, 57 (L_D = 50) cm x 50 µm 50 mM NaH_2PO_4 - 50 mM $Na_2B_4O_7$, pH 7.5, 20 mM SDS; hydrodyn. inj., U = 20 kV, T = 25 °C; UV at 214 nm; analysis time 6 min	MEKC with SDS	346
87	diosmin/hesperidin, isorhoifolin, hesperetin, diosmetin-7-O-glycoside, linarin and diosmetin	capillary, 70 cm x 50 µm 50 mM Tris - 46 mM HCl, pH 7.1, 50 mM SDS hydrodyn. inj., U = 25 kV, T = 60 °C UV at 280 nm; analysis time 13 min	MEKC with SDS	347
88	linarin, diosmin, isorhoifolin and hesperidin	capillary, 65 cm x 50 µm 200 mM boric acid, adj. to pH 10.5 with NaOH; hydrodyn. inj., U = 25 kV, T = 60°C UV at 270 nm; analysis time 12 min	borate complexation	347
89	peltatoside, isoquercetrin, hyperin, quercitrin and avicularin	capillary, 70 cm x 50 µm 200 mM boric acid, adj. to pH 10.5 with NaOH; hydrodyn. inj., U = 20 kV, T = 40°C UV at 275 nm; analysis time 15 min	borate complexation	348

Table 7.10. Catechols, catecholamines and other biogenic amines. Experiments are performed with normal polarity (detection at the cathodic end) if not otherwise stated.

Appl. No.	Compounds	Conditions	Remarks	Ref.
90	norepinephrine, epinephrine, 3,4-dihydroxybenzylamine, dopamine, L-DOPA, catechol and 4-methylcatechol	capillary, 64.3 cm x 26 µm 10 mM NaH_2PO_4 - 6 mM $Na_2B_4O_7$, pH 7, 10 mM SDS; electrokin. inj. 4 s, 20 kV, U = 20 kV; amperom. det. (0.7 V vs SCE); analysis time 11 min	MEKC with SDS; borate complexation	349
91	serotonine, norepinephrine, isoproterenol and 4-methylcatechol	capillary, 78.6 cm x 12.7 µm 25 mM MES, adj. to pH 5.7 with NaOH electrokin. inj. 2 s, 30 kV, U = 30 kV amperom. detection (0.7 V vs SCE); analysis time 18 min	$c \approx 10^{-7}$- 10^{-8} M	143
92	dopamine, norepinephrine, epinephrine and catechol	capillary, 87.9 cm x 26 µm 20 mM MES, adj. to pH 6.05 with NaOH electrokin. inj. 2 s, 5 kV, U = 25 kV amperom. detection (0.7 V vs SCE); analysis time 13 min	injected amount 36 fmol epinephrine and 47 fmol catechol	147
93	L-DOPA, norepinephrine, catechol, epinephrine, 4-methylcatechol, 3,4-dihydroxybenzylamine and dopamine	conditions as in Appl No. 90, but: 100 mM SDS instead of 10 mM analysis time 32 min	MEKC with SDS; borate complexation; $c \approx 10^{-4}$ M	246
94	L-DOPA, norepinephrine, catechol, epinephrine, 4-methylcatechol, 3,4-dihydroxybenzylamine and dopamine	conditions as in Appl No. 90, but: 100 mM SDS, 40 mM sodium octylsulfate analysis time 16 min	MEKC with mixed micelles; borate complexation	246

Table 7.10. (continued)

Appl. No.	Compounds	Conditions	Remarks	Ref.
95	serotonine, norepinephrine, epinephrine, L-DOPA, 5-hydoxyindoleacetic acid, homovanillic acid and dihydroxyphenylacetic acid	capillary, 85 cm x 5 µm 25 mM MES, adj. to pH 5.65 with NaOH, 10% (v/v) 2-propanol; electrokin. inj. 5 s, 25 kV, U = 25 kV; amperom. det. (0.7 V vs SCE); analysis time 30 min	inj.volume 36 pL	148
96	dopamine, 3-methoxytyramine, L-DOPA, homovanillic acid and dihydroxyphenylacetic acid	capillary, 81 cm x 2 µm 25 mM MES, adj. to pH 5.65 with NaOH, 10% (v/v) 2-propanol; electrokin. inj. 5 s, 25 kV, U = 25 kV; amperom. det. (0.7 V vs SCE); analysis time 20 min	inj. volume 4 pL	148
97	metanephrine, normetanephrine, deoxyepinephrine, dopamine, 3,4-dihydroxybenzylamine, isoproterenol, epinephrine, norepinephrine, L-DOPA and vanillylmandelic acid	capillary, 70 (L_D = 50) cm x 100 µm 100 mM boric acid, adj. to pH 9.1 with NaOH hydrodyn. inj., U = 10 kV UV at 217 nm; analysis time 25 min	c = 0.5 mM of each; addition of 100 mM TTAB improves separation (but: t=50 min)	350
98	ephedrine, norephedrine, 3,4-dihydroxyphenylacetic acid, 3,4-dihydroxyphenylglycol, 5-hydroxyindole-3-acetic acid, 5-hydroxytryptophan, 3,4-dihydroxybenzylamine and norepedrine·HCl	capillary, 45 (L_D) cm x 100 µm 100 mM $Na_2B_4O_7$ - 50 mM NaH_2PO_4, pH 7.0, 80 mM SDS; hydrodyn. inj., U = 15 kV UV at 210 nm; analysis time 63 min	MEKC with SDS; c = 1000 ppm of each in MeOH	351
99	putrescine, cadaverine, spermidine and spermine in rat tissues	capillary, 100 cm x 50 µm 5 mM sodium borate, pH 9, 2% SDS, 5% ethyleneglycol, 0.1% ethylenediamine hydrodyn. inj., U = 30 kV fluoresc. det.; analysis time 25 min	ethylenediamine suppresses adsorption at the wall	352

Table 7.10. (continued)

Appl. No.	Compounds	Conditions	Remarks	Ref.
100	putrescine, spermidine and spermine in the presence of Na^+, K^+, L-histidine, L-lysine and L-arginine	capillary, 60 (L_D = 35) cm x 75 µm 8 mM quinine sulfate, pH 5.9, 20% EtOH electrokin. inj. 3 s, 30 kV, U = 30 kV UVind. at 236 nm; analysis time 10 min	c = 100 µM of each; for determination of polyamines in tu-mor cell cultures	353

Table 7.11. Vitamins. Experiments are performed with normal polarity (detection at the cathodic end) if not otherwise stated.

Appl. No.	Compounds	Conditions	Remarks	Ref.
101	vitamin B_6, vitamin C, pantothenic acid, vitamin B_2, niacin and vitamin B_1	capillary, 60 cm x 75 µm 20 mM NaH_2PO_4, adj to pH 9.0 with $Na_2B_4O_7$, 50 mM SDS; hydrodyn. inj., U = 18 kV UV at 214 nm; analysis time 18 min	MEKC with SDS	321
102	niacinamide, vitamin B_{12}, vitamin B_6, niacin and vitamin B_2	capillary, 57 (L_D = 50) cm x 75 µm 60 mM $Na_2B_4O_7$, pH 8.92, 60 mM SDS, hydrodyn. inj., U = 30 kV UV at 214 nm; analysis time 10 min	MEKC with SDS	75
103	vitamin B_1, vitamin B_{12}, vitamin B_6, vitamin C, nicotinic acid and folic acid	capillary, 57 (L_D = 50) cm x 75 µm 10 mM Tris - 10 mM NaH_2PO_4, pH 7.56 hydrodyn. inj., U = 20 kV UV at 200 nm; analysis time 10 min	c = 60 $\mu g \cdot mL^{-1}$ of each, but vitamin C: 180 $\mu g \cdot mL^{-1}$	354
104	vitamin B_6 and its common metabolites	capillary, 100 cm x 75 µm 10 mM Na_2HPO_4 - 6 mM $Na_2B_4O_7$, 50 mM SDS; hydrodyn. inj., U = 30 kV UV at 200 nm and LIF; analysis time 40 min	MEKC with SDS; for analysis of these metabolites in human urine	355
105	different forms of vitamin B_{12} (hydroxo-cobalamin, cyano-cobalamin, 5'-deoxyadenosyl-cobalamin and methyl-cobalamin	capillary, 100 cm x 75 µm 10 mM NaH_2PO_4, adj. to pH 2.5 with H_3PO_4 hydrodyn. inj., U = 30 kV UV at 214 nm; analysis time 40 min		356
106	vitamin B_3, B_6, B_{12}, vitamin C, riboflavin phosphate and nicotinic acid	capillary, 80 cm x 100 µm 20 mM NaH_2PO_4, pH 9, 50 mM SDS; hydrodyn. inj., U = 30 kV; UV at 254 nm; time 25 min	MEKC with SDS	357
107	vitamin B_6 vitamers (pyridoxine, pyridoxamin and pyridoxal)	capillary, 80 cm x 100 µm 10 mM NaH_2PO_4, adj. to pH 4.6 with H_3PO_4, 10 mM SDS; hydrodyn. inj., U = 15 kV; amperomet. det. (1.2 V vs. Ag/AgCl); time 9 min		358

7.4 Neutral Substances

In principle, neutral substances like aliphatic and aromatic hydrocarbons cannot be separated electrophoretically. One can either give them a charge by complexation or solvophobic association, or one can use MEKC for the separation of the uncharged molecules. Although phenols are dissociated in the higher pH region, they are also mentioned in this section. Some examples are given in Table 7.12.

7.5 Herbicides

So far, only few examples can be found in literature treating the electrophoretic separation and determination of herbicides. Some of these are given in Table 7.13.

Table 7.12. Neutral compounds. Experiments are performed with normal polarity (detection at the cathodic end) if not otherwise stated.

Appl. No.	Compounds	Conditions	Remarks	Ref.
108	benzo[*g,h,i*]perylene, perylene, pyrene 9-methylanthracene and mesityle oxide	capillary, 100 cm x 75 µm 25 mM THA perchlorate, 50% CH_3CN electrokin. inj., U = 20 kV UV at 229 nm; analysis time 26 min	tetrahexylammonium (THA) interacts with analytes to give positively charged species	74
109	aromatic hydrocarbons (toluene, naphthalene, 9-fluorenone, fluorene, xanthene, dibenzyl, phenanthrene, stilbene and fluoranthene)	capillary, 65 (L_D = 50) cm x 50 µm 20 mM $Na_2B_4O_7$ - 20 mM NaH_2PO_4, pH 9.0, 50 mM SDS, 6 - 8 M urea hydrodyn. inj., U = 20 kV UV at 210 nm	MEKC with SDS	243
110	resorcinol, phenol, *p*-nitroaniline, nitrobenzene, toluene and 2-naphthol	capillary, 70 (L_D = 50) cm x 52 µm 100 mM $Na_2B_4O_7$ - 50 mM NaH_2PO_4, pH 9.0, 50 mM SDS, 6 - 8 M urea hydrodyn. inj., U = 20 kV UV at 210 nm	MEKC with SDS	243
111	aromatic hydrocarbons as in Appl. No. 109	capillary, 50 (L_D) cm x 50 µm 20 mM NaH_2PO_4, adj. to pH 9.0 with $Na_2B_4O_7$, 50 mM SDS, 4M urea, 30 mM ß-CD; hydrodyn. inj., U = 20 kV UV at 220 nm; analysis time 25 min	MEKC with SDS in the presence of ß-cyclodextrin (CD) and urea	249
112	8 nitroaromatics (1-chloro-2,4 dinitrobenzene, 2,4-dinitrotoluene, 1-chloro-2-nitrobenzene, 1-chloro-4-nitrobenzene, 4-nitrotoluene, 4-chloro-2-nitrotoluene, 2-chloro-6-nitrotoluene and 2-chlor-4-nitrotoluene)	capillary, 55 cm x 50 µm 100 mM $Na_2B_4O_7$ - 50 mM phosphate, pH 7.0, 30 mM SDS hydrodyn. inj., U = 20 kV UV at 254 nm; analysis time 16 min	MEKC with SDS	359

Table 7.12. (continued)

Appl. No.	Compounds	Conditions	Remarks	Ref.
113	10 polyaromatic hydrocarbons (naphthalene, acenaphthene, fluorene, phenanthren, pyrene, chrysene, benzo-[*b*]fluoranthene, benzo[*a*]pyrene and benzo[*g,h,i*]perylene)	capillary, 50 cm x 50 μm 12 mM $Na_2B_4O_7$, adj. to pH 8.2, 5 mM SUA, 35% CH_3CN hydrodyn. inj., E = 185 $V \cdot cm^{-1}$ UV at 275 nm; analysis time 32 min	MEKC with sodium 12-undecylenate (SUA)	360
114	chlorophenols and neutral phenols (2-CP, 2,4-DCP, 2,6-DCP, *o*-phenylphenol, 2,3,4,6-TCP, 4,5,6-TCP, penta-chloro-phenol, 4-CP and 2,4,6-TCP)	capillary, 65 cm x 25 μm 15 mM $Na_2B_4O_7$ - 45 mM NaH_2PO_4, adj. to pH 8.0; electrokin. inj., U = 20 kV; amperom. detection (1.4 V vs. SCE); analysis time 23 min	c = 5-8 $ng \cdot L^{-1}$ of each; for analysis of CP's in industrial waste water (LOD = 0.005 $ng \cdot L^{-1}$)	361
115	7 chlorophenols (2-CP, 3-CP, 2,3-DCP, 2,5-DCP, 2,4,5-TCP, 2,4,6-TCP and pentachlorophenol	capillary, 40 (L_D = 26.5) cm x 50 μm 50 mM phosphate, pH 7.0, 50 mM SDS electrokin. inj., U = 18 kV, T = 40 °C UV at 210 nm; analysis time 10 min	MEKC with SDS	233
116	phenol and 5 chlorophenols (2-CP, 2,5-DCP, 2,4,5-TCP, 2,3,4,5-TCP and pentachlorophenol)	capillary, 65 cm (L_D = 50)x 50 μm 25 mM $Na_2B_4O_7$ - 50 mM NaH_2PO_4, adj. to pH 7.0, 100 mM SDS electrokin. inj., U = 15 kV, T = 35 °C UV at 220 nm; analysis time 35 min	MEKC with SDS	231

Table 7.13. Herbicides. Experiments are performed with normal polarity detection at the cathodic end) if not otherwise stated.

Appl. No.	Compounds	Conditions	Remarks	Ref.
117	10 chloro-phenoxy acids	cap., 44 (L_D = 37) cm x 50 μm 20 mM phosphate, 100 mM SDS, 3 mM Brij 35 and 10% MeOH hydrodyn. inj., U = 15 kV; UV at 200-300 nm simultaneously; analysis time 9.5 min	MEKC with SDS in the presence of Brij 35 and organic modifier	240
118	atrazine and simazine	cap., 50 (L_D = 36) cm x 75 μm 10 mM NaH_2PO_4, 25 mM SDS, pH 8.0; hydrodyn. inj., U = 10 kV UV at 225 nm; time 10 min	MEKC with SDS; analysis in river water after extraction (LOD = 0.4 ppb)	362
119	prometryne and prometon	cap., 60 (L_D = 30) cm x 50 μm 10 mM NaH_2PO_4, pH 6.5, 50% CH_3CN; hydrodyn. inj., U = 15 kV;UV at 226 nm	CZE after "on-line" preconcentration by a factor of 10 (LOD ≈ $4 \cdot 10^{-7}$ M	363
120	paraquat and diquat	cap., 80 (L_D = 50) cm x 50 μm 100 mM NaH_2PO_4, adj. to pH 7.0 electrokin. inj., 5 s, 15 kV, U = 15 kV; UV at 254 nm; time 10 min	LOD ≈ 1.5 μM	364
121	prometon, prometryne, propazine and butachlor	cap., 80 (L_D = 50) cm x 50 μm 400 mM boric acid, adj. to pH 8.0 with NaOH, 50 mM OG; hydrodyn. inj., U = 15 kV; UV at 210 nm; analysis time 13 min	MEKC with an octylglucoside (OG) - borate micellar system	365

7.6 Amino Acids, Peptides and Proteins

Fundamentally, peptides and proteins are composed of amino acids linked by peptidic and disulphide bonds. This may justify the common discussion of these three classes of compounds in one section.

The main problem to be solved in the analysis of amino acids is the method of detection. The majority of natural amino acids show only little absorbance in the useful UV wavelength range. Hence, most amino acids are assayed after derivatization. To date, the most sensitive methods for amino acid determination are based on derivatization with fluorescent reagents which allow analysis in the subfemtomole range (see Sect. 4.2.4). Another promising detection mode that appears to be applicable is indirect UV absorbance or fluorescence detection (see

Sect. 4.2.6). By far the most applications of amino acid analysis are reported on chiral separations that are described in Sect. 7.9. Table 7.14. summarizes some separations of amino acid mixtures.

The chemical characteristics of peptides are determined not only by the type and number of amino acids but also by the sequence in the peptide chain. The electrophoretic properties of peptides are fixed by their amino acid sequence. Besides the amine terminus of the sequence, the amine and guanidine residues of lysine and arginine are the main carriers of positive charges whereas the negative contribution to the net charge is associated to the carboxylic acid terminus and the acidic groups of the aspartic and glutamic acids. An important characteristic in electrophoresis of peptides is their isoelectric or isoionic point. Although isoelectric and isoionic points are mostly similar, they are generally not identical. While the isoelectric point is determined by a given aqueous medium, the isoionic point is only related to interactions with protons. The relation of the electrophoretic mobility of peptides and their relative molecular weight and charge is described by Offord's equation (see Eq. 3-43). For large peptides calcu lation of the net charge cannot easily be done from the pK values of the acidic and basic groups. Additional factors such as conformational differences, primary sequence, chirality, etc., also effect the mobility. Tables 7.15. and 7.16. show some applications of CE for peptide analysis.

The high resolving power of CZE for peptide analysis is demonstrated by the peptide mapping of recombinant human interleukin 6 (IL-6) by means of reversed phase HPLC and CZE (Fig. 7.2.). Almost all of the 23 different protein fragments from a tryptic digest were resolved by CZE in 20 min, whereas HPLC analysis showed worse resolution even after 50 min. In Table 7.17. some more examples for the use of CZE in peptide mapping are given.

Analysis of proteins is traditionally done by SDS-PAGE and slab gel isoelectric focusing. Besides a purity profile of a protein sample, both methods provide informations about molecular weight, isoelectric point and microheterogeneity. Although these techniques give excellent results with respect to detection limit and resolution, they are labor-intensive, difficult to automate and only semi-quantitative. Thus CE supplements both techniques well.

Successful separation of proteins by CZE involves efficient suppression of adsorption to the fused silica wall (see Sect. 3.1.2). Basically, there are two approaches to prevent protein adsorption: modification of the fused silica surface by dynamic or static coating (see Sect. 5.1.2) or by performing analysis under experimental conditions that minimize adsorption. As in the case of peptides relative mobilities of proteins can be correlated to the isoelectric point, the net charge and the molecular weight according to Eq. 3-43. For instance, the relative mobilities of collagen proteins were found to be linear related to their pI values. These results were obtained in untreated fused silica capillaries in the pH range 6.9 - 10.5. Since the electrophoretic mobility is dependent on the ionic strength I of the buffer solution the isoelectric point also depends on I, where in many cases the pI is linear related to the square root of the ionic strength.

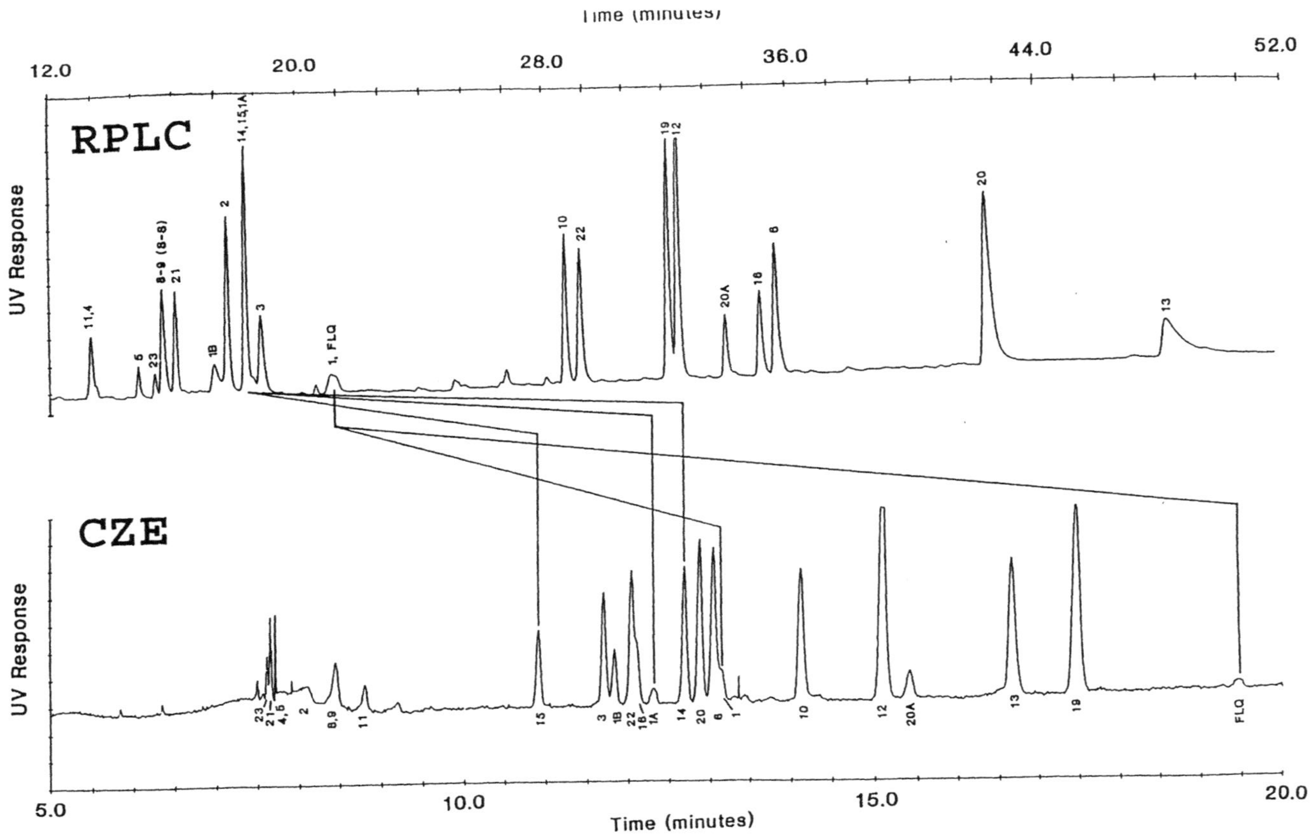

Fig. 7.2. Tryptic mapping of recombinant human interleukin 6 by RP-HPLC and CZE (submitted; by courtesy of Dr. B. Helk, Sandoz Pharma Ltd. Basel). Experimental: (**a**) HPLC, gradient elution using a Vydac 218TP54, 250 x 4.6 mm column, mobile phase: A: 0.1% CF_3CO_2H, B: 0.1% CF_3CO_2H in 90% CH_3CN, gradient: 0 - 60% B in 58 min, flow rate 1 $mL \cdot min^{-1}$, detection 215 nm. (**b**) CZE, fused silica capillary 50 cm (L_D) x 50 µm i.d., buffer 150 mM H_3PO_4, pH 1.6, UV detection at 200 nm, field strength 278 $V \cdot cm^{-1}$, temperature 30 °C

Table 7.14. Amino acids. Experiments are performed with normal polarity (detection at the cathodic end) if not otherwise stated.

Appl. No.	Compounds	Conditions	Remarks	Ref.
122	arg, pro, leu/ala, phe, ser/tyr, cys, glu and asp	capillary, 100 cm x 50 µm 1 mM sodium salicylate, 0.2 mM Na_2CO_3, adj. to pH 9.7 with NaOH electrokin. inj., 1 s, 45 kV, U = 45 kV ind. fluor. (325/405); analysis time 12 min	$c = 10^{-4}$ M of each; injected volume 2 nL	162
123	23 PTH amino acids	capillary, 50 (L_D = 30) cm x 50 µm 100 mM $Na_2B_4O_7$ - 50 mM NaH_2PO_4, adj. to pH 9.0, 100 mM SDS, 4.3 mM urea hydrodyn. inj., U = 10.5 kV UV at 210 nm; analysis time 30 min	MEKC with SDS in the presence of urea; aa's derivatized with phenylthiohydantoin (PTH)	243
124	15 Dns-amino acids	capillary, 60 cm x 50 µm 100 mM $Na_2B_4O_7$ - 50 mM NaH_2PO_4, adj. to pH 7.56, 40 mM SDS hydrodyn. inj., U = 15 kV fluorescence 325/550; analysis time 30 min	MEKC with SDS; amino acids derivatized with dansyl chloride	366
125	17 FTC amino acids	capillary, 100 cm x 50 µm 10 mM phosphate buffer, pH 7.0 electrokin. inj., 10 s, 0.5 kV, U = 30 kV LIF 488/525; analysis time 14 min	$c \approx 10^{-9}$ M of each; aa's deriv. with fluorescein isothiocyanate (FITC) to give thiocarbamyl (FTC) derivatives	113
126	18 FTC amino acids	capillary, 99 cm x 50 µm 5 mM carbonate buffer, pH 10.0 electrokin. inj., 10 s, 2 kV, U = 25 kV LIF 488/525; analysis time 13 min	see Appl. No. 123; injected amount 2 -7 amol of each, $c \approx 10^{-11}$ M	119

Table 7.14. (continued)

Appl. No.	Compounds	Conditions	Remarks	Ref.
127	17 amino acids derivatized with CBQCA	capillary, 104 (L_D = 73) cm x 50 µm 50 mM TES, pH 7.02, 50 mM SDS hydrodyn. inj., U = 25 kV LIF 442/550; analysis time 30 min	MEKC with SDS; $c = 8.7 \cdot 10^{-6}$ M	123
128	8 amino acids derivatized with CBQCA	20 mM $Na_2B_4O_7$ - 20 mM Na_2HPO_4, adj. to pH 9.5, 0.1 mM diaminopentane Beckman P/ACE 2050 CE-LIF system analysis time 8 min	addition of 1,4-diaminopentane to prevent adsorption on the wall	367
129	14 CBI amino acids	capillary, 70 (L_D = 50) cm x 50 µm 100 mM boric acid, adj. to pH 9.0 with NaOH, 50 mM SDS, 10 mM ß-CD hydrodyn. inj., U = 15 kV LIF 442/490; analysis time 26 min	MEKC with SDS and ß-CD; $c = 2.5 \cdot 10^{-7}$ M of each; aa's derivatized with naphthalene-2,3-dicarboxaldehyde (NDA) to give 1-cyano-2-substituted benz[*f*]isoindole (CBI)	132

Table 7.15. Synthetic and bioactive peptides consisiting of not more than 50 amino acid residues (MW< 5000 Da). Experiments are performed with normal polarity (detection at the cathodic end) if not otherwise stated.

Appl. No.	Compounds	Conditions	Remarks	Ref.
130	6 tyrosyl-x-dipeptides (x = glycine, alanine, valine, leucine, glutamic acid and tyrosine)	capillary, 110 (L_D = 75) cm x 50 µm 150 mM H_3PO_4, pH 1.5; electrokin. inj., 5 s, 0.5 kV; U = 0.5 - 25 kV over the first 15 min, then held constant at 25 kV UV at 190 nm; analysis time 33 min	gradient voltage programming	26
131	10 dipeptides (5 pairs of sequence isomers)	capillary, 57 (L_D = 50) cm x 75 µm 50 mM phosphate buffer, pH 2.5, hydrodyn. inj., U = 15 kV UV at 200 nm; analysis time 17 min	c = 1 - 5 $mg{\cdot}mL^{-1}$ of each peptide	53
132	di-, tri-, tetra-, penta- and hexalanine	conditions as in Appl. No. 131		53
133	10 synthetic peptides as CBQCA derivatives (4 tripeptides, 4 tetrapeptides and 2 octapeptides)	capillary, 90 (L_D = 60) cm x 50 µm 50 mM $Na_2B_4O_7$, adj. to pH 9.5, 20 mM α-CD; hydrodyn. inj., U = 20 kV LIF 442/550; analysis time 16 min	α-CD allows increased detection sensitivity and narrower peptide peaks; LOD ≈ 10^{-18} mol	123
134	peptide 8656 labeled with FITC (primary structure: arg-lys-arg-ala-arg-lys-glu)	capillary, 44 cm x 50 µm 5 mM $Na_2B_4O_7$, adj. to pH 9.2 electrokin. inj., 5 s, 2 kV, U = 20 kV LIF 488/525; analysis time 5 min	attachment of one single fluorescent label by taking the peptide through 1 cycle of Edman degradation reaction before derivatization	368
135	6 heptapeptides	capillary, 65 (L_D = 45) cm x 50 µm 20 mM citric acid, adj. to pH 2.5 hydrodyn. inj., E = 277 $V{\cdot}cm^{-1}$, T = 30 °C UV at 200 nm; analysis time 16 min		369

Table 7.15. (continued)

Appl. No.	Compounds	Conditions	Remarks	Ref.
136	6 heptapeptides as in Appl. No. 135	capillary, 120 (L_D = 100) cm x 50 µm 20 mM CAPS, adj. to pH 11.0 hydrodyn. inj., E = 250 $V \cdot cm^{-1}$, T = 30 °C UV at 200 nm; analysis time 15 min		369
137	6 synthetic octapeptides (angiotensin II homologues)	capillary, 110 (L_D = 75) cm x 53 µm 150 mM H_3PO_4, pH 1.5; electrokin. inj., 10s, 2.5 kV; U = 2.5 - 30 kV over the first 5 min, then held constant at 30 kV UV at 190 nm; analysis time 19 min	gradient voltage programming	26
138	2 angiotensin II homologues differing by a CH_2 group at position 5	capillary, 65 (L_D = 45) cm x 50 µm 10 mM phosphate buffer, pH 7.0, SDS hydrodyn. inj., E = 308 $V \cdot cm^{-1}$, T = 30 °C UV at 200 nm; analysis time 16 min	MEKC with SDS; unfortunately the SDS concentration is not given	370
139	5 nonapeptides	conditions as in Appl. No. 131, but 1) buffer: 25 mM Tris - 50 mM DoTAB, pH 7.0 and 2) reversed polarity analysis time 14 min	MEKC with dodecyltrimethylammonium bromide (DoTAB) forming positively charged micelles	53
140	9 peptide hormones (angiotensin II, α-MSH, TRH, LHRH, bradykinin, bombesin, leucine enkephalin, methionine enkephalin and oxytocin)	capillary, 75 cm x 50 µm 20 mM sodium citrate, adj. to pH 2.5 with HCl, 30 mM NaCl; hydrodyn. inj., U = 25 kV, T = 30 °C; UV at 200 nm; analysis time 16 min	c ≈ 0.01 $mg \cdot mL^{-1}$ of each peptide	371

Table 7.15. (continued)

Appl. No.	Compounds	Conditions	Remarks	Ref.
141	9 peptide hormones as in Appl. No. 140	capillary, 57 (L_D = 6.9) cm x 50 µm 100 mM phosphate buffer, pH 2.44 hydrodyn. inj. at the outlet end, U = 25 kV (reversed polarity), T = 20 °C UV at 200 nm; analysis time 2.5 min	injecting the sample at the "outlet end" (7 cm away from the detection window) and working under reversed polarity results in a very short analysis time	75
142	5 bradykinin peptide standards	capillary, 85 (L_D = 45) cm 59 mM phosphate buffer, pH 2.5 electrokin. inj., 10 s, 25 kV, U = 25 kV UV at 210 nm; analysis time 14 min	number of amino acid residues varies between 8 and 10	330
143	des-tyr'-met-enkephalin, met-enkephalin, leu-enkephalin, (val[5])-angiotensin II, angiotensin I and angiotensin III	capillary, 85 (L_D = 45) cm x 75 µm 10 mM $Na_2B_4O_7$ - 10 mM boric acid, pH 8.5, 50 mM SDS electrokin. inj., 2 s, E = 275 $V{\cdot}cm^{-1}$ UV at 215 nm; analysis time 11 min	MEKC with SDS	324
144	6 adrenocorticotropic hormone-related fragments consisting of 2 - 6 amino acid residues	capillary, 57 (L_D = 50) cm x 75 µm 20 mM EACA, adj. to pH 4.4 with CH_3CO_2H (alternative buffer: 40 mM imidazole, adj. to pH 7.5 with MOPS); hydrodyn. inj., U = 25 kV; UV at 214 nm; analysis time 14 min		372
145	bradykinin, neurotensin and angiotensin I	capillary, 57 (L_D = 50) cm x 75 µm 100 mM $Na_2B_4O_7$, pH 9.2; hydrodyn. inj., U = 25 kV; UV at 200 nm	sample: 1 $mg{\cdot}mL^{-1}$ of each dissolved in 0.1% CF_3CO_2H	373

Table 7.15. (continued)

Appl. No.	Compounds	Conditions	Remarks	Ref.
146	5 peptides consisting of 3 - 9 amino acid residues	capillary, 23 (L_D = 14) cmx 50 µm 10 mM NaH_2PO_4, adj. to pH 2.7 with HCl electrokin. inj., 10 s, 1 kV, U = 5.5 kV UV at 200 nm; analysis time 4 min	application of an external electric field across the separation capillary from outside to directly control the EOF	374
147	substance P and 7 of its fragments	coated capillary, Bio-Rad, 20 cm x 25 µm 100 mM phosphate buffer, pH 2.5 electrokin. inj., 5 s, 8 kV, U = 8 kV UV at 200 nm; analysis time 9 min	c = 50 $\mu g \cdot mL^{-1}$ of each; this system can also be used for separation of the 9 peptide hormone standards (Appl. No. 138) in 10 min	375
148	multiple antigen peptides	coated capillary, Bio-Rad, 20 cm x 25 µm 100 mM phosphate buffer, pH 3.5 electrokin. inj., 5 s, 4 kV, U = 8 kV UV at 206 nm; analysis time 20 min	crude samples	376
149	leucinostatins A, D, H and K	coated capillary, Bio-Rad, 20 cm x 25 µm 100 mM phosphate buffer, pH 2.5 electrokin. inj., 6 s, 7 kV, U = 8 kV UV at 206 nm; analysis time 10 min	c = $3.33 \cdot 10^{-2}$ $mg \cdot mL^{-1}$ of each	377
150	collagen CNBr-released peptides	capillary, 60 (L_D = 50) cm x 100 µm 2.5 mM $Na_2B_4O_7$, pH 10.5 electrokin. inj., 6 s, 10 kV, U = 20 kV UV at 220 nm; analysis time 8 min		378
151	4 endorphin peptides	coated capillary, Bio-Rad, 20 cm x 25 µm 50 mM phosphate buffer, pH 2.5 electrokin. inj., 5 s, 8 kV, U = 8 kV UV at 220 nm; analysis time 10 min	number of amino acid residues varies between 32 and 27	330

Table 7.15. (continued)

Appl. No.	Compounds	Conditions	Remarks	Ref.
152	5 endorphin peptides and fragments (2-17, 1-17, 1-31, N-Ac-1-31 and 2-31)	capillary, 65 (L_D = 45) cm x 50 µm 20 mM citric acid, pH 2.5; E = 277 $V \cdot cm^{-1}$, T = 30 °C; UV at 200 nm; analysis time 16 min		379
153	3 endorphin peptides (1-31, N-Ac-1-31 and 2-31)	conditions as in Appl. No. 152, but: E = 307 $V \cdot cm^{-1}$; analysis time 12 min		380
154	4 endorphin fragments (6-17, 7-17, 8-17 and 9-17)	conditions as in Appl. No. 152 analysis time 12 min		381
155	14 motilin fragments	capillary, 72 (L_D = 48) cm x 50 µm 20 mM sodium citrate, adj. to pH 2.5; hydrodyn. inj., U = 20kV; UV at 200 nm; time 14 min	the fragments consist of 6 - 12 amino acid residues	382
156	salmon and eel calcitonin	capillary, 75 cm x 50 µm 20 mM sodium citrate, adj. to pH 2.5; hydrodyn. inj., U = 20kV; UV at 200 nm; time 14 min	the calcitonins consist of 32 amino acid residues and differ in only 3	383
157	vasoactice intestinal peptide in rat brain	capillary, 72 cm x 50 µm 20 mM sodium citrate, adj. to pH 2.5 hydrodyn. inj., U = 25 kV, T = 30 °C UV at 200 nm; analysis time 14 min		384
158	synthetic peptide fragment of the HIV transmembrane glycoprotein gp 41	coated capillary, Bio-Rad, 20 cm x 25 µm 100 mM phosphate buffer, pH 2.56 electrokin. inj., 2 s, 12 kV, U = 8 kV UV at 215 nm; analysis time 4 min	for purity control of synthetic peptide mixtures; the peptide consists of 41 amino acid residue	385
159	6 synthetic somatostatin analog peptides	capillary, 60 cm x 50 µm 25 mM triethylammonium phosphate, adj. to pH 2.25, 13.2% (v/v) CH_3CN hydrodyn. inj., U = 30 kV; UV at 215 nm; analysis time 6 min		386

Table 7.16. Recombinant peptide and protein products. Experiments are performed with normal polarity (detection at the cathodic end) if not otherwise stated.

Appl. No.	Compounds	Conditions	Remarks	Ref.
160	human growth hormone (hGH) and one deaminated form	coated capillary, Bio-Rad, 20 cm x 25 µm 50 mM phosphate buffer, pH 8.0, 0.2% HMC, 0.1% G 3707; U = 8 kV UV at 200 nm; analysis time 25 min	coating: hydrophilic linear polymer; coating procedure see Sect. 5.1.2.1	387
161	comparison of recombinant and pituitary-derived human growth hormone	coated capillary, Bio-Rad, 50 cm x 50 µm 100 mM phosphate buffer, pH 2.56 electrokin. inj., 5 s, 8 kV, U = 8 kV UV at 200 nm; analysis time 25 min	coating: hydrophilic linear polymer; coating procedure see Sect. 5.1.2.1	388
162	human growth hormone (hGH) and 3 of its deaminated forms	capillary, 105 (L_D = 81.5) cm x 50 µm 10 mM TRICINE, 0.58 mM morpholine, 20 mM NaCl, pH 8.0 hydrodyn. inj., E = 300 V·cm^{-1}, T = 24 °C UV at 200 nm; analysis time 9 min		379
163	biosynthetic human insulin (BHI) and 3 of its derivatives	conditions as in Appl. No. 162, but: capillary, 95.5 (L_D = 81.5) cm x 50 µm analysis time 10 min	for purity control of synthetic BHI	379
164	recombinant insulin-like growth factor (r-IGF I) and a by-product	capillary, 120 (L_D = 105) cm x 75 µm 10 mM CAPS, 10 mM $Na_2B_4O_7$, 1 mM EDTA, pH 11.1 electrokin. inj., 10s, 10 kV, E = 250 V·cm^{-1} UV at 215 nm; analysis time 15 min	for purity control of r-IGF I; c = 0.1% of each	389

Table 7.16. (continued)

Appl. No.	Compounds	Conditions	Remarks	Ref.
165	(65 aa) r-hirudin, (64 aa) r-hirudin and (63 aa) r-hirudin	capillary, 10 (L_D = 75) cm x 75 µm 16.7 mM PIPES, 12 mM $Na_2B_4O_7$, 1 mM EDTA, pH 6.7; electrokin. inj., 10s, 10 kV, E = 300 $V{\cdot}cm^{-1}$; UV at 215 nm; time 38 min	for purity control of r-hirudin; c = 0.15% of each	389
166	fluorescamine derivatized recombinant leukozyte A interferon	capillary, 57 (L_D = 50) cm x 75 µm 50 mM sodium phosphate buffer, pH 7.0, 50 mM LiCl;hydrodyn. inj., U = 12 kV UV at 280 nm; analysis time 30 min	c = 232 $\mu g{\cdot}\mu L^{-1}$	390
167	recombinant leukozyte A interferon in a formulation mixture	capillary, 89 (L_D = 41) cm x 75 µm 50 mM $Na_2B_4O_7$, adj. to pH 8.3, 25 mM LiCl electrokin. inj., 12 s, 10 kV, U = 13 kV UV at 210 nm; analysis time 28 min		391
168	recombinant interleukin 1α in a formulation mixture	conditions as in Appl. No. 167		391
169	recombinant tissue plasminogen activator (rtPA)	coated capillary, Bio-Rad, 14 cm x 25 µm 2% ampholyte (pH 6-8), 2% CHAPS, 6 M urea; anolyte 10 mM H_3PO_4, catholyte 20 mM NaOH, mobilizer 10 mM NaOH, 80 mM NaCl; UV at 280 nm; focusing U = 12 kV, mobilization U = 8 kV	CIEF in a capillary coated with a linear hydrophilic polymer; coaing procedure see Sect. 5.1.2.1	392
170	desialylated rtPA	coated capillary, Bio-Rad, 14 cm x 25 µm 100 mM ammonium phosphate buffer, pH 4.6, 0.01% Triton-X 100, 200 mM EACA elelctrokin. inj. , 8 s, 8 kV, U = 6 kV UV at 200 nm; analysis time 20 min	studies of glycosylation forms of glycoproteins	392

Table 7.17. Peptide mapping of selected proteins. Experiments are performed with normal polarity (detection at the cathodic end) if not otherwise stated.

Appl. No.	Compounds	Conditions	Ref.
171	tryptic map of recombinant human growth hormone	10 mM TRICINE, 45 mM morpholine, 20 mM NaCl, pH 8.0 hydrodyn. inj., E = 316 $V \cdot cm^{-1}$ UV at 200 nm; analysis time 15 min	379
172	tryptic map of recombinant human growth hormone	coated capillary, Bio-Rad, 20 cm x 25 μm phosphate buffer, pH 2.5 electrokin. inj., 5 s, 8 kV, U = 8 kV UV at 200 nm; analysis time 12 min	388
173	tryptic map of recombinant human growth hormone	capillary, 100 (L_D = 80) cm x 50 μm 100 mM glycine, adj. to pH 2.35 with 1 M HCl; hydrodyn. inj., U = 30 kV, T = 30 °C UV at 200 nm; analysis time 33 min	63 393
174	tryptic map of human insulin-like growth factor II (IGF II)	conditions as in Appl. No. 173 UV at 200 nm; analysis time 32 min	63
175	tryptic map of human insulin-like growth factor II (IGF II)	conditions as in Appl. No. 173, but: 100 mM TRICINE - 20 mM morpholine, pH 8.15; UV at 200 nm; analysis time 30 min	63
176	CBQCA derivatized tryptic map of β-casein	capillary, 90 (L_D = 60) cm x 50 μm 50 mM borate buffer, pH 9.5, 20 mM α-CD electrokin. inj., 5 s, 3 kV, U = 20 kV LIF 442/550; analysis time 20 min	123
177	tryptic map of β-lactoglobulin	capillary, 75 cm x 50 μm 20 mM sodium citrate, adj. to pH 2.5 with HCl; hydrodyn. inj., U = 25 kV, T = 30 °C UV at 200 nm; analysis time 30 min	394
178	protease V8 map of β-lactoglobulin	capillary, 65 (L_D = 45) cm x 50 μm 20 mM citric acid, adj. to pH 2.5 hydrodyn. inj., E = 313 $V \cdot cm^{-1}$, T = 30 °C UV at 200 nm; analysis time 15 min	369
179	CBQCA derivatized tryptic map of β-lactoglobulin	capillary, 47 (L_D = 40) cm x 75 μm 10 mM $Na_2B_4O_7$ - 10 mM Na_2HPO_4, pH 9.5, 0.1 mM diaminopentane hydrodyn. inj., U not given, T = 25 °C LIF 448/560; analysis time 8 min	395

Table 7.17. (continued)

Appl. No.	Compounds	Conditions	Ref.
180	tryptic map of recombinant tissue plasminogen activator	capillary, 57 (L_D = 50) cm x 75 μm 100 mM phosphate buffer, pH 2.5, hydrodyn. inj., U = 12 kV, T = 25 °C UV at 200 nm; analysis time 60 min	396
181	tryptic map of bovine serum albumin	coated capillary, Bio-Rad, 32 cm x 25 μm 200 mM phosphate buffer, pH 2.56, U = 8 kV; UV at 190 nm; analysis time 27 min	387
182	tryptic map of bovine serum albumin	capillary, 27 (L_D = 20) cm x 50 μm 100 mM NaH_2PO_4, adj. to pH 2.5, 1.5 M urea, 30 mM β-CD hydrodyn. inj., U = 12 kV, T = 25 °C UV at 200 nm; analysis time 25 min	75
183	benzoin derivatized tryptic map of alkylated human serum albumin	capillary, 70 (L_D = 50) cm x 50 μm 50 mM CAPS, pH 9.1, 60 mM SDS, 10% CH_3CN; hydrodyn. inj., U = 25 kV LIF ; analysis time 30 min	397
184	cyanogen bromide map of human serum albumin with intact disulphide bridges and reduced and alkylated disulphide bridges	capillary, 65 (L_D = 50) cm x 50 μm, coated with linear polyacrylamide 50 mM glutamine - 50 mM triethylamine, pH 9.7; hydrodyn. inj., U = 25 kV UV at 215 nm; analysis time 20 min	397
185	tryptic and chymotryptic map of trypsinogen	capillary, 55 (L_D = 40) cm x 50 μm 50 mM CAPS, pH 9.5, 10% MeOH hydrodyn. inj., U = 25 kV UV at 215 nm; analysis time 15 min	397
186	tryptic map of oxidized lysozyme	capillary, 65 (L_D = 50) cm x 50 μm 50 mM sodium phosphate, pH 2.3 hydrodyn. inj., U = 27 kV UV at 215 nm; analysis time 20 min	397
187	tryptic, chymotryptic and peptic maps of reduced, carboxymethylated lysozyme	Superox-coated capillary, 80 (L_D = 74) cm x 75 μm; 25 mM TRIS, adj. to pH 4.8 with HCl electrokin. inj., 2 s, 10 kV, U = 15 kV UV at 215 nm; analysis time 85 min	398

Principally, separation of proteins can be performed under denaturing or non-denaturing (native) conditions. Denaturing conditions are useful for investigating protein subunits. Denaturation refers to molecular states in which proteins are partially or completely unfolded. It is favored by extreme pH, high temperature, ionic surfactants and denaturing agents such as urea or guanidine. However, not all denaturing conditions are useful for CE. Under certain conditions, e.g. elevated temperature, proteins tend to form aggregates or precipitates which interfere in CE. Urea has been found to be effective in high concentrations of 4 to 8 M, where it prevents aggregation and causes randomization of the protein structure.

To maintain the native structure of a protein during electrophoresis, experimental conditions must be chosen carefully depending on the nature of the protein. An interesting example of the separation of similar proteins is shown in Fig. 7.3. Hemoglobin A (HbA) subunits are separated at pH 2.5 where wall interactions are suppressed but the proteins are not denatured. Fig. 7.3.a shows the separation pattern of HbA from healthy human beings. HbA is separated to its subunits with the major component being HbA_0 (90 - 97%) and the more negatively charged glycosylated HbA_1 and HbA_2 fractions. Fig. 7.3.b shows the separation pattern of human HbA from a diabetes mellitus patient. In this sample the amount of HbA_{1c} is increased to ca. 20% due to the increased glucose concentration.

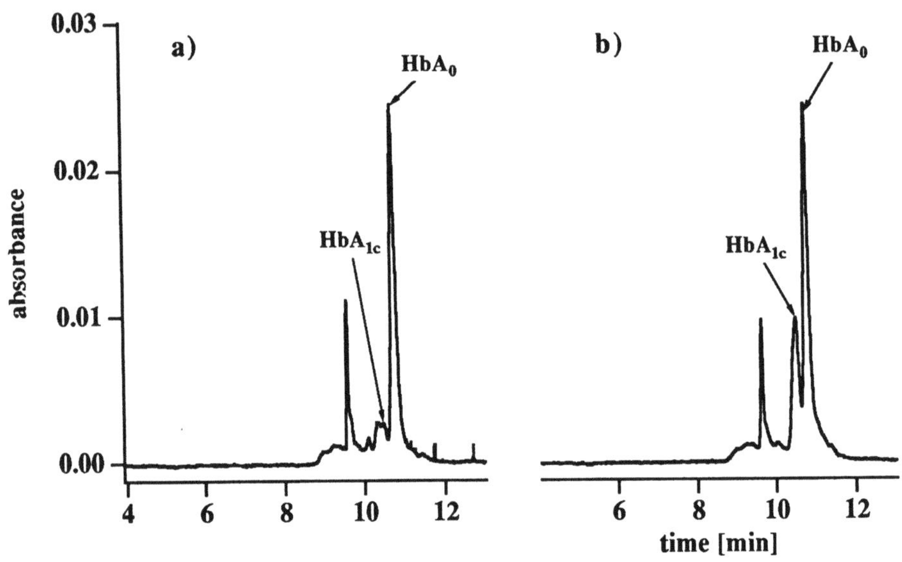

Fig. 7.3. Separation of human hemoglobin from (**a**) healthy people and (**b**) diabetes mellitus patients by CZE. Instrument: Beckman P/ACE 2000; experimental conditions: fused silica capillary, 57 cm x 50 µm i.d., hydrodynamic injection for 1 s, voltage 20 kV, temperature 25 °C, UV detection at 200 nm, electrolyte system 20 mM sodium citrate, adjusted to pH 2.5 with hydrochloric acid, 30 mM NaCl

Table 7.18 illustrates the different approaches to protein analysis by means of CE showing the separation of standard protein mixtures. Finally, Table 7.19. gives an overview of the analysis of some selected proteins.

Table 7.18. Standard protein mixtures. Experiments are performed with normal polarity (detection at the cathodic end) if not otherwise stated.

Appl. No.	Compounds	Conditions	Remarks	Ref.
188	lysozyme and cytochrome C	capillary, 100 (L_D = 63) cm x 52 µm 20 mM CAPS - 10 mM KCl, pH 11.0 electrokin. inj., 5 s, 16 kV, U = 30 kV UV at 230 nm; analysis time 8 min	separation of basic proteins by coulombic repulsion between proteins and silica wall by raising pH above pI; not universally useful because proteins with lower pI's can be denaturated at basic pH values	25
189	myoglobin (horse heart and whale skeletal muscle), carbonic anhydrase A and B, β-lactoglobulin A and B	capillary, 101 (L_D = 55) cm x 52 µm 10 mM TRICINE - 20 mM KCl, pH 8.22 electrokin. inj., 6 s, 2 kV, U = 20 kV UV at 230 nm; analysis time 12 min	see Appl. No. 188; all sample proteins are negatively charged at the chosen pH	25
190	β-lactoglobulin A, cytochrome C (horse), lysozyme (chicken), myoglobin (horse heart) and parvalbumin (rabbit)	capillary, 110 (L_D = 75) cm x 53 µm 150 mM H_3PO_4, pH 1.5 electrokin. inj., 10 s, 2.5 kV, U = 2.5 - 30 kV in 5 min, then 30 kV UV at 190 nm; analysis time 19 min	separation at low pH where all proteins are positively charged; deactivation of the silia surface by phosphate; drawback: diminished charge differences of the proteins due to full protonation deteriorating resolution	26
191	6 cytochrome C species	capillary, 125 (L_D = 75) cm x 53 µm 150 mM H_3PO_4, adj. to pH 5.0 electrokin. inj., 10 s, 2.5 kV, U = 2.5 - 30 kV in 5 min, then 30 kV UV at 190 nm; analysis time 19 min	separation of basic proteins at pH 5; whereas at pH < 5 mobility differences are too small, at a slightly higher pH (5.25), the proteins begin to adsorb at the wall	26

Table 7.18. (continued)

Appl. No.	Compounds	Conditions	Remarks	Ref.
192	17 standard proteins	capillary, 110 cm (L_D = 75) x 52 µm, coated with [(3-methacryloyl)propyl]trimethoxysilane and 1-vinyl-2-pyrolidinone; 38.5 mM H_3PO_4 - 20 mM NaH_2PO_4, pH 5.0; electrokin. inj., 5 s, 5 kV, U = 2.5 - 30 kV in 5 min, then 30 kV, UV at 190 nm; time 25 min	deactivation of the silica surface by static coating with polyvinylpyrolidinone (PVP); gradient voltage programming	26
193	lysozyme and α-chymotrypsinogen	capillary, 75 (L_D = 65) cm x 25 µm 40 mM phosphate buffer, pH 7.0, 250 mM K_2SO_4; electrokin. inj., 8 s, 8 kV, U = 10 kV fluor. det. 280/340; anal. time 70 min	separation of proteins in buffers containing high concentrations of salts; K_2SO_4 is superior to NaCl, LiCl, KCl, CsCl, KBr, KNO_3; drawback: the high ionic strength allows only low voltages resulting in long runs	28
194	lysozyme and α-chymotrypsinogen	conditions as in Appl. No. 193, but: 100 mM CHES, pH 9.0; anal. time 50 min	see Appl. No. 193	28
195	lysozyme and α-chymotrypsinogen	conditions as in Appl. No. 193, but: 40 mM phosphate buffer, 2 M betaine, 100 mM K_2SO_4, pH 7.6; U = 20 kV analysis time 25 min	separation of proteins in buffers containing high concentrations of zwitterionic salts combined with an ionic salt; in comparison to Appl. No. 193, run time is much shorter	28
196	5 basic proteins (lysozyme, cytochrome C, ribonuclease A, α-chymotrypsinogen and myoglobin)	capillary, 50 (L_D = 35) cm x 75 µm, coated with alkylsilane and Brij35 (Sect. 5.1.2.3); 10 mM phosphate buffer, pH 7.0, 0.001% (w/w) Brij35, E = 300 $V \cdot cm^{-1}$ UV at 200 nm; analysis time 17 min	separation of proteins using a nonionic surfactant coating. EOF is rather constant in the pH range 4 -11 allowing the use of the best pH to give optimum selectivity without changing EOF; detergents added to the buffer should be used below their CMC	201

Table 7.18. (continued)

Appl. No.	Compounds	Conditions	Remarks	Ref.
197	myoglobin, conalbumin, transferrin, β-lactoglobulin B and A, ovalbumin	conditions as in Appl. No. 196, but: L_D = 30 cm; analysis time 8 min	see Appl. No. 196	201
198	5 basic proteins as in Appl. No. 196	capillary, 85 (L_D = 65) cm x 50 µm, coated with base-catalyzed diol-epoxide (Sect. 5.1.2.4); 10 mM phosphate buffer, pH 7.0; E = 300 $V \cdot cm^{-1}$ UV at 200 nm; analysis time 32 min	separation of proteins using an epoxy polymer coating; also applicable to the analysis of proteins with lower pI values	202
199	cytochrome C, lysozyme, myoglobin, trypsin, ribonuclease, trypsinogen, chymotrypsinogen	capillary, 89 (L_D = 65) cm x 50 µm, coated with PEG (Sect. 5.1.2.5.2) 30 mM KH_2PO_4, adj. to pH 3.8 electrokin. inj., 10 s, 10 kV, U = 20 kV fluor. det. 280/340; analysis time 9 min	separation of proteins using polyethylene glycol (PEG) modified capillaries in the pH range of 3 - 5, at higher pH values peak deformation and decrease in resolution are observed	204
200	7 protein markers	capillary, 100 cm x 20 µm, deactivated with terminal arylpentafluoro groups; 200 mM ammonium phosphate - 100 mM KCl, pH 7; hydrodyn. inj., E = 200 $V \cdot cm^{-1}$ UV at 219 nm; analysis time 33 min	coating procedure: (1) silylate with 0.1% γ-aminopropyltrimethoxysilane, (2) rinse with 200 mM pentafluorobenzoyl chloride in toluene, (3) reequilibrate with toluene, MeOH and finally H_2O	205
201	4 basic proteins (lysozyme, cytochrome C, ribonuclease and α-chymotrypsinogen	capillary, 80 (L_D = 50) cm x 50 µm, coated with PEG 2000 100 mM phosphate buffer, pH 6.0 hydrodyn. inj., U = 17 kV UV at 210 nm; analysis time 40 min	separation of proteins using polyethylene glycol (PEG) modified capillaries in the pH range of 4 - 7.5; a procedure for restoring collapsed capillaries was developed by the authors	206

Table 7.18. (continued)

Appl. No.	Compounds	Conditions	Remarks	Ref.
202	lysozyme, trypsin and chymotrypsinogen	capillary, 72 (L_D = 50.5) cm x 50 µm, coated with epoxy-diol 50 mM phosphate buffer, pH 4.0 electrokin. inj., U = 20 kV UV at 205 nm; analysis time 15 min	separation of proteins using capillaries modified with γ-glycidoxypropyltrimethoxysilane ("epoxy-diol" coating); only effective at pH < 5; efficiency is worse than than on PEG coating of Appl. No. 199	207
203	lysozyme and cytochrome C	capillary, 39.5 (L_D = 20) cm x 50 µm, coated with maltose; 50 mM phosphate buffer, pH 6, 0.01% (w/v) NaN_3 electrokin. inj., 10 s, 10 kV, U = 20 kV UV at 205 nm; analysis time 12 min	separation of proteins using carbohydrate modified capillaries; good shielding of the surface up to pH 7.0, but lower efficiency that PEG coating of Appl. No. 199	207
204	horse heart myoglobin, bovine ribonuclease, bovine chymotrypsinogen, horse heart cytochrome and hen egg lysozyme	PEI 200-EDGE coated capillary, 50 (L_D = 35) cm x 75 µm 20 mM NH_3, adj. to pH 7.0 with HCl hydrodyn. inj., U = 12.5 kV UV at 200 nm; analysis time 32 min	separation of proteins using positively charged polyethylene-imine (PEI) coated capillaries with ethyleneglycol diglycidyl ether (EDGE) as crosslinking agent; re-versed polarity with detection at the anode	208
205	12 standard proteins	capillary, 60 (L_D = 50) cm x 100 µm 2.5 mM $Na_2B_4O_7$, adj. to pH 10.5 with 0.1 M NaOH; electrokin. inj., 6 s, 10 kV, U = 20 kV; UV at 220 nm; an. time 18 min	see Appl. No. 188	378
206	horse heart myoglobin, human and bovine erythrozyte carbonic anhydrase B and β-lactoglobulin A	capillary, 57 (L_D = 50) cm x 75 µm, coated with linear PAA; 4% ampholyte (pH 3.5-10), anolyte 150 mM H_3PO_4, catholyte 50 mM NaOH; mobilizer 50 mM NaCl - 50 mM NaOH; hydrodyn. inj., IEF at U = 25 kV for 20 min; UV at 280 nm	CIEF in a capillary coated with a linear hydrophilic polymer; coating procedure see Sect. 5.1.2.1	412

Table 7.18. (continued)

Appl. No.	Compounds	Conditions	Remarks	Ref.
207	β-lactoglobulin A and B, lactalbumin and hemoglobin A and S	coated capillary, Bio-Rad, 20 cm x 25 µm 300 mM $Na_2B_4O_7$, adj. to pH 8.5 UV; analysis time 10 min	between every run, capillary is rinsed with 100 mM phosphate buffer, pH 2.5, followed by H_2O to eliminate adsorbed or precipitated proteins; reversed polarity with detection at anode; coating procedure see Sect. 5.1.2.1	400
208	horse heart myoglobin, ribonuclease A, cytochrome C3 and lysozyme	capillary, 100 (L_D = 90) cm x 50 µm 10 mM phosphate buffer, pH 7, 50 µg·mL^{-1} FC 134 electrokin. inj., 10 s, 20 kV, U = 30 kV UV at 230 nm; analysis time 20 min	dynamic coating of the silica wall with the fluorosurfactant Fluorad FC 134 from 3M Company; reversed polarity with detection at the anode	401
209	carbonic anhydrase, urease, ovalbumin, α-lactalbumin and bovine serum albumin	capillary, 37.5 (L_D = 30.5) cm x 75 µm 50 mM $Na_2B_4O_7$, adj. to pH 10.0 with NaOH; hydrodyn. inj., U = 10 kV UV at 200 nm; analysis time 9 min	40 µg·mL^{-1} of each protein is dissolved in 20 mM boric acid, pH 4.0, containing 20% ethylene glycol	402
210	7 model proteins	capillary, 100 (L_D = 63.5) cm x 50 µm 50 mM $Na_2B_4O_7$, adj. to pH 9.5 hydrodyn. inj., U = 22 kV UV at 200 nm; analysis time 14 min	see Appl. No. 188; after each run, the capillary is rinsed with 1 M NaOH, followed by reconditioning with buffer	403
211	8 model proteins	capillary, 37 (L_D = 30) cm x 25 µm 100 mM $Na_2B_4O_7$, adj. to pH 11.5 hydrodyn. inj., U = 12 kV UV at 200 nm; analysis time 6 min	see Appl. No. 210	404

Table 7.18. (continued)

Appl. No.	Compounds	Conditions	Remarks	Ref.
212	β-casein, α-lactalbumin, α-casein/β-lactoglobulin B and β-lactoglobulin A in milk	capillary, 23 cm x 21 μm 50 mM sodium phosphate buffer, pH 7, 4 M urea; hydrodyn. inj., U = 10 kV UV at 200 nm; analysis time 6 min		405
213	5 basic model proteins	capillary, 70 (L_D = 57) cm x 75 μm 20 mM sodium phosphate buffer, pH 3.0, 30 mM NaCl, 0.05% (w/w) PVA (MW 15000); hydrodyn. inj., U = 25 kV UV at 200 nm; analysis time 16 min	deactivation of silica by dynamic coating with non-ionic polyvinylalcohols (PVA); for the separation of basic proteins	406
214	5 basic proteins	capillary, 65 cm x 50 μm, coated with p[HPMA]; 50 mM TRIS, adj. to pH 4.7 with HCl, electrokin. inj., 1 s, 15 kV, U = 20 kV UV at 215 nm; analysis time 25 min	poly(2-hydroxypropylmethacrylate) (p[HPMA]) is used as hydrophilic polymer coating; coated capillaries can be used in the pH range of 4 - 8 (also for acidic proteins)	407
215	6 protein molecular mass standards (from 14400 to 78000 Da)	capillary, 45 (L_D = 25) cm x 75 μm TRIS - borate, pH 8.1, 0.1% SDS, 6% acrylamide, 0.5% APS, 0.04% TEMED hydrodyn. inj., U = 12 kV UV at 230 nm; analysis time 20 min	SDS-CGE of proteins using non-crosslinked PAA in uncoated capillaries; if T > 4%, no apprec. gel displacement is observed in the uncoated cap., because EOF is decreased significantly; reversed polarity with detection at the anode	408
216	6 protein molecular mass standards (from 29000 to 205000 Da)	conditions as in Appl. No. 215, but: 4% acrylamide; analysis time 19 min	see Appl. No. 215; lower gel concentrations can be used for wider molecular mass range	408

Table 7.18. (continued)

Appl. No.	Compounds	Conditions	Remarks	Ref.
217	α-lactalbumin, β-lactalbumin, trypsinogen and pepsin	capillary, 20 (L_D) cm x 75 μm; gel composition 10% T, 3.3% C, buffer 90 mM TRIS, adj. to pH 8.6 with NaH_2PO_4, 0.1% SDS, 8 M urea; electrokin. inj. 10 s, 6 μA, E = 400 $V \cdot cm^{-1}$; UV at 230 nm; analysis time 55 min	SDS-CGE using conventional crosslinked PAA gel in a capillary pretreated with a bifunctional agent acc. to Karger (Sect. 5.2.2.2.1); reversed polarity with detection at the anode	197
218	6 protein molecular mass standards (from 14400 to 97400 Da)	capillary, 24 (L_D = 7) cm x 75 μm, coated with linear PAA; gel composition 5.1% T, 2.6% C, buffer 375 mM TRIS, adj. to pH 8.8 with NaH_2PO_4, 0.1% SDS, ethylene glycol (1.8-2.7 M); electrokin. inj. 10 s, 2.5 kV, E = 83 $V \cdot cm^{-1}$; UV at 214 nm; time 21 min	see Appl. No. 217	409
219	myoglobin, ovalbumin, bovine serum albumin and conalbumin	capillary, 57 (L_D = 50) cm x 100 μm 50 mM H_3PO_4, adj. to pH 5.5 with NaOH, 0.5% SDS, gel, T = 10%; hydrodyn. inj., U = 20 kV; UV at 254 nm; anal. time 60 min	see Appl. No. 215	410
220	lysozyme, carbonic anhydrase, ovalbumin, bovine serum albumin and phosphorylase B	coated capillary, 15 (L_D) cm x 75 μm 120 mM TRIS - 120 mM histidine, pH 8.8, 0.1% SDS, gel, 12% T; electrokin. inj. 6 s, 400 $V \cdot cm^{-1}$, E = 560 $V \cdot cm^{-1}$ UV at 280 nm; analysis time 10 min	see Appl. No. 215, but: the capillary is coated with linear PAA gel; coating procedure see Sect. 5.1.2.1	229
221	6 protein molecular mass standards (from 14400 to 94000 Da)	capillary, 47 (L_D = 40) cm x 100 μm, coated with linear PAA; gel composition 5.1% T, 2.6% C, buffer 100 mM TRIS - 100 mM CHES, pH 8.8, 0.1% SDS, 3% (w/v) PEG 100000; hydrodyn. inj., E = 300 $V \cdot cm^{-1}$ UV at 214 nm; analysis time 17 min	SDS-CGE using PEG as UV-transparent linear hydrophilic polymer network	229 411

Table 7.19. Monoclonal antibodies, serum proteins, hemoglobins, histones, selected enzymes and glycoproteins. Experiments are performed with normal polarity (detection at the cathodic end) if not otherwise stated.

Appl. No.	Compounds	Conditions	Remarks	Ref.
222	iron-free transferrin isoforms after incubation with neuraminidase	capillary, 18.5 (L_D = 20) x 100 μm 18 mM TRIS - 18 mM boric acid, pH 8.4, 0.3 mM EDTA; hydrodyn. inj., U = 8 kV UV at 280 nm; analysis time 6 min	separation of glycoprotein isoforms differing in their carbohydrate content (microheterogeneity studies)	412
223	unpurified alkaline phosphatase - IgG conjugate	capillary, 27 (L_D = 20) x 75 μm 100 mM $Na_2B_4O_7$, adj. to pH 10.0, 0.5% MC, 0.5 mM SDS; hydrodyn. inj., U = 5 kV, T = 15 °C; UV at 280 nm; analysis time 6 min	methyl cellulose (MC) is added as molecular sieving agent; model for characterization of enzyme - antibody conjugates	413
224	comparison of the IEF patterns of 2 different murine IgG preparations	coated capillary, Bio-Rad, 12 cm x 25 μm ampholyte pH 3-10 and pH 5-8, anolyte 10 mM H_3PO_4, catholyte 20 mM NaOH; mobilizer 80 mM NaCl - 20 mM NaOH; hydrodyn. inj., IEF and mobilization at 8 kV; UV at 280 nm	CIEF in a capillary coated with a linear hydrophilic polymer; coating procedure see Sect. 5.1.2.1; sample is mixed with the carrier solution	414
225	humanized anti-TAC monoclonal antibody	capillary, 37 (L_D = 30) cm x 75 μm, coated with linear PAA; ampholyte pH 3-10, anolyte 20 mM H_3PO_4, catholyte 20 mM NaOH; mobilizer 80 mM NaCl - 20 mM NaOH; hydrodyn. inj., IEF at 6 kV (10 min) , mobilization at 8 kV (40 min); UV at 280 nm	CIEF in a capillary coated with a linear hydrophilic polymer; coating procedure see Sect. 5.1.2.1; sample is mixed with the carrier solution	415
226	IgG monoclonal antibody chimeric L6 (cL6)	coated capillary, Bio-Rad, 20 (L_D = 17.2) cm x 25 μm; sodium phosphate buffer, pH 5.6 electrokin. inj., U = 12 kV UV at 200 nm; analysis time 15 min	cL6 is separated into its isoelectrotypes; coating procedure see Sect. 5.1.2.1	416

Table 7.19. (continued)

Appl. No.	Compounds	Conditions	Remarks	Ref.
227	human serum proteins	capillary, 37.5 (L_D = 30.5) cm x 75 µm 50 mM $Na_2B_4O_7$; hydrodyn. inj., U = 10 kV UV at 200 nm; analysis time 10 min	serum diluted 40:1 with 1 mM boric acid, pH 4.5, containing 20% ethylene glycol	402
228	human serum proteins	capillary, 100 (L_D = 63.5) cm x 50 µm 50 mM $Na_2B_4O_7$, adj. to pH 9.5; hydrodyn. inj., U = 30 kV; UV at 200 nm; time 10 min		403
229	human serum proteins	capillary, 25 cm x 25 µm 150 mM $Na_2B_4O_7$, adj. to pH 10.0; hydrodyn. inj., U = 20 kV; UV at 200 nm; time 90 s	serparation pattern comparable to conventional serum electrophoresis on cellulose acetate	404
230	assay of bovine serum albumin	capillary, 27 (L_D = 20) cm x 75 µm 150 mM $Na_2B_4O_7$, adj. to pH 8.5; hydrodyn. inj., U = 12 kV; UV at 214 nm; time 2.5 min	coefficient of variation 7.59% at BSA concentrations of 25 - 1000 $\mu g \cdot mL^{-1}$	417
231	globin chains of hemoglobins from different subjects	capillary, 42 cm x 75 µm 25 mM sodium phosphate buffer, pH 11.8 hydrodyn. inj., I = 40 µA UV at 214 nm; analysis time 7 min		418
232	comparison of the IEF patterns of different hemoglobins (A, A+S, A+F and A+C)	capillary, 17 cm x 25 µm, coated with linear PAA; 2% ampholyte pH 3-10, anolyte 20 mM H_3PO_4, catholyte 40 mM NaOH; mob. 80 mM NaCl - 20 mM NaOH; hydrodyn. inj., IEF at 7kV (5 min) , mob. at 8 kV (13 min); UV at 280 nm	CIEF in a capillary coated with a linear hydrophilic polymer; coating procedure see Sect. 5.1.2.1; sample is mixed with the carrier solution	419
233	chains of different hemoglo-bins	conditions as in Appl. No. 232 analysis time 25 min	globin chains are prepared by treatment with acidic acetone	419

Table 7.19. (continued)

Appl. No.	Compounds	Conditions	Remarks	Ref.
234	hemoglobin reference standard	capillary, 35 cm x 25 µm, coated with linear PAA; 100 mM sodium phosphate buffer, pH 3.2, 7 M urea, 1% reduced Triton X-100 electrokin. inj., 3 s, 8 kV, U = 12 kV UV at 210 nm; analysis time 14 min	coating procedure see Sect. 5.1.2.1	419
235	multi-acetylated cuttlefish testis histone H4	capillary, 70 (L_D = 48) cm x 50 µm, coated with polymeric coating agent (ABI); 10 mM sodium acetate, pH 6.6; hydrodyn. inj., E = 215 $V \cdot cm^{-1}$ UV at 200 nm; analysis time 12 min	coating of the capillary with a positively charged polymer; reversed polarity with detection at the anode	420
236	whole histones in its 5 fractions	coated capillary, Bio-Rad, 35 x 50 µm 100 mM phosphate buffer, pH 2.5, electrokin. inj., 10 s, 10 kV, U = 10 kV, UV at 200 nm; analysis time 9 min		421
237	phosphorylated histone H1 variants	capillary, 57 (L_D = 50) cm x 75 µm 100 mM sodium phosphate buffer, pH 2.0, 0.03% HPMC; hydrodyn. inj., U = 16 kV UV at 210 nm; analysis time 15 min	hydroxypropyl methyl cellulose (HPMC) is added as molecular sieving agent	422
238	myoglobin subunits	capillary, 47 (L_D = 40) cm x 50 µm, coated with linear PAA; 2.5% ampholyte pH 3-10, anolyte 75 mM H_3PO_4, catholyte 25 mM NaOH; mobilizer 80 mM NaCl - 20 mM NaOH; hydrodyn. inj., IEF at 25 kV (10 min) , mobilization at 25 kV (15 min); UV at 280 nm	CIEF in a capillary coated with a linear hydrophilic polymer; coating procedure see Sect. 5.1.2.1; sample is injected separately	423

Table 7.19. (continued)

Appl. No.	Compounds	Conditions	Remarks	Ref.
239	bovine pancreatic ribonuclease B	capillary, 57 (L_D = 50) cm x 75 µm 20 mM CAPS, pH 11.0; hydrodyn. inj., U = 10 kV; UV at 214 nm; analysis time 15 min	separation of glycoprotein isoforms (microheterogeneity studies)	424
240	recombinant glycoprotein (MW 40000, pI 4.5-5.0) with different sialic acid residues	capillary, 27 (L_D = 20) cm x 75 µm 100 mM NaOAc, adj. to pH 4.0 with 100 mM H_3PO_4; hydrodyn. inj., U = 10 kV UV at 214 nm; analysis time 12 min	separation of glycoprotein isoforms (microheterogeneity studies)	425
241	ribonuclease A, B1 and B2	capillary, 120 (L_D = 100) cm x 50 µm 20 mM CAPS, pH 11.0; hydrodyn. inj. E = 250 V·cm^{-1}; UV at 200 nm; time 10 min	separation of glycoprotein isoforms (microheterogeneity studies)	379
242	reconbinant RNase T1 and site-directed mutants	capillary, 70 (L_D = 48) cm x 50 µm 25 mM sodium phosphate, pH 7.0; hydrodyn. inj., E = 430 V·cm^{-1}; UV at 200 nm; time 9 min	separation of a protein in its isoenzymes	426
243	glutathione peroxidase in rat liver extract	capillary, 27 (L_D = 20) cm x 50 µm 100 mM $Na_2B_4O_7$, adj. to pH 8.2, 100 mM SDS; hydrodyn. inj., U = 12 kV, T = 30 °C UV at 214 nm; analysis time 4 min	MEKC with SDS	427
244	ovalbumin	capillary, 87 (L_D = 80) cm x 50 µm 100 mM $Na_2B_4O_7$, adj. to pH 8.5, 5 mM 1,4-diamino butane; hydrodyn. inj., U = 25 kV, T = 30 °C; UV at 214 nm; time 30 min	separation of glycoprotein isoforms in the presence of 1,4-diamino butane to prevent adsorption; borate complexation	428
245	Ca^{2+} and Zn^{2+} binding proteins (calmodulin, parvalbumin, thermolysin and carbonic anhydrase)	capillary, 122 (L_D = 100) cm x 50 µm 100 mM TRIS - 100 mM TRICINE, pH 8.3, various amounts of Ca^{2+} and Zn^{2+} and EDTA hydrodyn. inj., U = 30 kV; UV at 200 nm	binding-shift assay	429

7.7 Carbohydrates and Their Derivatives

Carbohydrates play an important role in many diverse research and industrial domains. The considerable number of stereoisomers, the immense combination possibilities of carbohydrate monomers, and the lack of chromophores make sugar analysis a critical task. For sugar analysis, electrophoresis on supporting media has been applied as an alternative method to paper and thin-layer chromatography for the rapid identification of sugars occurring in foodstuffs and hydrolysates of polysaccharides. The separated carbohydrates are visualized by staining procedures after electrophoresis. CE represents an alternative to these time-consuming and only qualitative electrophoretic methods.

Since carbohydrates show only low UV absorptivities, UV detection of underivatized sugars is restricted to the nanogram range. A significant increase in sensitivity can be achieved by derivatizing reducing sugars with 2-amino pyridine to *N*-pyridylglucosamines which can be determined by UV detection at 240 nm. Two other derivatizing agents, ethyl *p*-aminobenzoate and *p*-aminobenzoic acid, do not only react with reducing sugars and allow therefore simultaneous UV detection of aldoses, ketoses and uronic acids. Aminosugars can be derivatized with CBQCA to highly fluorescent isoindol derivatives which can be detected by LIF. If reductive amination is performed before the derivatization step, also reducing sugars which are converted to amino sugars by the amination step are accessible to this highly sensitive detection method. Another approach to increase sensitivity is the use of indirect detection methods.

Another problem arising in CE of carbohydrates is the neutral character of many of the interesting species. However, they can be converted in situ in anionic borate complexes in carriers containing borate (see Sect. 3.3.6.1). The use of borate buffers has the additional advantage that absorbance in the low UV range becomes increased significantly. One reason for this fact is the shift of the equilibrium between carbonyl or open-chain and annular form toward the carbonyl form. Figure 7.4. shows the separation of eight underivatized monosaccharides using borate buffer at elevated temperature (see Sect. 3.2.3).

Instead of using borate buffers, electrophoretical separation of sugars can be performed at pH values higher than 11, where the sugars exist in their deionized form. Carbohydrates possessing acidic or basic groups can be electrophoretically separated without the need of borate buffers.

Whereas mono- and oligosaccharides consisting of not more than about 15 monomer units can easiliy be separated in free solution, higher oligosaccharides possess unfavorable mass-to-charge ratios preventing their effective resolution in open tubes. Hence, homologous series of oligo- and polysaccharides involve the use of a gel matrix to allow their size based separation. Table 7.20 gives an overview of model separations of sugar mixtures by means of CE. In Table 7.21. some selected application are summarized.

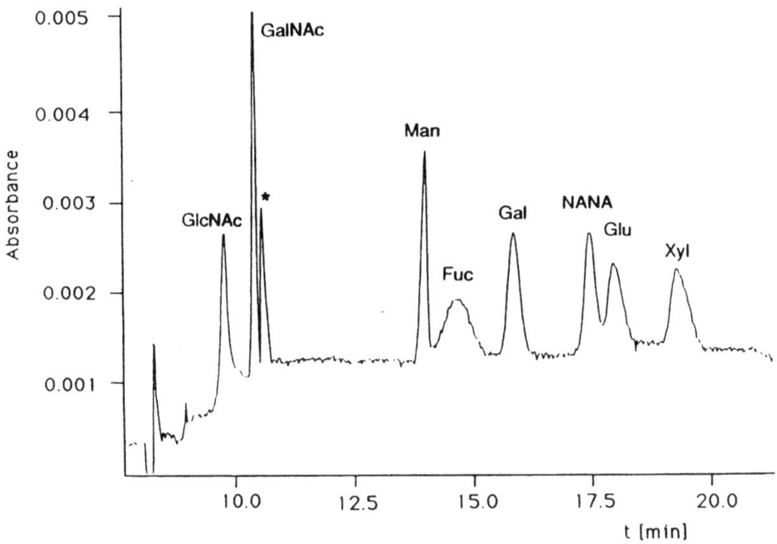

Fig. 7.4. CE of a mixture of 8 underivatized monosaccharides. Instrument: Spectra-Phoresis 1000, Spectra-Physics; experimental conditions: fused silica capillary, 94 cm x 75 µm i.d., hydrodynamic injection for 1 s, volatge 20 kV, UV detection at 195 nm, temperature 60 °C, electrolyte system 60 mM $Na_2B_4O_7$, pH 9.3; sample: *N*-acetylglucosamine (GlcNAc) and *N*-acetylgalactosamine (GalNAc), each 0.75 mM, mannose (Man), fucose (Fuc), galactose (Gal), glucose (Glu), and xylose (Xyl), each 7.5 mM, and sialic acid (NANA), 1.5 mM. (Reproduced from Ref. 57 with permission of the American Chemical Society)

Table 7.20. Model mixtures of carbohydrates. Experiments are performed with normal polarity (detection at the cathodic end) if not otherwise stated.

Appl. No.	Compounds	Conditions	Remarks	Ref.
246	11 reducing monosaccharides as N-2-pyridylglycamines	capillary, 65 (L_D = 50) x 50 µm 200 mM boric acid, adj. to pH 10.5 with KOH hydrodyn. inj., U = 15 kV UV at 240 nm; analysis time 30 min	precolumn derivatization of reducing monosaccharides with 2-aminopyridine; borate complexation; c = 10 - 100 mM	430
247	maltotetraose, -pentaose, hexaose and heptaose as N-2-pyridylglycamines	capillary, 80 (L_D = 50) cm x 50 µm 100 mM sodium phosphate buffer, pH 4.0 electrokin. inj., 15 s, 18 kV, U = 20 kV UV at 240 nm; analysis time 25 min	precolumn derivatization of malto-oligosaccharides with 2-aminopyridine; (tetrabutylammonium bromide (TBA) improves resolution)	431
248	sucrose, glucose and fructose	capillary, 90 cm x 18 µm 1 mM coumarin 343, adj. to pH 11.5 electrokin. inj., 1 s, 30 kV, U = 30 kV ind. fluorescence; analysis time 6.5 min	separation of sugars in their ionized form; c = 1 mM; also applicable to other mono- and disaccharides	164
249	8 monosaccharides	capillary, 94 (L_D = 87) cm x 75 µm 60 mM $Na_2B_4O_7$, pH 9.3 hydrodyn. inj., U = 20 kV, T = 60 °C UV at 195 nm; analysis time 20 min	detection of underivatized monosaccharides in the low UV range as their borate complexes	57
250	4 disaccharides	conditions as in Appl. No. 249, but: 50 mM $Na_2B_4O_7$, pH 9.3 analysis time 12 min	detection of underivatized disaccharides in the low UV range as their borate complexes	57
251	4 aldopentoses as their MPP derivatives	capillary, 78 (L_D = 63) x 50 µm 200 mM boric acid, adj. to pH 9.5 with KOH hydrodyn. inj., U = 15 kV UV at 245 nm; analysis time 22 min	precolumn derivatization of reducing monosaccharides with 3-methyl-1-phenyl-2-pyrazolin-5-one (MPP); borate complexation	432

Table 7.20. (continued)

Appl. No.	Compounds	Conditions	Remarks	Ref.
252	8 aldohexoses as their MPP derivatives	conditions as in Appl. No. 251 analysis time 25 min	see Appl. No. 251	432
253	isomaltose oligomers with 1 - 10 glucose units as their MPP derivatives	conditions as in Appl. No. 251 analysis time 20 min	see Appl. No. 251; also applicable to the separation of cellulose oligomers with 1 - 6 glucose units	432
254	6 model amino sugars derivatized with CBQCA	capillary, 80 (L_D = 50) cm x 50 µm 20 mM Na_2HPO_4 - 20 mM $Na_2B_4O_7$, pH 9.12, 50 mM SDS; hydrodyn. inj., U = 16 kV; LIF 442/550 ; time 25 min	MEKC with SDS; sensitive determination of derivatized amino sugars; mass LOD = 10^{-18} mol	124
255	glucosamin oligomers with 1 - 7 monomer units derivatized with CBQCA	capillary, 85 (L_D = 55) cm x 50 µm 10 mM Na_2HPO_4 - 30 mM $Na_2B_4O_7$, pH 9.4, 50 mM SDS, 30% MeOH; hydrodyn. inj., U = 20 kV; LIF 442/550; time 30 min	MEKC with SDS; sensitive determination of derivatized amino sugars	124
256	partially hydrolyzed dextrin 15 derivatized with CBQCA after reductive amination	capillary, 88 (L_D = 58) cm x 50 µm 10 mM Na_2HPO_4 - 10 mM $Na_2B_4O_7$, pH 9.4 hydrodyn. inj., U = 20 kV LIF 442/550; analysis time 15 min	sensitive determination of derivatized amino sugars; maltose oligomers with 1 to 10 glucose units are resolved	433
257	partially hydrolyzed dextrin 15 derivatized with CBQCA after reductive amination	capillary, 26 (L_D = 19) cm x 50 µm, coated with linear PAA; gel composition 10% T, 3% C, buffer 100 mM TRIS - 250 mM $Na_2B_4O_7$, pH 8.33, 7 M urea; hydrodyn. inj., E = 269 $V \cdot cm^{-1}$ LIF 442/550; analysis time 30 min	CGE in highly concentrated cross-linked PAA gel (Sect. 5.2.2.2.3); sensitive determination of derivatized amino sugars; maltose oligomers with 1 to 18 glucose units are resolved; reversed polarity	433

Table 7.20. (continued)

Appl. No.	Compounds	Conditions	Remarks	Ref.
258	enzymatically digested chondroitin sulfate derivatized with CBQCA after reductive amination	capillary, 32 (L_D = 23) cm x 50 µm, coated with linear PAA; gel composition 18% T, 3% C, buffer 100 mM TRIS - 250 mM $Na_2B_4O_7$, pH 8.48, 2 M EDTA; hydrodyn. inj., E = 178 $V \cdot cm^{-1}$; LIF 442/550; analysis time 60 min	see Appl. No. 257; separation of acidic oligosaccharides such as glucosaminoglycans; also applicable to enzymatically digested hyaluro-nic acid [24]	433 (24)
259	partially hydrolyzed poly-(galacturonic acid) derivatized with CBQCA after reductive amination	conditions as in Appl. No. 258, but: electrokin. inj., 25 s, 5 kV, E = 234 $V \cdot cm^{-1}$; analysis time 80 min	see Appl. No. 258; oligomers with 1 to 60 monomer units are resolved	125
260	ribose, lyxose, arabinose and xylose as their MPP derivatives	capillary, 53 cm x 50 µm 100 mM $M(OAc)_2$; hydrodyn. inj. U = 10 kV; UV at 245 nm analysis time 9 - 15 min, depending on M^{2+}	see Appl. No. 251; M = Ca^{2+}, Ba^{2+} or Sr^{2+}; cations absorb at the wall and reverse EOF (detection at anode!); Ba^{2+} gave best results	434
261	6 monosaccharides and 5 uronic acids	capillary, 122 (L_D = 100) cm x 50 µm 6 mM sorbate, pH 12.1; hydrodyn. inj. U = 28 kV; $UV_{ind.}$ at 256 nm, time 21 min	c = 0.95 - 2.66 mM	152
262	8 monosacharides, 3 uronic acids, 2 disaccharides and maltotriose derivatized with *p*-aminobenzoic acid	capillary, 72 (L_D = 50) cm x 50 µm 175 mM boric acid, adj. to pH 10.5 with 2 M NaOH; hydrodyn. inj., U = 25 kV, T = 30 °C; UV at 305 nm, time 18 min	precolumn derivatization of aldoses, ketoses and uronic acids with *p*-aminobenzoic acid; borate complexation	435
263	mixture of 14 aldoses, ketoses and uronic acids derivatized with ethyl *p*-aminobenzoate	capillary, 72 (L_D = 50) cm x 50 µm 150 mM boric acid, adj. to pH 10.0 with 2 M NaOH; hydrodyn. inj., U = 28 kV UV at 285 nm, analysis time 22 min	precolumn derivatization of aldoses, ketoses and uronic acids with ethyl *p*-aminobenzoate; borate complexation	436

Table 7.20. (continued)

Appl. No.	Compounds	Conditions	Remarks	Ref.
264	N-acetylchitooligosaccharides consisting of 1 - 6 monomers as their 2-aminopyridyl derivatives	capillary, 80 (L_D = 50) cm x 50 µm 100 mM phosphate buffer, 50 mM TBA bromide, pH 5.0; hydrodyn. inj., U = 18 kV UV at 240 nm, analysis time 30 min	precolumn derivatization of reducing sugars with 2-aminopyridine; tetrabutylammonium (TBA) bromide serves as ion pairing agent	437
265	N-acetylchitooligosaccharides consisting of 1 - 6 monomers as their 6-aminoquinolyl derivatives	see Appl. No. 264 analysis time 32 min	precolumn derivatization of reducing sugars with 6-aminoquinoline; TBA serves as ion pairing agent	437
266	xyloglucan oligosaccharides as their 2-aminopyridyl derivatives	see Appl. No. 264, but: pH 4.75, U = 20 kV analysis time 32 min	see Appl. No. 264; CZE analysis of branched oligosaccharides	437

Table 7.21. Selected application of CE in sugar analysis. Experiments are performed with normal polarity (detection at the cathodic end) if not otherwise stated.

Appl. No.	Compounds	Conditions	Remarks	Ref.
267	monosaccharides occuring in the polysaccharides of Flos matriccariae derivatized with *p*-aminobenzoate	capillary, 72 (L_D = 50) x 50 µm 200 mM boric acid, adj. to pH 10.5 with NaOH; hydrodyn. inj., U = 25 kV, T = 30 °C UV at 305 nm; analysis time 24 min	the same separation system can be used for monosaccharide analysis of other plant extracts	435
268	saccharose, glucose and fructose in in apple and orange juice	capillary, 112 (L_D = 90) x 50 µm 6 mM sorbic acid, adj. to pH 12.1 with NaOH hydrodyn. inj., U = 24 kV $UV_{ind.}$ at 256 nm; analysis time 15 min	sample diluited 1:25 with bidistilled water; separation of monosaccharides in their ionized form	436
269	α–, β- and γ-cyclodextrin	capillary, 50 (L_D = 14) x 75 µm 30 mM benzoic acid - 30 mM TRIS, pH 6.2 hydrodyn. inj., U = 20 kV $UV_{ind.}$ at 254 nm; analysis time 7 min	cyclodextrins form inclusion complexes with benzoic acid serving also as visualizing agent; c = 10^{-3} mM of each	299
270	β-cyclodextrin and piroxicam	capillary, 50 (L_D = 14) x 75 µm 10 mM benzoic acid - 10 mM TRIS, pH 6.2 hydrodyn. inj., U = 25 kV $UV_{ind.}$ at 254 nm; analysis time 9 min	simultaneous determination of β-CD and a drug in a tablet; tablet containing 191.2 mg β-CD and 20 mg piroxicam is dissolved in 50 mL water	299
271	oligosaccharides derived from ovalbumin as N-2-pyridylglycamins	coated capillary, Bio-Rad, 20 x 25 µm 100 mM phosphate buffer, pH 2.5, electrokin. inj., 30 s, 8 kV, U = 8 kV, UV at 240 nm; analysis time 20 min	linear maltose oligomers separated on the basis of polymerisation degree: number of glucose units increases whereas charge remains constant	438

Table 7.21. (continued)

Appl. No.	Compounds	Conditions	Remarks	Ref.
272	see Appl. No. 271	capillary, 95 cm x 50 μm 200 mM boric acid, adj. to pH 10.5 with NaOH; electrokin. inj., 30 s, 8 kV, U = 20 kV fluorescence 395/316; analysis time 24 min	branched oligosaccharides separated on the basis of the structures of outer monosaccharide residues due to borate complexation; complementary mode to Appl. No. 272	438
273	α_1-acid glycoprotein oligosaccharides as N-2-pyridylglycamins	capillary, 80 (L_D = 45) cm x 50 μm, modified with a hydrophilic inert coating [206] 100 mM sodium phosphate buffer, pH 5.0, 50 mM TBA bromide; electrokin. inj., 2 s, 18 kV, U = 18 kV; UV at 240 nm; time 40 min	tetrabutylammonium (TBA) serves as ion pairing agent; method can be used to compare patterns of α_1-acid glycoprotein oligosaccharides of different species	439
274	ganglioside micelles including G_{M1}, G_{D1b} and G_{T1b}	capillary, 72 (L_D = 50) cm x 50 μm 2.5 mM potassium phosphate buffer, pH 7.4 hydrodyn. inj., U = 30 kV UV at 195 nm; analysis time 10 min	three peaks are only observed shortly after mixing the ganglioides; than mixed micelle formation occurs resulting in a single peak (not at 0 °C!)	440
275	chondroitin sulfate and dermatan sulphate derived disaccharides	capillary, 68 cm x 75 μm 10 mM $Na_2B_4O_7$ - 50 mM boric acid, pH 8.8 hydrodyn. inj., U = 10 kV UV at 232 nm; analysis time 40 min	separation of non-sulfated and sulfated disaccharides having net charges from -1 to -4 derived from glucosaminoglycans (GAGs)	441
276	heparin and heparan sulphate derived disaccharides	capillary, 68 cm x 75 μm 10 mM $Na_2B_4O_7$ - 50 mM boric acid, pH 8.8 hydrodyn. inj., U = 10 kV UV at 232 nm; analysis time 40 min	see Appl. No. 275; mass LOD = 50 fmol	442

Table 7.21. (continued)

Appl. No.	Compounds	Conditions	Remarks	Ref.
277	9 heparin disaccharides	capillary, 57 (L_D = 50) cm x 75 µm 200 mM NaH_2PO_4, adj. to pH 3.0 with H_3PO_4 hydrodyn. inj., U = 7.5 kV UV at 214 nm; analysis time 100 min	see Appl. No. 275; c = 0.1 mg·mL^{-1} of each disaccharide; EOF mobility is smaller than electrophoretic mobilities; reversed polarity with detection at the anode	443
278	colominic acid hydrolysate	conditions see Appl. No. 277, but: pH 4.0 analysis time 55 min		443
279	unsaturated disaccharides derived from glucosaminoglycans as their MPP derivatives (see Appl. No. 251)	capillary, 51 cm x 50 µm 100 mM boric acid, adj. to pH 9.0 with KOH U = 25 kV, T = 30 °C UV at 214 nm; analysis time 10 min	see Appl. No. 275; applicable to the analysis of GAG fractions of a urine sample digestion with chodroitinase ABC	444
280	11 glucosinolates	capillary, 72 (L_D = 50) x 50 µm 18 mM $Na_2B_4O_7$ - 30 mM NaH_2PO_4, adj. to pH 7.0, 50 mM CTAB; hydrodyn. inj.; U = 20 kV, T = 30 °C; UV at 235 nm; time 15 min	MEKC with CTAB; reversed polarity with detection at the anode	445
281	8 inositol phosphates	capillary, 55 x 50 µm; 2.5 mM K_2CrO_4, 5 mM boric acid, pH 7.3, 0.5 mM TTAB electrokin. inj., 2 s, 5 kV, U = 20 kV $UV_{ind.}$ at 270 nm; analysis time 7 min	tetradecyltrimethylammonium bromide (TTAB) is used as EOF modifier; reversed polarity with detection at the anode	446
282	7 inositol phosphates	conditions as in Appl. No. 281, but: 5 mM potassium hydrogenphthalate, 2 mM $Na_2B_4O_7$, pH 5.9, 0.5 mM TTAB	see Appl. No. 281	446

7.8 Nucleotides, Oligonucleotides and Nucleic Acids

A nucleotide is composed of three parts: a purin or pyrimidine base, a sugar and one or more phosphate groups. The sugar can either be a ribose or a deoxyribose. The sugar and the base build together a nucleoside, the nucleotide being the phosphate ester of the nucleoside. Natural occurring nucleotides are phosphorylated in position 5 of the sugar molecule. Since nucleotides are composed of one or more phosphate residues behaving like strong acids and a weak base, they are negatively charged over a wide pH range and can be easily analyzed by conventional CZE or MEKC (see Table 7.22.).

Table 7.22. Nucleotides. Experiments are performed with normal polarity (detection at the cathodic end) if not otherwise stated.

Appl. No.	Compounds	Conditions	Ref.
283	12 ribonucleosides (the mono-, di- and triphosphate nucleosides of cytidine, uridine, guanosine and adenosine)	capillary, 45 (L_D = 25) x 50 µm 10 mM TRIS - 10 mM Na_2HPO_4, pH 7.05, 100 mM dodecyltrimethylammonium bromide (DTAB) for MEKC; electrokin. inj., 2 s, 5 kV, U = 18 kV (reversed polarity); UV at 254 nm; analysis time 3 min	447
284	ATP, dATP, CTP, dCTP, GTP, dGTP, dTTB, UTP and ITP	capillary, 69.5 (L_D = 62.8) x 75 µm, coa-ted with linear PAA; 50 mM phosphate buffer, pH 2.7, 2 mM EDTA hydrodyn. inj., U = 20 kV UV at 200 nm; analysis time 17 min	448
285	4 deoxynucleoside-5'-monophosphates as their fluorescein-ethylenediamine conjugates	capillary, 95 (L_D = 50) cm x 75 µm 10 mM TRIS - 10 mM boric acid, adj. to pH 10.4 with 2 M NaOH, 10% CH_3CN hydrodyn. inj., U = 15 kV fluorescence ; analysis time 17 min	449
286	nucleotides from human peripheral blood lymphozytes (ATP, GTP, GDP, AMP, ADP, UMP, CDP and UDP)	capillary, 57 (L_D = 50) x 75 µm 140 mM borate buffer, pH 9.4 hydrodyn. inj., U = 16 kV UV at 254 nm; analysis time 25 min	450
287	cyclic nucleotides (c-AMP, c-GMP and c-IMP)	capillary, 100 (L_D = 55) x 75 µm 50 mM $Na_2B_4O_7$, adj. to pH 8.3 with 1 N HCl; electrokin. inj., 15 s, 10 kV, U = 22 kV, UV at 210 nm; analysis time 23 min	451

Table 7.22. (continued)

Appl. No.	Compounds	Conditions	Ref.
288	5 nucleosides and 4 bases	capillary, 80 (L_D = 60) x 50 µm 25 mM $Na_2B_4O_7$ - 50 mM NaH_2PO_4, pH 7.0, 100 mM SDS U = 14 kV, UV at 210 nm; analysis time 35 min	248 452
289	uracil, cytosine, thymine and adenine	conditions as in Appl. No. 288, but: capillary, 65 (L_D = 50) x 50 µm	248

The polycondensation products of nucleotides are oligonucleotides and nucleic acids. Synthetic oligonucleotides can be distinguished in homooligonucleotides consisting of identical monomer units all having the same base and heterooligonucleotides with monomer units of altering base composition. In nature, two different types of heteronucleic acids exist: ribonucleic acid (RNA) with ribose and deoxyribonucleic acid (DNA) with deoxyribose as sugar unit. Whereas RNA consists of only one strain (single-stranded), DNA is composed of two complementary polynucleotide strains (double-stranded). There are 4 different bases occuring in a DNA chain, namely adenine, cytosine, guanine and thymidine. In RNA, thymidine is exchanged against uracil.

Since each monomer residue contributes to the polymer chain with a strongly acidic free hydroxyl group at the phosphate and one weak base, polynucleotides behave like fully dissociated polyanions at neutral and weakly alkaline pH values. Consequently, nucleic acids of different chain length possess very similar or even identical charge-to-mass ratios. Hence, they involve the use of a molecular sieving medium for high resolution separations. Besides a few applications dealing with MEKC (Table 7.23.) most separations are performed in chemical or physical gels (see Sect. 5.2).

Most gel separations are carried out at pH values of 8-9. The migration properties of oligonucleotides are dependent on their base composition. In denaturing gels, the homooligomer mixtures $pd(A)_{12\text{-}18}$, $pd(C)_{12\text{-}18}$, $pd(G)_{12\text{-}18}$ and $pd(T)_{12\text{-}18}$ have parallel plots of relative migration time as a function of the chain length with the migration order A > C > G > T. For the influence of temperature on the separation see Sect. 3.2.3.

The mobility of DNA fragments larger than a few thousands base pairs in length is strongly field dependent in the way, that larger fragments migrate much faster as expected. This leads to a comigration of larger fragments with smaller ones, if the field strength is too high. This effect is even more pronounced in gels of high %T. Thus, manipulation of the applied field and proper selection of gel composition are important parameters in the optimization of nucleic acid separations.

For further applications of CE in DNA sequencing see Refs. 129, 222, 464 - 468 and further recommended reading.

Table 7.23. Oligonucleotides and nucleic acids. Experiments are performed with reversed polarity (detection at the anodic end) if not otherwise stated.

Appl. No.	Compounds	Conditions	Remarks	Ref.
290	12 oligothymidines with 2 - 4, 6 10 and 13 - 18 monomers	capillary, 65 (L_D = 45) x 50 µm 5 mM TRIS - 5 mM Na_2HPO_4, pH 7.0, 7M urea, 50 mM SDS, 0.3 mM Cu^{2+}; U = 20 kV UV at 260 nm; analysis time 38 min	MEKC with SDS in the presence of metal cations; normal polarity with detection at the cathode	248
291	6 oligonucleotides, each with eight bases	conditions as in Appl. No. 290, but: 20 mM TRIS, 3 mM Zn^{2+}; analysis time 22 min	see Appl. No. 290; normal polarity with detection at the cathode	248
292	DRIgest (λ DNA-Hind III/ΦX174 DNA-Hae III)	capillary, 30 (L_D = 15) x 75 µm 100 mM TRIS - borate, pH 8.1, 7 M urea, 0.1% SDS, 2.5 mM EDTA; hydrodyn. inj., U = 15 kV; UV at 260 nm; time 10 min	MEKC with SDS; sample heated for 20 min at 60 °C and injected hot; 20 peaks are observed; normal polarity with detection at the cathode	453
291	DRIgest II (λ DNA-Hind III/ΦX174 DNA-Hinc II)	see Appl. No. 292, but: capillary, 50 (L_D = 25) x 75 µm; analysis time 25 min	MEKC with SDS; as expected a total of 21 peaks are observed; normal polarity with detec-tion at the cathode	453
294	DNA size standard-low range consisting of 9 DNA fragments ranging from 88 - 1746 bases	coated capillary, Bio-Rad, 50 x 50 µm 89 mM TRIS - borate, pH 8.0, 7 M urea, 0.1% SDS, 0.5% HPMC (4000 cP); U = 8 kV UV at 260 nm; analysis time 25 min	c = 0.2 $\mu g \cdot \mu L^{-1}$ total; hydroxypropylmethylcellulose (HPMC) added as molecular sieving agent; coating procedure see Sect. 5.1.2.1	86 454
295	DNA 123 base pair ladder, double-stranded	conditions as in Appl. No. 294, but: 0.5% MC (4000 cP); analysis time 30 min	methylcellulose (MC) added as molecular sieving agent	454
296	10 oligocytidines with 1 - 10 monomers	capillary, 50 (L_D = 35) x 50 µm, coated with linear PAA; 200 mM histidine - MES, pH 6.0, 5 mM spermine · 4 HCl; U = 15 kV, UV at 254 nm; analysis time 24 min	coating procedure see Sect. 5.1.2.1; spermine associates with native DNA to neutralize its charge partially	455

Table 7.23. (continued)

Appl. No.	Compounds	Conditions	Remarks	Ref.
297	chain-termination sequencing reaction products with different Templates (TEM80.1, TEM80.2, TEM 80.3)	capillary, 68 (L_D = 54) x 75 µm; gel composition 3% T, 5% C in 100 mM TRIS - borate, pH 7.6, 7 mM urea, 2.5 mM EDTA; electrokin. inj. 30 s, 10 kV, E = 300 $V \cdot cm^{-1}$; LIF; analysis time 32 min	CGE in crosslinked PAA gel according to Karger (see Sect. 5.2.2.2.1); primer molecules are tagged at the 5' end with the fluorescent dye JOE	456
298	chain-termination sequencing reaction products with single-stranded M13mp18 as template	capillary, 65 (L_D = 50) x 75 µm; gel composition 3% T, 5% C in 100 mM TRIS - borate, pH 8, 7 mM urea, 2.5 mM EDTA; electrokin. inj. 15 s, 10 kV, E=350 $V \cdot cm^{-1}$; LIF; time 70 min	see Appl. No. 297; the fragments differ in length from 18 to > 330 bases; single base resolution is achieved	456
299	oligodeoxyadenylic acids $pd(A)_{40-60}$	capillary, 27 x 75 µm; gel composition 7.5% T, 3.3% C in 100 mM TRIS - 250 mM boric acid, pH 8.3, 7 mM urea; electrokin. inj. 10 s, E = 400 $V \cdot cm^{-1}$; UV at 260 nm; analysis time 7.5 min	see Appl. No. 297; sample is heated to 60 °C for 20 min before injection	198
300	oligodeoxyadenylic acids $pd(A)_{40-60}$	capillary, 60 (L_D = 40) x 75 µm; gel composition 4% T, 3.3% C in 100 mM TRIS - 100 mM boric acid, pH 8.6, 7 mM urea, 2 mM EDTA; electrokin. inj. 3 s, 5 kV, E = 250 $V \cdot cm^{-1}$; UV at 260 nm; analysis time 34 min	see Appl. No. 297; c = 0. 2 $\mu g \cdot \mu L^{-1}$; separation of phosphorylated and dephosphorylated species	213
301	crude and purified samples of specific heterooligonucleotides	conditions as in Appl. No. 300	see Appl. No. 297; for purity control of synthetic heterooligonucleotides (here a 60mer, 48mer or 37mer)	213
302	oligothymidylic acids $pd(T)_{20-160}$	capillary, 75 (L_D = 50) x 100 µm; gel composition 2.5% T, 3.3% C in 89 mM TRIS - 69 mM boric acid, pH 8.6, 7 mM urea, 2 mM EDTA; electrokin. inj. 10 s, 10 kV, E = 400 $V \cdot cm^{-1}$; UV at 260 nm; analysis time 28 min	see Appl. No. 297; linear relationship between base number versus migration time	213

Table 7.23. (continued)

Appl. No.	Compounds	Conditions	Remarks	Ref.
303	oligodeoxythymidylic acids $pd(T)_{12-30}$	capillary, 40 (L_D = 20) x 75 µm; gel composition 7.5% T, 3.3% C in 50 mM TRIS - 50 mM boric acid, 7 mM urea; electrokin. inj. 1 s, 4 kV, E = 375 $V \cdot cm^{-1}$; UV at 260 nm; analysis time 12 min	see Appl. No. 297; gel contains 3% PEG; also applicable to purity control of synthetic heterooligonucleotides	457
304	polyuridine 5'-phosphate consisting of 431 bases	conditions see Fig. 5.8.	see Appl. No. 297; but: layer of non-crosslinked linear PAA placed between bifunct. agent and gel (Sect. 5.1.2.1)	215
305	11 DNA restriction fragments (1353, 1078, 872, 603, 310, 281/271, 234, 194, 118 and 72 base pairs in length)	capillary, 50 (L_D = 35) cm x 50 µm 89 mM TRIS - 89 mM boric acid, 5 mM EDTA, 0.25% HEC; hydrodyn. inj., E = 301.3$V \cdot cm^{-1}$, T = 30 °C; UV at 260 nm; anal. time 17 min	hydroxyethylcellulose (HEC) is added as molecular sieving agent	209
306	M13mp18 reaction fragments	capillary, 31 cm x 50 µm; gel composition 6% T, 5% C, 7 M urea, 20% (v/v) formamide; electrokin. inj., E = 150 $V \cdot cm^{-1}$, LIF 543.5/590; time 35 min	see Appl. No. 297; dye-labeled primer: ABI 21M13 TAMRA, 1.6 pM); sequencing rate 70 bases per hour	458
307	polydeoxyadenylic acids ranging from ca. 40 - 450 bases	capillary, 100 (L_D = 70) x 150 µm; gel composition 3% T, 3.3% C in 90 mM TRIS - borate, 2.5 mM EDTA, 7 mM urea; electrokin. inj. 3 s, 6 kV, U = 15 kV; UV at 260 nm; analysis time 130 min	see Appl. No. 297	214
308	M13T track sequencing product	capillary, 85 (L_D = 50) x 150 µm; gel composition 3% T, 3.3% C in 90 mM TRIS - borate, 2.5 mM EDTA, 7 mM urea; electrokin. inj. 10 s, 12 kV, U = 12 kV, T = 60 °C, UV at 260 nm; time 180 min	see Appl. No. 297; run at elevated temperature	214

Table 7.23. (continued)

Appl. No.	Compounds	Conditions	Remarks	Ref.
309	oligodeoxyadenylic acids $pd(A)_{40-80}$	UV transparent capillary, 23.3 (L_D = 13.7) x 53 µm; gel composition 7.5% T, 3.3% C in 100 mM TRIS - 100 mM boric acid, 2 mM EDTA, 7 mM urea; electrokin. inj. 5 s, 1.9 kV, E = 236 $V \cdot cm^{-1}$; UV at 260 nm; time 18 min	CGE in crosslinked PAA gel prepared by photopolymerization (see Sect. 5.2.2.2.2)	219
310	Hae III restriction digest of ϕX-174 DNA (see Appl. No. 305)	OV-17 coated capillary, 57 (L_D = 50) x 100 µm 89 mM TRIS - 89 mM boric acid, pH 8.5, 2 mM EDTA, 0.5% HPMC (4000 cP); electrokin. inj. 5 s, 2 kV, E = 236 $V \cdot cm^{-1}$; UV at 260 nm; analysis time 27 min	HPMC is added as molecular sieving agent; OV-17 (polysiloxane) coated capillary is purchased from J&W Scientific; addition of ethidium bromide allows separation of 271/281	459
311	13 DNA size standards	capillary, 40 (L_D = 30) x 75 µm; gel composition 3% T, 0.5% C in 100 mM TRIS - 100 mM boric acid, pH 8.3, 2 mM EDTA; electro-kin. inj. 0.5 s, 10 kV, E = 250 $V \cdot cm^{-1}$, UV at 260 nm; time 18 min	see Appl. No. 297; fragments differ in length from 72 to 7253 base pairs; low %C gel	216
312	Hae III restriction digest of ϕX-174 DNA (see Appl. No. 305)	capillary, 20 (L_D = 10) x 75 µm; gel composition 12% T in 100 mM TRIS - 100 mM boric acid, pH 8.3, 2 mM EDTA; electrokin. inj. 3 s, 150 $V \cdot cm^{-1}$, E = 300 $V \cdot cm^{-1}$, UV at 260 nm; time 12 min	CGE in non-crosslinked linear PAA gel; capillary is pretreated with bifunctional agent: coating of the wall with the linear PAA occurs automatically, when the gel is filled into the cap.	216
313	$pd(A)_{20}$ and $pd(A)_{40-60}$	capillary, 60 (L_D = 45) x 75 µm; gel composition 9% T in 100 mM TRIS - 100 mM boric acid, pH 8.3, 2 mM EDTA, 7 M urea; electrokin. inj. 2 s, 10 kV, E = 308 $V \cdot cm^{-1}$, UV at 260 nm; time 34 min	see Appl. No. 312	216
314	$pd(A)_{15-20}$ and $pd(A)_{40-60}$	capillary, 50 x 100 µm; gel composition 10% T in 100 mM TRIS - 100 mM boric acid, pH 8.0; electrokin. inj. 10 s, 8 kV, U = 10 kV, UV at 254 nm; analysis time 34 min	see Appl. No. 312	228

Table 7.23. (continued)

Appl. No.	Compounds	Conditions	Remarks	Ref.
315	pBR322 DNA Hae III restriction fragments ranging from 51 to 587 base pairs	capillary, 65 (L_D = 45) x 75 μm, coated with linear PAA; 100 mM TRIS - 100 mM TRICINE, pH 8.1, 2% galactomannan, 7 M urea; electro-kin. inj. 10 s, 5 kV, U = 19.5 kV, UV at 260 nm; analysis time 32 min	galactomannan (Synergel, from Diversified BioTech, Newton, MA) is added as molecular sieving agent	460
316	pBR322 DNA Hae III restriction fragments ranging from 26 to 622 base pairs	capillary, 47 (L_D = 40) x 100 μm, %T not given; 100 mM TRIS - 100 mM boric acid, pH 8.35, 2 mM EDTA; hydrodyn. inj., E = 100 (0-40 min), 200 (40-70 min) and 200 $V \cdot cm^{-1}$ (70-100 min) UV at 254 nm; analysis time 32 min	CGE in replacable non-crosslinked linear PAA using an uncoated capillary enhanced separation of a wide size range of DNA fragments by using an increasing stepwise gradient field	461
317	DNA size standards ranging from 8 - 2176 base pairs	capillary, 72 (L_D = 50) x 50 μm, coated with linear PAA; 100 mM TRIS - 100 mM boric acid, pH 8.7, 0.1 mM EDTA, 25 mM NaCl, 0.5% HEC (η of 2% aqueous solution 0.3 Pas), 1.27 μM ethidium bromide; hydrodyn. inj., U = 15 kV, T = 35 °C, UV at 260 nm; analysis time 24 min	hydroxyethylcellulose (HEC) is added as molecular sieving agent; NaCl and ethidium bromide enhance resolution of DNA fragments ($N = 2 - 4 \cdot 10^5$); normal polarity with detection at the cathode	462
318	Hae III restriction digest of φX-174 DNA (see Appl. No. 305)	capillary, 72 (L_D = 50) x 150 μm; gel 1.7% agarose in 89 mM TRIS - 89 mM boric acid, 2.5 mM EDTA; electrokin. inj. 16 s, 1 kV, E = 185 $V \cdot cm^{-1}$, T = 40 °C, UV at 260 nm; time 12 min	CGE in liquified agarose above its gelling temperature; for gel preparation see Sect. 5.2.3.1	225
319	Hae III restriction digest of φX-174 DNA (see Appl. No. 305)	capillary, 72 (L_D) x 75 μm; gel 2% agarose in 10 mM Na_2HPO_4 - 10 mM NaH_2PO_4; electrokin. inj. 5 s, 20 kV, E = 100 $V \cdot cm^{-1}$, T = 25 °C, UV at 254 nm; analysis time 40 min	CGE in agarose; for gel preparation see Sect. 5.2.3.1	224

7.9 Chiral Molecules

Principally, separation of optical isomers can be performed by two different strategies called direct and indirect chiral separation. Direct chiral separation involves the formation of diastereomeric molecule complexes of the two enantiomers with a chiral selector. For indirect enantio-separation both enantiomers undergo a derivatization reaction with a chiral selector to form diastereomers with disparate physico-chemical behavior. After the derivatization step the diastereomers can be separated by using non-chiral chromatographic or electrophoretic systems. However, careful validation of the reaction is necessary to avoid racemization and artifacts. This section will focus on techniques dealing with direct chiral separation in CE.

In general, enantioseparation of optical isomers is accomplished by the complex formation with a chiral selector resulting in diastereomeric complexes with different formation constants. According to the "three-point-interaction" rule of Dalglish [469] chiral recognition depends on a minimum of three simultaneous interactions between selector and selectands. At least one of these interactions has to be stereo-selective in order to discriminate between the enantiomers.

In capillary electrophoresis the chiral selector is usually added to the buffer solution. The complexation reaction between selectand and selector can be compared with the partition of a solute between a mobile and a pseudophase. Optimal separation conditions are found for the pseudophase moving into the opposite direction to the selectands. Alternatively, the chiral selector is immobilized in a gel matrix or is a constituent of a micellar system in MEKC. According to the principle of operation, three different separation modes are known:

> Host-guest complexation (see also Sect. 3.3.6.3)
> Ligand exchange complexation (see also Sect. 3.3.6.4)
> Solubilization by optically active micelles (see also Sect. 5.3)

Host-guest complexation

Complexes in which an analyte (guest molecule) is spatially enclosed by a ligand (host molecule) are called host-guest complexes or inclusion complexes. In CE two classes of compounds are used for host-guest complexation with enantiomers, (i) cyclodextrins (see Fig. 3.23.) and their derivatives and (ii) a chiral crown ether (see Fig. 3.25.).

Enantiorecognition of analytes with cyclodextrins is based on the inclusion of an aromatic or alkyl functionality into the cyclodextrin cavity and additional interactions between the secondary hydroxyl groups of the cone opening and substituents of the guest molecule. Crown ethers form stable complexes with potassium, ammonium and primary alkylamine cations. Ammonium or alkylamines form host-guest complexes by three $^{+}$NH···O hydrogen bonds in a tripod arrangement. Two different mechanisms are found to be responsible for chilic acid act like chiral barriers dividing the space availabe for the substituents of the chiral carbon atom adjecant to the amine function into two cavities. According to size and spatial arrangement of theses substituents diastereomeric complexes with different formation constants are formed. A second mechanism is given by the carboxylic acids which may show electrostatic interactions with polar guest substituents.

The separation of five dansylated amino acids with γ-cyclodextrin and 4 amino acids using a chiral crown ether is depicted in Fig. 7.5.

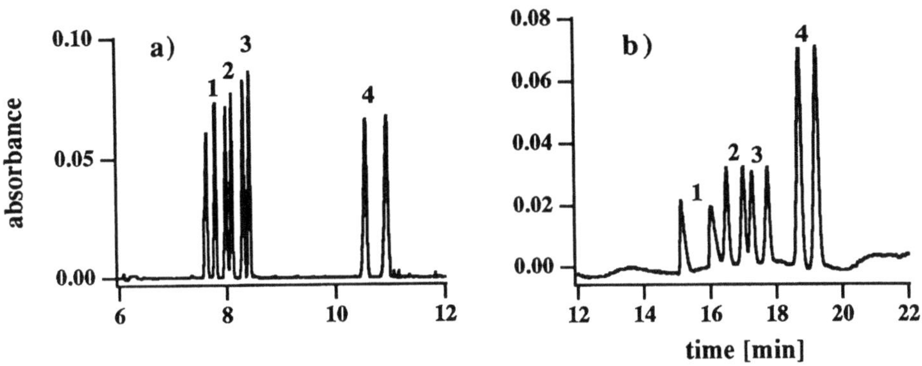

Fig. 7.5. (**a**) Chiral separation of DNS-D,L-amino acids by using γ-cyclodextrin. Instrument: Beckman P/ACE 2000; experimental conditions: fused silica capillary, 57 cm x 75 μm i.d., hydrodynamic injection for 1 s, voltage 15 kV for 8.5 min than 25 kV, temperature 25 °C, UV detection at 214 nm, electrolyte system 50 mM sodium tetraborate/10 mM γ-cyclodextrin, pH 9.0. Elution order: (**1**) leucine, (**2**) methionine, (**3**) threonine, (**4**) glutamic acid. (**b**) Chiral separation of D,L-amino acids by using 18-crown-6 tetracarboxylic acid ($18C6H_4$). Experimental conditions as in (**a**) but 15 kV and electrolyte system 10 mM TRIS/10 mM $18C6H_4$/citric acid, pH 2.2. Elution order: (**1**) (±)-3-amino-3-phenylpropionic acid, (**2**) D,L-tryptophan, (**3**) D,L-phenylalanine, (**4**) D,L-Dopa

Ligand exchange complexation

Enantioseparation by ligand exchange electrophoresis is based on multicomponent chelate complexes consisting of a central ion (i.e. Cu^{2+}, Ni^{2+}) and at least two chiral bifunctional ligands (chelators). The chelator concentration is chosen so, that all coordination positions of the central ion are saturated. The analyte enantiomers replace one chelator by forming a ternary complex (see Sect. 3.3.6.4).

Solubilization by optically active micelles

Chiral separation by micellar electrokinetic chromatography (MEKC) involves the use of surfactants above the critical micelle concentration (see Sect. 5.3). For enantiorecognition a chiral center has to be incorporated into the micelle. Several techniques have been described for this purpose. Mixed micelles composed of dodecyl-L-alanine (SDAla) and dodecylsulfate (SDS) together with copper ions or SDVal and SDS successfully separate amino acids. Another approach involves the use of chiral bile salts as surfactant. Particular useful for this purpose is sodium taurodeoxycholate.

Reviews of chiral separation by capillary electrophoresis were presented by Snopek et al. [470] and Kuhn and Hoffstetter-Kuhn [76]. Applications of chiral separations for pharmaceutical drug substances are summarized in Table 7.24., for underivatized and derivatized amino acids in Table 7.25. and for various other chiral compounds in Table 7.26.

7.10 Complex Samples

Because of its high peak capacity and efficiency, CE has a high potential for the analysis of complex samples such as fermentation broths, biological fluids and food samples. Unlike in the analysis of single compounds where a complex sample matrix can interfere with the separation, the electropherogram of a complex sample mixture serves as a fingerprint of the material and gives information about the production process or the quality of the product. With its simple automated instrumentation and short analysis times, CE is very well suited to be used as an on-line technique to monitor production processes. Additionally, sample preparation can normally be reduced to centrifugation and dilution. So far, their exist only a few applications of CE in complex sample analysis. Some of them are summarized in Table 7.27.

Table 7.24. Chiral separations of drugs. Experiments are performed with normal polarity (detection at the cathodic end) if not otherwise stated.

Appl. No.	Compounds	Conditions	Remarks	Ref.
320	adrenaline, epinephrine	coated capillary (Bio-Rad), 20 cm x 25 μm; 100 mM phosphate, pH 2.5, 20 mM dimethyl-β-CD, electrokin. inj., 4 s, 7 kV, U = 8 kV UV absorbance wavelength not given analysis time 6 min	host-guest complexation with dimethyl-β-CD; for quantitative analysis a correction factor is introduced to take into account UV absorbance shift caused by complexation with CD.	471
321	sympathomimetic drugs: ephedrine, norephedrine, epinephrine, norepinephrine and isoproterenol	coated capillary (Bio-Rad), 20 cm x 25 μm; 10 mM TRIS-H_3PO_4, pH 2.4, 18 mM dimethyl-β-CD, electrokin., 6 s, 6 kV, U = 8 kV UV absorbance wavelength not given; analysis time 5 min	host-guest complexation with dimethyl-β-CD; effect of CD concentration is studied in the range 4-36 mM showing that selectivity improves with increasing CD concentration	472
322	epinephrine	capillary, 50 (L_D = 45) cm x 75 μm; 10 mM TRIS, adj. to pH 2.4 with H_3PO_4, 18 mM methyl-β-CD; hydrodyn. inj., U = 15 kV, UV at 206 nm, analysis time 13 min	host-guest complexation with methyl-β-CD; validation of the method is presented; analysis by internal standard method	473
323	barbiturates: thiopental, pentobarbital, phenobarbital and barbital	capillary, 65 (L_D = 50) x 50 μm; 20 mM phosphate - borate, pH 9.0, 50 mM SDS and 30 mM γ-CD; hydrodyn. inj., U = 20 kV; UV at 220 nm; analysis time 20 min	MEKC with SDS in the presence of γ-cyclodextrin; effect of different CDs on separation factor is studied; in several experiments 4 M urea is added to the buffer	474
324	ergot alkaloids: isolysergic acid, terguride, meluol, nicergoline and lisuride	coated capillary (Bio-Rad) 20 cm x 25 μm resp. 50 cm x 50 μm; 100 mM phosphate, pH 2.5, 60 mM γ-CD, electrokin. inj., 7 s, 7 kV, U = 8 kV, UV at 206 nm; analysis time 10 min	host-guest complexation with γ-CD; effect of γ-CD concentration in the range 0-80 mM on migration time is studied; dimethyl-β-CD is also useful for this purpose; c = 5 μg/mL of each	475

Table 7.24. (continued)

Appl. No.	Compounds	Conditions	Remarks	Ref.
325	terbutaline and propranolol	coated capillary, (Bio-Rad) 20 cm x 25 µm; 100nM NaH_2PO_4, adj. to pH 2.5 with H_3POP_4, 5 mM dimethyl-β-CD, electrokin. inj., 10 s, 8 kV, U = 9 kV; UV at 206 nm; time 8 min	host-guest complexation with dimethyl-β-CD; 15 µM β-cyclodextrin is also useful for this purpose, propranolol is not baseline separated	476
326	clenbuterol and picumeterol	capillary, 57 (L_D = 50) cm x 75 µm 100 mM citric acid - 200 mM Na_2HPO_4, pH 4.0, 16 mM β-CD; hydrodyn. inj. 4 s, U = 13 kV; UV 214 nm, analysis time 35 min	host-guest complexation with β-CD; effect of β-cyclodextrin concentration, buffer I and T is studied. High ionic strength was essential in achieving chiral resolution; c = 0.1 mg/mL of each	477
327	verapamil, fluoxetine, bupivacaine, mepivacaine, carvedilol and pindolol	capillary, 57 (L_D = 50) cm x 75 µm; 18 mM TRIS, adj. to pH 2.8-3.0 with H_3POP_4, 10 mM trimethyl-β-CD or dimethyl-β-CD, 0.1% (w/w) methylhydroxyethylcellulose; hydrodyn. inj., U = 20 kV UV at 220 nm; analysis time 25 min	host-guest complexation with β-CDs; the influence of cationic detergents such as hexadecyltrimethylammonium bromide and cetylpyridinium chloride on resolution is studied;	478
328	chloramphenicol and related drugs	capillary, 50 cm x 49 µm; 20 mM TRIS - citric acid, pH 3.5, 10 mM dimethyl-β-CD, 0.1% (w/w) methylhydroxyethylcellulose; hydrodyn. inj., U = 18 kV UV at 254 nm; analysis time 30 min	host-guest complexation with β-CDs; CZE and ITP are compared with both having distinct advantages, CZE preferable in terms of resolution, ITP better suited for minute concentrations in a large excess of other components	479

Table 7.24. (continued)

Appl. No.	Compounds	Conditions	Remarks	Ref.
329	dopamine agonist: quinagolide	capillary, 57 (L_D = 50) cm x 75 µm; 50 mM phosphate, pH 2.5, 30 mM β-CD; hydrodyn. inj. 1 s, U = 15 kV UV at 214 nm; analysis time 40 min	influence of CD concentration, T and phosphate concentration on resolution is studied, the complex formation constant for both enantiomers is calculated	480
330	trimetoquinol, diltiazem and related compounds	capillary, 65 (L_D = 50) cm x 50 µm; 20 mM phosphate-borate, pH 7.0 or 9.0, 50 mM sodium taurodeoxycholate; U = 25 kV; UV at 210 nm; analysis time 20 min	MEKC with bile salts; various bile salts are compared, sodium taurodeoxycholate has evolved to be best suited	481 482 483
331	cicletanine	capillary, 57 (L_D = 50) cm x 75 µm; 0.1 M sodium borate, pH 8.6, 50 mM SDS, 50 mM β-cyclodextrin; hydrodyn. inj. 3 s; U = 10 kV; UV at 214 nm; analysis time 18 min	MEKC with SDS in the presence of β-cyclodextrin	484
332	non-steroidal anti-inflammatory drugs (NSAIDs): flurbiprofen, ibuprofen, ketoprofen	capillary, 50 - 90 cm x 75 µm; 10 mM sodium phosphate, pH 7.1, 2.5 - 10% of linear oligosaccarides; hydrodyn. inj., U = 30 kV; UV at 214 nm;	CZE with linear oligosaccharides (maltodextrins and corn syrups) up to 10 % as chiral selector.	485
333	aromatic primary amines (Dopa, ephedrine, norepinephrine, norephedrine)	capillary, 57 (L_D = 50) cm x 75 µm 10 mM TRIS, adj. to pH 2.2 with citric acid, 10 mM 18C6H_4; hydrodyn. inj., U = 15 kV UV at 254 nm; analysis time 50 min	host-guest complexation with crown ether (18-crown-6 tetracarboxylic acid)	155

Table 7.25. Chiral separations of amino acids. Experiments are performed with normal polarity (detection at the cathodic end) if not otherwise stated.

Appl. No.	Compounds	Conditions	Remarks	Ref.
334	Trp, Dopa, Phe and Tyr	capillary , 57 (L_D = 50) cm x 75 µm; 30 mM 18C6H4, pH 2.2, hydrodyn. inj., U = 15 kV; UV at 254 nm; analysis time 30 min	host-guest complexation with 18-crown-6 tetracarboxylic acid ($18C6H_4$); comparison between α-CD and $18C6H_4$ in terms of R and N is presented	480
335	tryptophan and derivatives	coated capillary (Bio-Rad), 20 cm x 25 µm; 0.1 M phosphate, pH 2.5, 40 mM dimethyl-β-CD, electrokin. inj., 7 s, 7 kV, U = 8 kV UV at 206 nm; analysis time 60 min	host-guest complexation with dimethyl-β-cyclodextrin	486
336	a) DNS-Leu, DNS-Met and DNS-Thr; b) Trp, Phe and Dopa	capillary, 57 (L_D = 50) cm x 75 µm; a) 50 mM $Na_2B_4O_7$, pH 9.0, 10 mM γ-CD; hydrodyn. inj. 1 s; U = 15 kV, UV at 214 nm, time 9 min;b) 10 mM TRIS - citric acid, pH 2.5, 10 mM $18C6H_4$; hydrodyn. inj.; U = 15 kV, UV at 254 nm, time 20 min	host-guest complexation with cyclodextrin and crown ether; a synergistic effect of both chiral selectors is demonstrated	76
337	5 dansylamino acids: DNS-Phe, DNS-Val, DNS-Leu, DNS-Met and DNS-Glu	capillary, 57 (L_D = 50) cm x 75 µm; 100 mM borate, pH 8.3, 100 mM SDS, 60 mM γ-CD; U = 12 kV; UV at 200 nm; time 24 min	MEKC with SDS and γ--CD	252
338	5 dansylamino acids: DNS-Phe, DNS-Nva, DNS-Leu, DNS-Met and DNS-Nle	capillary, 70 (L_D = 50) cm x 50 µm; 50 mM phosphate, pH 3.0, 50 mM taurodeoxycholate; hydrodyn. inj., T = 40 °C U = 8 kV, UV at 210 nm; analysis time 70 min	MEKC with bile salts	487
339	DNS-Glu, DNS-Ser and DNS-Leu	capillary, 30 (L_D = 15) cm x 75 µm; 100 mM TRIS - 250 mM boric acid, pH 8.3, 7 M urea, 75 mM β-CD in a PAA gel (T = 5%, C = 3.3%) electrokin. inj., 30 s, 250 $V \cdot cm^{-1}$, U = 9 kV; UV at 206 nm; time 8 min	host-guest complexation with β-CD incorporated in a polyacrylamide gel	488

Table 7.25. (continued)

Appl. No.	Compounds	Conditions	Remarks	Ref.
340	10 dansylamino acids	capillary, 75 cm x 75 µm 10 mM ammonium acetate, pH 7-8, 5 mM L-histidine, 10 mM $CuSO_4$; electrokin. inj. 5 s, 6 kV; E = 300 $V \cdot cm^{-1}$; fluorescence, time 11 min	ligand exchange complexation	489
341	18 dansylamino acids	capillary, 100 (L_D = 75) cm x 75 µm 10 mM $NH_4CH_3CO_2$, pH 7-8, 5 mM aspartame, 2.5 mM $CuSO_4$; electrokin. inj. 6 s,10 kV; U = 30 kV; fluorescence 325/550, time 11 min	ligand exchange complexation	490
342	7 cyano-benz[f]isoindole (CBI)-amino acids	capillary, 70 cm x 50 µm 100 mM borate, pH 9.0, 50 mM SDS, 10 mM β-CD; U = 15 kV, LIF 442/490 nm; time 28 min	host-guest complexation with β-cyclodextrin; injection volume 2.5 nL	491
343	6 phenylthiohydantoin (PTH)-amino acids	capillary, 63 (L_D = 49) cm x 50 µm 25 mM digitonin - 50 mM SDS, pH 3.0, U = 20 kV; UV at 260 nm; time 100 min	MEKC with mixed micelles composed of SDS and digitonin	492
344	6 phenylthiohydantoin (PTH)-amino acids	capillary, 65 (L_D = 50) cm x 50 µm; 30 mM SDS, 50 mM SDVal, 0.5 M urea, pH 9.0, 10 % (v/v) MeOH; U = 20 kV; UV at 260 nm; analysis time 50 min	MEKC with mixed micelles composed of SDS and SDVal	245
345	aromatic amino acids, di- and tripeptides	capillary, 57 (L_D = 50) cm x 75 µm 10 mM TRIS, adj. to pH 2.2 with citric acid, 10 mM $18C6H_4$; hydrodyn. inj., U = 15 kV UV at 254 nm; analysis time 50 min	host-guest complexation with crown ether (18-crown-6 tetracarboxylic acid)	155
346	aliphatic amino acids	capillary, 57 (L_D = 50) cm x 75 µm 5 mM TRIS, adj. to pH 2.2 with citric acid, 6 mM BTA chloride, 15 mM $18C6H_4$, hydrodyn. inj., U = 15 kV; $UV_{ind.}$ at 214 nm; time 50 min	host-guest complexation with crown ether (18-crown-6 tetracarboxylic acid); benzyltrimethylammonium (BTA) is used as visualizing agent	155

Table 7.26. Chiral separations of selected compounds. Experiments are performed with normal polarity (detection at the cathodic end) if not otherwise stated.

Appl. No.	Compounds	Conditions	Remarks	Ref.
347	Tröger's base	capillary , 57 (L_D = 50) cm x 75 µm; 50 mM phosphate, pH 2.5, 10 mM β-CD, hydrodyn. inj., U = 15 kV; UV at 214 nm; analysis time 55 min	host-guest complexation with β–cyclodextrin	76
348	1-phenylethanol, 1.1'-binaphthyl-2,2'-diylhydrogenphosphate	capillary, 97 (L_D = 80) cm x 50 µm; 20 mM borate-phosphate, pH 7, hydrodyn. inj., U = 20 kV; UV at 220 nm; analysis time 26 min	host-guest complexation with β-cyclodextrin immobilized to the silica surface forming a stationary phase	493
349	Binaphthyl derivatives: 1,1'-bi-2-naphthol, 1,1'-binaphthyldiylhydrogenphosphate, 1,1'-binaphthyldicarboxylic acid, 1,1'-biphenanthrene dihydroxide	capillary, 75 (L_D = 65) cm x 50 µm; 50 mM sodium deoxycholate with 12 % methanol; hydrodyn. inj.; U = 15 - 20 kV, LIF, analysis time 30 min	MEKC with bile salts	494
350	enantiomeric Co(III) complexes with ethylenediamine and amino acid ligands	capillary, 43 (L_D = 35) cm x 100 µm; 15 mM L-(+)-tartaric acid - 15 mM TRIS, pH 5.25; hydrodyn. inj., U = 10 kV; UV at 240 nm; analysis time 7 min	ligand exchange complexation with L-(+)-tartaric acid	84
351	aminoalcohols: phenylalaninol, phenylglycinol	capillary, 84 (L_D = 52) cm x 50 µm; 30 mM $18C6H_4$, pH 2.07; electrokin. inj. 5s, 5 kV, U = 15 kV, UV at 254 nm; time 50 min	host-guest complexation with 18-crown-6 tetracarboxylic acid ($18C6H_4$)	495

Table 7.27. Complex samples. Experiments are performed with normal polarity (detection at the cathodic end) if not otherwise stated.

Appl. No.	Compounds	Conditions	Remarks	Ref.
352	crude fermentation broth of *Aspergillus oryzae*	capillary , 57 (L_D = 50) cm x 75 µm; 25 mM phosphate buffer, pH 9.5, hydrodyn. inj., U = 15 kV UV at 200 nm; analysis time 10 min	fermentation broth diluted 1:3 with deionized water; main components identified as α-amylase, alkaline protease and subtilisin-like protease	496
353	urine from healthy controls and from patients with various metabolic disorders	capillary , 47 (L_D = 40) cm x 75 µm; 50 mM sodium phosphate buffer, pH 2.5, hydrodyn. inj., U = 20 kV UV at 214 nm; analysis time 20 min	urine sample are only filtered through 0.22 µm filters before injection	497
354	process control of the production of recombivax HB hepatitis B vaccine	capillary , 72 (L_D = 50) cm x 50 µm; 25 mM sodium phosphate buffer, pH 7.25, hydrodyn. inj., U = 27 kV UV at 200 nm; analysis time 10 min	monitoring the purification of a recombinant substance by analyzing each step in the process thus establishing a fingerprint for the entire process run	498
355	normal human urine	capillary, 47 (L_D = 50) cm x 75 µm; 50 mM $Na_2B_4O_7$, pH 8.3 electrokin. inj. 10 s, 5 kV, U = 15 kV UV at 210 nm; analysis time 90 min	urine derived from a normal individuum reveals about 10 major components	499
356	analysis of different kinds of beer	capillary, 72 (L_D = 50) cm x 50 µm; 20 mM sodium citrate, pH 2.5 hydrodyn. inj., E = 278 $V \cdot cm^{-1}$, T = 30 °C; UV at 200 nm; analysis time 40 min	each beer shows a characteristic profile; only neutral and positively charged species can be seen using this buffer system	500
357	process control of the production of Savinase	capillary, 45 (L_D = 25) cm x 50 µm; 25 mM sodium phosphate buffer, pH 7.2, 50 mM SDS hydrodyn. inj., U = 9 kV, T = 30 °C; UV at 200 nm; analysis time 14 min		501

8 Appendix

8.1 Buffer Tables

Table 8.1. Chemical buffer systems with their pK values and their useful pH ranges

Buffer system	pK value	Useful pH range
$CH_3CO_2H/CH_3CO_2^-$	4.76	3.6 - 5.8
$H_3PO_4/H_2PO_4^-$	2.12	1.5 - 3
$H_2PO_4^-/HPO_4^{2-}$	7.21	5 - 8
HPO_4^{2-}/PO_4^{3-}	12.67	11.5 - 13
NH_3/NH_4^+	9.25	8 - 10
citric acid/citrate	3.13/5.93/6.38	2 - 6
borate - HCl	9.14/(12.73/13.8)	8.0 - 9.1
borate - NaOH	9.14/(12.73/13.8)	9.2 - 11

Table 8.1. Biological buffers with their pK values and amino acids with their pI values, respectively, and their useful pH ranges

Abbreviation	Substance	pK	pI	Useful pH range
Ala	alanine		6.0	5.0 - 7.0
β-Ala	β-alanine		6.9	5.9 - 7.9
AMMEDIOL	2-amino-2-methyl-1,3-propandiol	8.86		7.9 - 9.9
AMPSO	2-hydroxypropane sulphonic acid	9.10		8.1 - 10.1
Arg	arginine		10.76	9.8 - 11.8
Asn	asparagine		5.41	4.4 - 6.4
Asp	asparagic acid		2.77	1.8 - 3.8
BES	2-aminoethane sulphonic acid	7.1		6.1 - 8.1
BICIN	*N,N'*-bis(2-hydroxyethyl)-glycine	8.3		7.3 - 9.3
BISTRIS	2-(bis(2-hydroxyethyl)-imino)-2-(hydroxymethylene)-1,3-propanediol	6.5		5.5 - 7.5

Table 8.1. (continued)

Abbreviation	Substance	pK	pI	Useful pH range
CAPS	3-(cyclohexylamino)propane-3-sulphonic acid	10.4		9.4 - 11.4
CHES	2-(cyclohexylamine)ethane-2-sulphonic acid	9.5		8.5 - 10.5
DIPSO	3-(*N*-bis(hydroxyethyl)-amino)-2-hydroxypropane sulphonic acid	7.5		6.5 - 8.5
EACA	6-aminocaproic acid	7.59		6.6 - 8.6
GABA	4-aminobutyric acid	7.33		6.3 - 8.3
Gln	glutamine		5.65	4.7 - 6.7
Glu	glutamic acid		3.22	2.2 - 4.2
Gly	glycine		5.97	5.0 - 7.0
GlyGly	glycylglycine		5.65	4.7 - 6.7
HEPES	*N*-(2-hydroxyethyl)piperazine-*N'*-2-ethane sulphonic acid	7.5		6.5 - 8.5
HEPPS	*N*-(2-hydroxyethyl)piperazine-*N'*-3-propane sulphonic acid	8.0		7.0 - 9.0
HEPPSO	*N*-(2-hydroxyethyl)piperazine-*N'*-2-propane sulphonic acid	7.8		6.8 - 8.8
His	histidine		7.47	6.5 - 8.5
Imi	imidazole	6.95		6.0 - 8.0
Lys	lysine		9.74	8.7 - 10.7
MES	morpholinoethane sulphonic acid	6.1		5.1 - 7.1
MOPS	morpholinopropane sulphonic acid	7.2		6.2 - 8.2
MOPSO	3-(*N*-morpholino)-2-hydroxypropane sulphonic acid	6.9		5.9 - 7.9
	morpholine	8.6		7.6 - 9.6
	ornithine		9.7	8.7 - 10.7
Pro	proline		6.3	5.3 - 7.3
Sar	sarcosine		6.12	5.1 - 7.1
Ser	serine		5.68	4.7 - 6.7
TAPS	*N*-tris(hydrohymethyl)methyl-3-aminopropane sulphonic acid	8.4		7.4 - 9.4
TAPSO	3-(*N*-trishydroxymethyl)-methyl-amino)-2-hydroxy-propane sulphonic acid	7.6		6.6 - 8.6
TES	*N*-tris(hydroxymethyl)-methyl-2-aminoethane sulphonic acid	7.4		6.4 - 8.4
TRICIN	*N*-tris(hydroxymethyl)-methyl glycine	8.2		7.2 - 9.2
TRIS	Tris(hydroxy)aminomethane	8.0		7.0 - 9.0

8.2 Derivatization Procedures

8.2.1 3-(4-Carboxybenzoyl)-2-quinoline Carboxaldehyde (CBQCA) [123]

> Dissolve the reagent in methanol to give a concentration of 3g/L.
> Dissolve potassium cyanide in water to give a 10 mM solution.
> Mix aliquots of the analyte (typical concentration ranging from 10^{-4} - 10^{-6} M) with 10 - 20 µL of potassium cyanide solution and 5 - 10 µL of CBQCA solution.
> Allow the mixture to stand at room temperature for at least 1 h prior to the sample injection.

8.2.2 Dansyl Chloride (Dns-Cl) [124]

> Prepare a stock solution of Dns-Cl in HPLC grade acetone to give a concentration of 3 $g \cdot L^{--1}$.
> If possible, bring the sample solution to a concentration of 10^{-4} - 10^{-6} M by adding 0.1 M sodium bicarbonate buffer.
> Mix 100 µL of each the reagent and the sample solution and let the mixture react at 37 - 50 °C for 15 - 60 min (or until the yellow color of Dns-Cl disappears.

8.2.3 4-Phenylspiro[furan-2(3H),1'-phthalan]-3,3'-dione (Fluorescamine) [128]

> Dissolve the reagent in acetone to give a concentration of 3 $g \cdot L^{-1}$ and add 20 $\mu L \cdot mL^{-1}$ pyridine to the solution.
> Transfer solutions of the analyte samples to a 500 µL microcentrifuge tube and adjust their total volume to 70 µL by addition of 0.1 M sodium tetraborate buffer, pH 9.0.
> Add 30 µL of fluorescamine solution to the sample while continuously and vigorously vortexing for 2 min. The concentration of the analyte samples should range from 2 - 1250 µg/100µL reaction mixture.

8.2.4 9-Fluorenylmethyl Chloroformate (FMOC) [112]

> Prepare a stock solution by dissolving FMOC in HPLC grade acetone to give a concentration of 15 mM.
> If possible, bring the sample solution to a concentration of 10^{-4} - 10^{-6} M by adding an appropriate amount of alkaline separation buffer.

> Mix 100 µL of each the reagent and the sample solution and let the mixture react for 1 min.

8.2.5 Fluorescein Isothiocyanate (FITC)

8.2.5.1 Preparation of Fluorescein Thiocarbamyl Derivatives [119]

> Prepare a stock solution of $5.5 \cdot 10^{-4}$ M FITC-isomer I in HPLC grade acetone, add a trace of pyridine and store it at 4 °C.
> If possible, bring the sample solution to a concentration of 10^{-4} - 10^{-6} M by adding 0.2 M sodium carbonate buffer, pH 9.0.
> Mix 1 mL of the sample solution with 10 µL of reagent solution and allow the mixture to react 2 - 4 hours at room temperature in the dark.

8.2.5.2 Preparation of Fluorescein Thiohydantoin Derivatives [113]

> Prepare a stock solution of $1.3 \cdot 10^{-3}$ M FITC in HPLC grade acetone and store it at 4 °C.
> If possible, bring the sample solution to a concentration of $5 \cdot 10^{-3}$ - 10^{-5} M by adding 0.2 M sodium carbonate buffer, pH 9.1.
> Mix 1 mL of the sample solution with 100 µL of reagent solution in a 1.5 mL vial and allow the mixture to react for 4 hours at room temperature in the dark.
> Mix 0.5 mL of each thiocarbamyl solution with 0.5 mL of trifluoroacatic acid and allow the mixture to react for 15 h at room temperature in the dark to form the fluorescein thiohydantoin derivative.
> Dilute the samples at least 4000-fold (for an original analyte concentration of $5 \cdot 10^{-3}$ M) before injection.

8.2.6 Naphthalene-2,3-dicarboxaldehyde (NDA) [132]

> Prepare a stock solution of reagent by dissolving 4.6 mg NDA in 25 mL HPLC grade acetonitrile to give a 1 mM solution. The stock solution can be used for 1 month.
> Prepare a 100 mM sodium tetraborate solution, pH 9.5 and transfer 700 µL into a 1.5 mL vial.
> Add 100 µL of a 10 mM sodium cyanide solution.
> Bring the sample solution to a concentration in the range from 1 nM - 1 µM for each analyte.
> Add 100 µL of the sample solution to the vial and mix.
> Add 100 µL of NDA solution and cap the vial. Gently shake the mixture and allow the reaction to proceed at 25 °C for 30 min.

8.2.7 *o*-Phthaldialdehyde (OPA) [112]

> Prepare a solution of 3.7 mM OPA in 20 mM sodium tetraborate buffer, pH 9.5 and add 2 % of methanol.

> Transfer sample solutions to a 500 µL microcentrifuge tube. The sample concentration should be in the range of 10^{-4} - 10^{-6} M.

> Mix the reagent solution and a 0.5 % solution of mercaptoethanol (or mercaptopropionic acid) with aliquots of sample solution and let react the mixture for 1 min by vortexing.

8.3 Glossary

A	ampere
ACE	affinity capillary electrophoresis
AU	absorbance unit
α	dissociation degree
α	separation factor
BIS	*N,N'*-methylenebisacrylamide
ß	thickness of the diffuse double layer
c	electrolyte concentration
C	coulomb [A·s]
[C]	concentration of the component C
CAE	capillary affinity electrophoresis
CE	capillary electrophoresis
CEC	capillary electrochromatography
CGE	capillary gel electrophoresis
CIA	capillary ion analysis
CIEF	capillary isoelectric focusing
CITP	capillary isotachophoresis
CMC	critical micellar concentration
CTAB	hexadecyl(cetyl)trimethylammonium bromide
CZE	capillary zone electrophoresis
d	thickness of the rigid layer
D	diffusion coefficient
Da	Dalton
E	electric field strength
E_A	activity energy
ε_0	permittivity or dielectric constant of free space [$8.854 \cdot 10^{-12}$ $F \cdot m^{-1}$]
ε_r	relative dielectric constant
EDTA	ethylenediamine tetraacetic acid (tetraacetate)
ε	permittivity or dielectric constant of the medium
ε	absorptivity

E	extinction
EKC	electrokinetic chromatography
EOF	electroosmotic flow
F	Faraday constant [96 485 $C{\cdot}mol^{-1}$]
F	Farad [$As{\cdot}V^{-1}$]
F_e	electric force
F_d	drag force
H	plate height
HPLC	high-performance liquid chromatography
η	Newtonian viscosity
I	electric current
I	ionic strength
I	light intensity
i	current density
i.d.	inner diameter
ITP	isotachophoresis
IEF	isoelectric focusing
J	joule [N·m]
k	Boltzman's constant [$1.38{\cdot}10^{-23}$ $J{\cdot}K^{-1}$]
K	Kelvin
K	cell constant
K_c	dissoziation constant
K	thermal conductance
κ	specific conductance
l	sample plug width
l_c	path length of the light through the detection cell
L	conductivity
L	liter [dm^3]
L_D	effective capillary length
L_T	total capillary length
LIF	laser-induced fluorescence detection
Λ	equivalent or molar conductance
Λ	limiting equivalent conductance
λ	ionic eqivalent conductance
μ_i	effective electrophoretic mobility
μ_i^0	limiting electrophoretic mobility
μ_{eo}	electroosmotic mobility
M	molecular mass
M	molar [$mol{\cdot}L^{-1}$]
MBE	moving-boundary electrophoresis
MEKC	micellar electrokinetic chromatography
MS	mass spectrometry
MW	molecular weight
N	theoretical plate number
N	newton [$kg{\cdot}m^{-1}{\cdot}s^{-2}$]

N_A	Avogadro's constant [$6.022 \cdot 10^{23}$ mol^{-1}]
n	peak capacity
o.d.	outer diameter
p	pressure
P	partition coefficient
P	power
Pa	pascal [$N \cdot m^{-2}$]
PAA	polyacrylamide
PAGE	polyacrylamide gel electrophoresis
pI	isoelectric point
pH	negative decadic logarithm of the proton concentration
pK	negative common logarithm of the dissociation constant K
PMT	photomultiplier tube
Q	heat output
Q_{eff}	effective charge
Ω	Ohm [$V \cdot A^{-1}$]
r	radius
R	molar gas constant [8.314 $J \cdot mol^{-1} \cdot K^{-1}$]
R	electric resistance
R	resolution
RI	refractive index
RSD	relative standard deviation [%]
S	Siemens [Ω^{-1}]
SDS	sodium lauryl (dodecyl) sulphate
σ^2	spatial variance
T	temperature
TEMED	*N,N,N'N'*-tetramethylene ethylenediamine
t	migration time
t_{eo}	migration time of the electroosmotic flow
U	electric potential, voltage
v	migration velocity
V	volt
V_d	detector dead volume
V_c	volume of the capillary column
V_i	injection volume
v_{eo}	electroosmotic flow velocity
w	temporal peak width at the baseline
$w/2$	temporal width at half of the peak height
w_s	spatial peak width
ω	regulating or omega function
z	charge number
Z	relative total net charge
ZE	zone electrophoresis
ζ	zeta potential

8.4 Manufacturers' Directory

ALDRICH CHEMICAL CO INC
Box 355, Milwaukee, Wisconsin 53201, USA
Phone: +1 414 273 3850; Fax: +1 414 273 4979

APPLIED BIOSYSTEMS
850 Lincoln Center Drive, Foster City, California 94404, USA
Phone: +1 415 570 6667; Fax: +1 415 572 2743

BECKMAN INSTRUMENTS INC
2500 Harbor Boulevard, Fullerton, California 92634, USA
Phone: +1 714 871 4848; Fax: +1 714 773 8898

BECKMAN INSTRUMENTS GmbH
Frankfurter Ring 115, Postfach 400 248, 8000 München 50, FRG
Phone: +49 89 3887 1; Fax: +49 89 3887 490

BERTAN High Voltage
121 New South Road, Hicksville, New York 11801, USA
Phone: +1 516 433 3110; Fax: +1 516 935 1766

BIO-RAD LABORATORIES INC
19 Dreve Du Senechal, B-1180 Brussels, Belgium
Phone: +132 2 375 5970; Fax: +32 2 374 6162

BIO-RAD LABORATORIES LIFE SCIENCE GROUP
2000 Alfred Nobel Drive, Hercules, California 94547, USA
Phone: +1 510 741 1000; Fax: +1 520 879 2289

CHROMPACK INTERNATIONAL BV
Herculesweg 8, Box 8033, 4330 EA Middleburg, The Netherlands
Phone: +31 1180 71000; Fax: +31 1180 33118

DIONEX CORP
Box 13603, Sunnyvale, California 94088-3603, USA
Phone: +1 408 737 0700; Fax: +1 408 730 9403

E. MERCK
Frankfurter Strasse 250, Postfach 4119, D-6100 Darmstadt, FRG
Phone: +49 6151 720; Fax: +49 6151 722000

EUROPHOR INSTRUMENTS
10 Avenue de L'Europe, 31520 Ramonville-Toulouse, France
Phone: +33 61 28 56 74; Fax: +33 61 28 56 00

FLUKA CHEMIE AG
Industriestrasse 25, CH-9470 Buchs, Switzerland
Phone: +41 85 6 95 11; Fax: +41 85 6 54 49

HEWLETT-PACKARD GmbH
Hewlett-Packard-Straße 8, D-7517 Waldbronn 2, FRG
Phone: +49-7243-602-0; Fax: 49-7243-602-666

ISCO INC
Box 5347, Lincoln, Nebraska 68505, USA
Fax: +1 402 464 0318

ISCO EUROPA AG
Brüschstrasse 17, CH-8708 Männedorf, Switzerland
Phone: +41 1 920 2425; Fax: +41 1 920 6208

J&W SCIENTIFIC
91 Blue Ravine Road, Folsom, California 95630, USA
Phone: +1 916 985 7888; Fax: +1 916 985 1101

JASCO
8649 Commerce Drive, Easton, Maryland 21601, USA
Phone: +1 410 822 1220; Fax: +1 410 822 7526

KONTRON INSTRUMENTS SPA
Via Fantoli 16/15, 20138 Milan, Italy
Phone: +39 2 50721; Fax: +39 2 506 0918

LAUERLABS BV
PO Box 2194, Kapiten Nemostraat BA, 7801 CD Emmen, The Netherlands
Phone: +31 5910 44088; Fax: +31 5910 43876

LC PACKINGS
Baarskesweg 154, 1057 HM Amsterdam, The Netherlands
Phone: +31 20 683 9768; Fax: +31 20 685 3452

LC PACKINGS (USA) INC
80 Carolina Street, San Francisco, California 94103, USA
Phone: +1 415 552 1855; Fax: +1 415 552 1859

PIERCE EUROPE BV
PO Box 1512, 3260 BA Out-Beijerland, The Netherlands
Phone: +31 1860 19277; Fax: +31 1860 19179

POLYMICRO TECHNOLOGIES INC
3035 North 33rd Drive, Phoenix, Arizona 85017, USA

Phone: +1 602 272 7437; Fax: +1 602 278 1776

SERVA FEINBIOCHEMICA GMBH & CO
PO Box 105260, D-6900 Heidelberg 1, FRG
Phone: +49 6221 5020; Fax: +49 6221 502188

SGE INC
2007 Kramer Lane, Austin, Texas 78758, USA
Phone: +1 512 836 9159

SIEMENS AG
Dept. Process Analytics AUT V351, Postfach 211262, 75 Karlsruhe 21, FRG
Phone: +49 721 595 6148; Fax: +49 721 595 6375

SIGMA CHEMICAL CO
PO Box 14508, St. Louis, Missouri 63178, USA

SIGMA CHEMICAL CO, LTD
Fancy Road Poole, Dorset BH17 7NH, UK
Phone:+44 202 733 114; Fax:+44 202 715 460

SPECTRA-PHYSICS ANALYTICAL
Box 5116, 45757 Northport Loop West, Fremont, California 94537, USA
Phone: +1 510 657 1100; Fax: +1 510 490 8182

SPECTRA-PHYSICS LTD
Boundary Way, Hemel Hempstead, Hertfordshire HP2 7Sh, UK
Phone: +44 442 232 322; Fax: +44 442 68538

STAGROMA AG
Alte Winterthurerstrasse 51, CH-8304 Wallisellen, Switzerland
Phone: +41 1 830 1175; Fax: +41 1 830 4088

SUPELCO INC
Supelco Park, Bellefonte, Pensylvania 16823, USA
Phone: +1 814 359 3041; Fax: +1 814 359 3044

UNICAM ANALYTICAL SYSTEMS
1001 Fourir Drive, Madison, Wisconsin 53717
Fax: +1 608 831 5156

WATERS CHROMATOGRAPHY DIVISION OF MILLIPORE
34 Maple Street, Milford, Massachusetts 01757, USA
Phone: +1 508 478 2000; Fax: +1 508 478 5839

WILMAD GLASS CO
Buena, New Jersey, USA

8.5 Further Recommended Reading

This collection of publications gives an overview of some interesting and comprehensive contributions to the corresponding topics which have not been mentioned in the context.

General Information:

W.G. Kuhr, Capillary Electrophoresis. *Anal. Chem.* **62** (1990) 403R-414R

W.G. Kuhr and C.A. Monnig, Capillary Electrophoresis. *Anal. Chem.* **64** (1992) 389R-407R

Influences:

H. Poppe, Overloading and Interaction Phenomena in Electrophoretic Separations. *Anal. Chem.* **64** (1992) 1908-1919

S.L. Delinger and J.M. Davis, Influence of Analyte Plug Width on Plate Number in Capillary Electrophoresis. *Anal. Chem.* **64** (1992) 1947-1959

E. Grushka, Effect of Hydrostatic Flow on the Efficiency in Capillary Electrophoresis. *J. Chromatogr.* **559** (1991), 81-94

Detection:

L. N. Amankwa, M. Albin and W.G. Kuhr, Fluorescence Detection in Capillary Electrophoresis. *Trends Anal. Chem.* **11** (1991) 114-120

P.D. Curry, C.E. Engstrom-Silverman and A.G. Ewing, Electrochemical Detection for Capillary Electrophoresis. *Electroanalysis* **3** (1991) 587-596

E.S. Yeung and W. G. Kuhr, Indirect Detection Methods for Capillary Separations. *Anal. Chem.* **63** (1991) 275A-282A

T. Keough, R. Takigiku, M.P. Lacey and M. Purdon, Matrix-Assisted Laser Desorption Mass Spectrometry of Proteins Isolated by Capillary Zone Electrophoresis. *Anal. Chem.* **64** (1992) 1594-1600

Techniques:

P.D. Grossman, T. Hino and D.S. Soane, Dynamic Light Scattering Studies of Hydroxyethyl Cellulose Solutions Used as Sieving Media for Electrophoretic Separations. *J. Chromatogr.* **608** (1992) 79-84

S. Terabe and N. Matsubara, Microemulsion Electrokinetic Chromatography: Comparison with Micellar Electrokinetic Chromatography. *J. Chromatogr.* **608** (1992) 23-30

A. Guttman, J. Horvath and N. Cooke, Influence of Temperature on the Sieving Effect of Different Polymer Matrices in Capillary SDS Gel Electrophoresis of Proteins. *Anal. Chem.* **65** (1993) 199-203

Applications:

Y. Ma and R. Zhang, Optimization of Indirect Photometric Detection of Anions in High-Performance Capillary Electrophoresis. *J. Chromatogr.* **625** (1992) 341-348

W.R. Jones and P. Jandik, Various Approaches to Analysis of Difficult Sample Matrices of Anions Using Capillary Ion Elecrophoresis. *J. Chromatogr.* **608** (1992) 385-393

J.S. Stamler and J. Loscalzo, Capillary Zone Electrophoresis Detection of Biological Thiols and Their S-Nitrosated Derivatives. *Anal. Chem.* **64** (1992) 779-785

J.C. Kraak, S. Busch and H. Poppe, Study of Protein-Drug Binding Using Capillary Zone Electrophoresis. *J. Chromatogr.* **608** (1992) 257-264

H. Swerdlow, J.Z. Zhang, D.Y. Chen, H.R. Harke, R. Grey, S. Wu and N.J. Dovichi, Three DNA Sequencing Methods Using Capillary Gel Electrophoresis and Laser-Induced Fluorescence. *Anal. Chem.* **63** (1991) 2835-2841

X.C. Huang, M.A. Quesada and R.A. Mathies, DNA Sequencing Using Capillary Array Electrophoresis. *Anal. Chem.* **64** (1992) 2149-2154

S.C. Smith and M. G. Khaledi, Optimization of pH for the Separation of Organic Acids in Capillary Zone Electrophoresis. *Anal. Chem.* **65** (1993) 193-198

References

[1] F. Kohlrausch, Ueber Concentrations-Verschiebungen durch Electrolyse im Inneren von Lösungen und Lösungsgemischen. *Ann. Phys. Chem.* **62** (1897) 209-239

[2] S. Hjertén, Free Zone Electrophoresis. *Chromatogr. Rev.* **9** (1967) 122-239

[3] F.M. Everaerts, J.L. Beckers and Th.P.E.M. Verheggen (eds.), Isotachophoresis - Theory, Instrumentation and Applications. J. Chromatogr. Library, Elsevier, Amsterdam - Oxford - New York (1976)

[4] R. Virtanen, Zone Electrophoresis in a Narrow-Bore Tube Employing Potentiometric Detection. *Acta Polytechnica Scand.* **123** (1974) 1-67

[5] F.E.P. Mikkers, F.M. Everaerts and Th.P.E.M. Verheggen, High-Performance Zone Electrophoresis. *J. Chromatogr.* **169** (1979) 11-20

[6] J.W. Jorgenson and K.D. Lukacs, High Resolution Separations Based on Electrophoresis and Electroosmosis. *J. Chromatogr.* **218** (1981) 209-216

[7] J.W. Jorgenson and K.D. Lukacs, Zone Electrophoresis in Open-Tubular Glass Capillaries. *Anal. Chem.* **53** (1981) 1298-1302

[8] J.W. Jorgenson and K.D. Lukacs, Capillary Zone Electrophoresis. *Science* **222** (1983) 266-272

[9] M. Bier, O.A. Palusinski, R.A. Mosher and D.A. Saville, Electrophoretic Mathematical Modeling and Computer Simulation. *Science* **219** (1983) 1281-1287

[10] O. Vesterberg, History of Electrophoresis. *J. Chromatogr.* **480** (1989) 3-19

[11] P. G. Righetti (ed.), Isoelectric Focusing: Theory, Methodology and Applications. Elsevier, Amsterdam - Oxford - New York (1983)

[12] G. Wedler (ed.), Lehrbuch der Physikalischen Chemie. Verlag Chemie, Weinheim (1982) 193

[13] G.R. Wiese, R.O. James and T.W. Healy, Discreteness of Charge and Solvation Effects in Cation Adsorption at the Oxide/Water Interface. *Discuss. Faraday Soc.* **52** (1971) 302-311

[14] K. Salomon, D.S. Burgi and J.C. Helmer, Evaluation of Fundamental Properties of a Silica Capillary in Capillary Electrophoresis. *J. Chromatogr.* **559** (1991), 69-80

[15] B.J. Wanders, A.A.A.M. Van De Goor and F.M. Everaerts, Methods for On-Line Determination and Control of Electroendosmosis in Capillary Electrochromatography and Electrophoresis. *J. Chromatogr.* **470** (1989) 89-93

[16] K.D. Altria and C.F. Simpson, The Effect of Electrolyte Chain Length on Electroendosmotic Flow in High Voltage Capillary Zone Electrophoresis. *Anal. Proceed.* **25** (1988) 85

[17] K.D. Altria and C.F. Simpson, High Voltage Capillary Zone Electrophoresis: Operating Parameters Effects on Electroendosmotic Flows and Electrophoretic Mobilities. *Chromatographia* **24** (1987) 527-532

[18] T. Kaneta, S. Tanaka and H. Yoshida, Improvement of Resolution in the Capillary Electrophoretic Separation of Catecholamines by Complex Formation with Boric Acid and Control of Electroosmosis with Cationic Surfactant. *J. Chromatogr.* **538** (1991) 385-391

[19] J.C. Giddings, Generation of Variance, "Theoretical Plates", Resolution, and Peak Capacity in Electrophoresis and Sedimentation. *Sep. Sci.* **4** (1969) 181-189

[20] J.C. Giddings, Harnessing Electrical Forces for Separation, Capillary Zone Electrophoresis, Isoelectric Focusing, Field-Flow Fractionation, Split-Flow Thin-Cell Continuous-Separation and Other Techniques. *J. Chromatogr.* **480** (1989) 21-33

[21] J. Knox, Thermal Effects and Band Spreading in Capillary Electro-Separation. *Chromatographia* **26** (1988) 329-337

[22] A.L. Lehninger (ed.), Biochemie. Verlag Chemie, Weinheim-New York (1979) p140

[23] Y. Walbroehl and J.W. Jorgenson, Capillary Zone Electrophoresis for the Determination of Electrophoretic Mobilities and Diffusion Coefficients of Proteins. *J. Microcol. Sep.* **1** (1989) 41-45

[24] J. Liu, V. Dolnik, Y.-Z. Hsieh and M. Novotny, Experimental Evaluation of the Separation Efficiency in Capillary Electrophoresis Using Open Tubular and Gel-Filled Columns. *Anal. Chem.* **64** (1992) 1328-1336

[25] H.H. Lauer and D.McManigill, Capillary Zone Electrophoresis of Proteins in Untreated Fused Silica Tubing. *Anal. Chem.* **58** (1986) 166-170

[26] R. McCormick, Capillary Zone Electrophoresis Separation of Peptides and Proteins Using Low pH Buffers in Modified Silica Capillaries. *Anal. Chem.* **60** (1988) 2322-2338

[27] J.S. Green and J.W. Jorgenson, Minimizing Adsorption of Proteins on Fused Silica in Capillary Zone Electrophoresis by the Addition of Alkali Metal Salts to the Buffer. *J. Chromatogr.* **478** (1989) 63-70

[28] M.M. Bushey and J.W. Jorgenson, Capillary Electrophoresis of Proteins in Buffers Containing High Concentrations of Zwitterionic Salts. *J. Chromatogr.* **480** (1989) 301-310

[29] S. Hjertén, High-Performance Electrophoresis - Elimination of Electroendosmosis and Solute Adsorption. *J. Chromatogr.* **347** (1985) 189-198

[30] Y.-F. Maa, K.J. Hyver and S.A. Swedberg, Impact of Wall Modification on Protein Elution in High Performance Capillary Zone Electrophoresis. *HRC&CC* **14** (1991) 65-67

[31] S. Hjertén, Zone Broadening in Electrophoresis with Special Reference to High-Performance Electrophoresis in Capillaries: An Interplay between Theory and Practice. *Electrophoresis* **11** (1990) 665-690

[32] D.S. Burgi, K. Salomon and R.-L. Chien, Methods for Calculating the Internal Temperature of Capillary Columns during Capillary Electro-phoresis. *J. Liq. Chromatogr.* **14** (1991) 847-867

[33] R.J. Nelson, A. Paulus, A.S. Cohen and B.L. Karger, Use of Peltier Thermoelectric Devices to Control Column Temperature in High-Performance Capillary Electrophoresis., *J. Chromatogr.* **480** (1989) 111-127

[34] E. Grushka, R.M. McCormick and J.J. Kirkland, Effect of Temperature Gradients on the Efficiency of Capillary Zone Electrophoresis. *Anal. Chem.* **61** (1989) 241-246

[35] A.E. Jones and E. Grushka, Nature of Temperature Gradients in Capillary Zone Electrophoresis. *J. Chromatogr.* **466** (1989) 219-225

[36] Y. Kurosu, K. Hibi, T. Sasaki and M. Saito, Influence of Temperature Control in Capillary Electrophoresis. *HRC & CC* **14** (1991) 200-203

[37] A. Vinther and H. Søeberg, Temperature Elevations of the Sample Zone in Free Solution Capillary Electrophoresis Under Stacking Conditions. *J. Chromatogr.* **559** (1991) 27-42

[38] H. Wätzig, The Measurement of Temperature Inside Capillaries for Electrophoresis Using Thermochromic Solutions. *Chromatographia* **33** (1992) 445-448

[39] M.S. Bello and P.G. Righetti, Unsteady Heat Transfer in Capillary Zone Electrophoresis. I. Mathematical Model *J. Chromatogr.* **606** (1992) 95-102

[40] M.S. Bello and P.G. Righetti, Unsteady Heat Transfer in Capillary Zone Electrophoresis. II. Computer Simulations. *J. Chromatogr.* **606** (1992) 103-111

[41] F.E.P. Mikkers, F.M. Everaerts and Th.P.E.M. Verheggen, Concentration Distributions in Free Zone Electrophoresis. *J. Chromatogr.* **169** (1979) 1-10

[42] R.A. Mosher, D.A. Saville and W. Thormann, The Dynamics of Electrophoresis, in: B.J. Radola (ed.), Electrophoresis Library. Verlag Chemie, Weinheim (1982)

[43] V. Sustacek, F. Foret and P. Bocek, Selection of the Background Electrolyte Composition with Respect to Electromigration Dispersion and Detection of Weakly Absorbing Substances in Capillary Zone Electrophoresis. *J. Chromatogr.* **545** (1991) 239-248

[44] R.-L. Chien and D.S. Burgi, Sample Stacking of an Extremely Large Injection Volume in High-Performance Capillary Electrophoresis. *Anal. Chem.* **64** (1992) 1046-1050

[45] K. Kleparniak and P. Bocek, Theoretical Background for Clinical and Biomedical Applications of Electromigration Techniques. *J. Chromatogr.* **569** (1991) 3-42

[46] X. Huang, W.F. Coleman and R. Zare, Analysis of Factors Causing Peak Broadening in Capillary Zone Electrophoresis. J. *Chromatogr.* **480** (1989) 95-110

[47] A. Vinther and H. Søeberg, Mathematical Model Describing Dispersion in Free Solution Capillary Electrophoresis Under Stacking Conditions. *J. Chromatogr.* **559** (1991) 3-26

[48] F. Foret, M. Deml and P. Bocek, Capillary Zone Electrophoresis - Quantitative Study of the Effects of Some Dispersive Processes on the Separation Efficiency. *J. Chromatogr.* **452** (1988) 601-613

[49] T.T. Lee and E.S. Yeung, Facilitating Data Transfer and Improving Precision in Capillary Zone Electrophoresis with Migration Indices. *Anal. Chem.* **63** (1991) 2842-2848

[50] T. Tsuda, Electroosmotic Flow and Electric Current in Capillary Electrophoresis. *J. Liq. Chromatogr.* **12** (1989) 2501-2514

[51] J. Colburn, B. Black, S.-M. Chen, D. Demorest, J. Wiktorowicz and K. Wilson, Factors Affecting Capillary Electrophoresis Analysis. *Applied Biosystems Research News,* Issue 1 (1990)

[52] J.W. Jorgenson, Capillary Zone Electrophoresis. *ACS Symp. Ser.* **335** (1987) 182-198

[53] H.J. Issaq, G.M. Janini, I.Z. Atamna, G.M. Muschik and J. Lukszo, Capillary Electrophoresis Separation of Small Peptides: Effect of pH, Buffer Additives, and Temperature. *J. Liq. Chromatogr.* **15** (1992) 1129-1142

[54] D. Demorest and R. Dubrow, Factors Influencing the Resolution and Quantification of Oligonucleotides Separated by Capillary Electrophoresis on a Gel-Filled Capillary. *J. Chromatogr.* **559** (1991), 43-56

[55] A. Guttman and N. Cooke, Effect of Temperature on the Separation of DNA Restriction Fragments in Capillary Gel Electrophoresis. *J. Chromatogr.* **559** (1991), 285-294

[56] R.S. Bush, A.S. Cohen and B.L. Karger, Influence of Column Temperature on the Electrophoretic Behavior of Myoglobin and α-Lactalbumin in High-Performance Capillary Electrophoresis. *Anal. Chem.* **63** (1991) 1346-1350

[57] S. Hoffstetter-Kuhn, A. Paulus, E. Gassmann and H.M. Widmer, Influence of Borate Complexation on the Electrophoretic Behavior of Carbohydrates in Capillary Electrophoresis. *Anal. Chem.* **63** (1991) 1541-1547

[58] C-W. Wang and E.S. Yeung, Temperature Programming in Capillary Zone Electrophoresis. *Anal. Chem.* **64** (1992) 502-506

[59] P.G. Righetti and T. Caravaggio, Isoelectric Points and Molecular Weights of Proteins - A Table -. *J. Chromatogr.* **127** (1967) 1-28

[60] P.G. Righetti, G. Tudor and K. Ek, Isoelectric Points and Molecular Weights of Proteins - A New Table -. *J. Chromatogr.* **220** (1981) 115-194

[61] A. Tiselius and H. Svensson, The Influence of Electrolyte Concentration on the Electrophoretic Mobility of Egg Albumin. *Trans. Faraday Soc.* **36** (1940) 16-22

[62] R.E. Offord, Electrophoretic Mobilities of Peptides on Paper and Their Use in the Determination of Amide Groups. *Nature* **5049** (1966) 591-593

[63] E.C. Rickard, M.M. Strohl and R.G. Nielsen, Correlation of Electrophoretic Mobilities from Capillary Electrophoresis with Physicochemical Properties of Proteins and Peptides. *Anal. Biochem.* **197** (1991) 197-207

[64] R. Consden, A.H. Gordon and A.J.P. Martin, Ionophoresis in Silica Jelly - Method for the Separation of Amino-Acids and Peptides. *Biochem.* **40** (1946) 33-37

[65] S. Wren, Optimization of pH in the Electrophoretic Separation of 2-, 3- and 4-Methylpyridines. *J. Microcol. Sep.* **3** (1991) 147-154

[66] J. Cai, J.T. Smith and Z.E. Rassi, Determination of the Ionization Constants of Weak Electrolytes by Capillary Zone Electrophoresis. *HRC & CC* **15** (1992) 30-32

[67] B.B. VanOrman, G.G. Liversidge, G.L. McIntire, T.M. Olefirowicz and A.G. Ewing, Effects of Buffer Composition on Electroosmotic Flow in Capillary Electrophoresis. *J. Microcol. Sep.* **2** (1990) 176-180

[68] H.J. Issaq, I.Z. Atamna, G.M. Muschik and G.M. Janini, The Effect of Electric Field Strength, Buffer Type and Concentration on Separation Parameters in Capillary Zone Electrophoresis. Chromatographia **32** (1991) 155-161

[69] D.C. Henry, *Proc. Roy. Soc., Series A* **133** (1931) 106-129

[70] R.J. Wieme, in: E. Heftman (ed.), Chromatography, A Laboratory Handbook of Chromatographic and Electrophoretic Methods, 3rd Edition. Van Nostrand Reinhold, New York (1975)

[71] I.Z. Atamna, C.J. Metral, G.M. Muschik and H.J. Issaq, Factors that Influence Mobility, Resolution and Selectivity in Capillary Zone Electrophoresis. III. The Role of Buffer's Anion. *J. Liq. Chromatogr.* **13(16)** (1990) 3201-3210

[72] I.Z. Atamna, C.J. Metral, G.M. Muschik and H.J. Issaq, Factors that Influence Mobility, Resolution and Selectivity in Capillary Zone Electrophoresis. II. The Role of the Buffer's Cation. *J. Liq. Chromatogr.* **13** (1990) 2517-2527

[73] H.J. Issaq, I.Z. Atamna, C.J. Metral and G.M. Muschik, Factors that Influence Mobility, Resolution and Selectivity in Capillary Zone Electrophoresis. I. Sodium Phosphate vs. Potassium Phosphate. *J. Liq. Chromatogr.* **13** (1990) 1247-1259

[74] Y. Walbroehl and J.W. Jorgenson, Capillary Zone Electrophoresis of Neutral Organic Molecules by Solvophobic Association with tetraalkylammonium Ion. *Anal. Chem.* **58** (1986) 479-481

[75] G.M. McLaughlin, J.A. Nolan, J.L. Lindahl, R.H. Palmieri, K.W. Anderson, S.C. Morris, J.A. Morrison and T.J. Bronzert, Pharmaceutical Drug Separations by HPCE: Practical Guidelines. *J. Liq. Chromatogr.* **15** (1992) 961-1021

[76] R. Kuhn and S. Hoffstetter-Kuhn, Chiral Separation by Capillary Electrophoresis. *Chromatographia* **34** (1992) 505-512

[77] D.Y. Pharr, Z.S. Fu, T.K. Smith and W.L. Hinze, Solubilization of Cyclodextrins for Analytical Applications. *Anal. Chem.* **61** (1989) 275-279
[78] Technical Brochure of Encapsin HPB, Janssen Biotech N.V., Belgium, 1992
[79] M. Tagvaei and G. Stewart, β-Cyclodextrin Solubility in Reversed-Phase High-Performance Liquid Chromatographic Eluents. *Anal. Chem.* **63** (1991) 1902-1904
[80] A.K. Chatjigakis, C. Donzé and A.W. Coleman, Solubility Behavior of β-Cyclodextrin in Water/Cosolvent Mixtures. *Anal. Chem.* **64** (1992) 1632-1634
[81] S. Terabe, H. Ozaki, K. Otsuka and T. Ando, Electrokinetic Chromatography with 2-O-Carboxymethyl-β-Cyclodextrin as a Moving "Stationary" Phase. *J. Chromatogr.* **332** (1985) 211-217
[82] A. Nardi, S. Fanali and F. Foret, Capillary Zone Electrophoretic Separation of Cyclodextrins with Indirect UV Photometric Detection. *Electrophoresis* **11** (1990) 774-776
[83] R.A. Mosher, The Use of Metal Ion-Supplemented Buffers to Enhance the Resolution of Peptides in Capillary Zone Electrophoresis. *Electrophoresis* **11** (1990) 765-769
[84] S. Fanali, L. Ossicini, F. Foret and P. Bocek, Resolution of Optical Isomers by Capillary Zone Electrophoresis: Study of Enantiomeric and Diastereomeric Co(III) Complexes with Ethylenediamine and Amino Acid Ligands. *J. Microcol. Sep.* **1** (1989) 190-194
[85] S. Fujiwara and S. Honda, Effect of Addition of Organic Solvent on the Separation of Positional Isomers in High-Voltage Capillary Zone Electrophoresis. *Anal. Chem.* **59** (1987) 487-490
[86] M. Zhu, D.L. Hansen, S. Burd and F. Gannon, Factors Affecting Free Zone Electrophoresis and Isoelectric Focusing in Capillary Electrophoresis. *J. Chromatogr.* **480** (1989) 311-319
[87] A.D. Tran, S. Park, P.J. List, O.T. Huynh, R.R. Ryall and P.A. Lane, Separation of Carbohydrate-Mediated Microheterogeneity of Recombinant Human Erythropoietin by Free Solution Capillary Electrophoresis. Effects of pH, Buffer Type and Organic Additives. *J. Chromatogr.* **542** (1991) 459-471
[88] A.S. Cohen, D.R. Najarian and B.L. Karger, Separation and Analysis of DNA Sequence Reaction Products by Capillary Gel Electrophoresis. *J. Chromatogr.* **516** (1990) 49-60
[89] R. Kuhn and H. Wagner, Application of Free-Flow Electrophoresis to the Preparative Purification of Basic Proteins from an E. Coli Cell Extract. *J. Chromatogr.* **481** (1989) 343-351
[90] H.H. Lauer, J.B. Ooms, B.H. Nieland and P. Waardenburg, Prince: A Variable Pressure Injection Device for Modular CE-Systems. Poster at the HPCE '91, Amsterdam, 1991
[91] R.S. Rush and B.L. Karger, Sample Injection with P/ACE System 2000. Importance of Temperature Control with Respect to Quantitation. *Beckman Instruments, Technical Information 104* (1990)
[92] X. Huang, M.J. Gordon and R.N. Zare, Bias in Quantitative Capillary Zone Electrophoresis Caused by Electrokinetic Sample Injection. *Anal. Chem.* **60** (1988) 375-377
[93] E.V. Dose and G.A. Guiochon, Internal Standardization Technique for Capillary Zone Electrophoresis. *Anal. Chem.* **63** (1991) 1154-1158
[94] R.J. Nielson, Reduction of Solvent Evaporation from Sample Microvials in Capillary Electrophoresis during Unattended Operation of the P/ACE System 2000. *Beckman Instruments, Technical Information 101* (1990)
[95] J. Teherani, R. Macomber and L. Day, Capillary Electrophoresis: An Integrated System with a Unique Split-Flow Sample Introduction Mechanism. *HRC&CC* **14** (1991) 10-14

[96] T.M. Olefirowicz and A.G. Ewing, Capillary Electrophoresis for Sampling Single Nerve Cells. *Chimia* **45** (1991) 106-108
[97] D.J. Rose and J.W. Jorgenson, Characterization and Automation of Sample Introduction Methods for Capillary Electrophoresis. *Anal. Chem.* **60** (1988) 642-648
[98] M. Yu and N.J. Dovichi, Sub-Femtomole Determination of DABSYL-Amino Acids with Capillary Zone Electrophoresis Separation and Laser-Induced Thermo-Optical Absorbance Detection. *Mikrochim. Acta* **III** (1988) 27-40
[99] Y. Wahlbroehl and J.W. Jorgenson, On-Column UV Absorption Detector for Open Tubular Capillary Zone Electrophoresis. *J. Chromatogr.* **315** (1984) 135-143
[100] J.S. Green and J.W. Jorgenson, Design of a Variable Wavelength UV Absorption Detector for On-Column Detection in Capillary Electrophoresis and Comparison of its Performance to a fixed Wavelength UV Absorption Detector. *J. Liq. Chromatogr.* **12** (1989) 2527-2561
[101] S. Kobayashi, T. Ueda and M. Kikumotop, Photodiode Array Detection in High-Performance Liquid Chromatography. *J. Chromatogr.* **480** (1989) 179-184
[102] G.J.M. Bruin, G. Stegeman, A.C. van Asten, X. Xu, J.C. Kraak and H. Poppe, Optimization and Evaluation of the Performance of Arrangements for UV Detection in High Resolution Separations Using Fused-Silica Capillaries. *J. Chromatogr.* **559** (1991), 163-182
[103] T. Wang, R.A. Hartwick and P.B. Champlin, Adjustable Aperture-Width Detector Cell for On-Column Detection in Capillary Zone Electrophoresis. *J. Chromatogr.* **462** (1989) 147-154
[104] F. Foret, M. Deml, V. Kahle and P. Bocek, On-Line Fiber Optic UV Detection Cell and Conductivity Cell for Capillary Zone Electrophoresis. *Electrophoresis* **7** (1986) 430-432
[105] A.E. Bruno, E. Gassmann, N. Pericles and K. Anton, On-column Capillary Flow Cell Utilizing Optical Waveguides for Chromatographic Applications. *Anal. Chem.* **61** (1989) 876-883
[106] J.P Chervet, R.E.J. van Soest and M. Ursem, Z-shaped Flow Cell for UV Detection in Capillary Electrophoresis. *J. Chromatogr.* **543** (1991) 439-449
[107] J.P. Chervet, LC Packings, Eur. Pat. No. 891067001990
[108] T. Tsuda, J.V. Sweedler and R.N. Zare, Rectangular Capillaries for Capillary Zone Electrophoresis. *Anal. Chem.* **62** (1990) 2149-2152
[109] T. Wang, J.H. Aiken, C.W. Huie and R.A. Hartwick, Nanoliter-Scale Multireflection Cell for Absorption Detection in Capillary Electrophoresis. *Anal. Chem.* **63** (1991) 1372-1376
[110] J.A. Taylor and E.S. Yeung, Axial-beam Absorbance Detection for Capillary Electrophoresis. *J. Chromatogr.* **550** (1991) 831-837
[111] X. Xi and E.S. Yeung, Axial-Beam Absorption Detection for Capillary Electrophoresis with a Conventional Light Source. *Appl. Spectroscopy* **45** (1991) 1199-1203
[112] M. Albin, R. Weinberger, E. Sapp and S. Moring, Fluorescence Detection in Capillary Electrophoresis: Evaluation of Derivatizing Reagents and Techniques. *Anal. Chem.* **63** (1991) 417-422
[113] S. Wu and N.J. Dovichi, Capillary Zone Electrophoresis Separation and Laser-Induced Fluorescence Detection of Zeptomole Quantities of Fluorescein Thiohydantoin Derivatives of Amino Acids. *Talanta* **39** (1992) 173-178
[114] T. Higashijima, T. Fuchigami, T. Imasaka and N. Ishibashi, Determination of Amino Acids by Capillary Zone Electrophoresis Based on Semiconductor Laser Fluorescence Detection. *Anal. Chem.* **64** (1992) 711-714

[115] B. Nickerson and J.W. Jorgenson, High Speed Capillary Zone Electrophoresis with Laser Induced Fluorescence Detection. *HRC&CC* **11** (1988) 533-534
[116] J.S. Green and J.W. Jorgenson, Variable-Wavelength On-Column Fluorescence Detector for Open-Tubular Zone Electrophoresis. *J. Chromatogr.* **352** (1986) 337-343
[117] S. Wu and N.J. Dovichi, High-Sensitivity Fluorescence Detector for Fluorescein Isothiocyanate Derivatives of Amino Acids Separated by Capillary Zone Electrophoresis. *J. Chromatogr.* **480** (1989) 141-155
[118] Y. Kurosu, T. Sasaki and M. Saito, Fluorescence Detection with an Immersed Flow Cell in Capillary Electrophoresis. *HRC&CC* **14** (1991) 186-190
[119] Y.-F. Cheng and N.J. Dovichi, Subattomole Amino Acid Analysis by Capillary Zone Electrophoresis and Laser-Induced Fluorescence. *Science* **242** (1988) 562-564
[120] L. Hernandez, R. Marquina, J. Escalona and N.A. Guzman, Detection and Quantification of Capillary Electrophoresis Zones by Fluorescence Microscopy. *J. Chromatogr.* **502** (1990) 247-255
[121] L. Hernandez, J. Escalona, N. Joshi and N. Guzman, Laser-Induced Fluorescence and Fluorescence Microscopy for Capillary Electrophoresis Zone Detection. *J. Chromatogr.* **559** (1991), 183-196
[122] C. Toulas and L. Hernandez, Applications of a Laser-Induced Fluorescence Detector for Capillary Electrophoresis to Measure Attomolar and Zeptomolar Amounts of Compounds. *LC-GC* **10** (1992) 471-476
[123] J. Liu, Y. Hsieh, D. Wiesler and M. Novotny, Design of 3-(4-Carboxybenzoyl)-2-quinolinecarboxaldehyde as a Reagent for Ultrasensitive Determination of Primary Amines by Capillary Electrophoresis Using Laser Fluorescence Detection. *Anal. Chem.* **63** (1991) 408-412
[124] J. Liu, O. Shirota and M. Novotny, Capillary Electrophoresis of Amino Sugars with Laser-Induced Fluorescence Detection. *Anal. Chem.* **63** (1991) 413-417
[125] J. Liu, O. Shirota and M. Novotny, Sensitive, Laser-Assisted Determination of Complex Oligosaccharide Mixtures Separated by Capillary Gel Electrophoresis at High Resolution. *Anal. Chem.* **64** (1992) 973-975
[126] J.F. Lawrence and R.W. Frei (eds.), Chemical Derivatization in Liquid Chromatography. J. Chromatogr. Library - Volume 7, Elsevier, Amsterdam - Oxford - New York (1976)
[127] P. Gozel, E. Gassmann, H. Michelson and R.N. Zare, Electrokinetic Resolution of Amino Acids Enantiomers with Copper(II)-Aspartame Support Electrolyte. *Anal. Chem.* **59** (1987) 44-49
[128] N.A. Guzman, J. Moschera, C.A. Bailey, K. Iqbal and A.W. Malick, Assay of Protein Drug Substances Present in Solution Mixtures by Fluorescamine Derivatization and Capillary Electrophoresis. *J. Chromatogr.* **598** (1992) 123-131
[129] H. Swerdlow, S. Wu, H. Harke and N.J. Dovichi, Capillary Gel Electrophoresis for DNA Sequencing. Laser-Induced Fluorescence Detection with the Sheath Flow Cuvette. *J. Chromatogr.* **516** (1990) 61-67
[130] DNA Modification Reagents for Use in Automated DNA Synthesis: A User's Manual, CLONTECH Laboratories, Palo Alto, CA.
[131] P. de Montigny, J.F. Stobaugh, R.S. Givens, R.G. Carlson, K. Srinivasachar, L.A. Sternson and T. Higuchi, Naphthalene-2,3-dicarboxaldehyde/Cyanide Ion: A Rationally Designed Fluorogenic Reagent for Primary Amines. *Anal. Chem.* **59** (1987) 1096-1101
[132] T. Ueda, R. Mitchell, F. Kitamura, T. Metcalf, T. Kuwana and A. Nakamoto, Separation of Naphthalene-2,3-dicarboxaldehyde-Labeled Amino Acids by

High-Performance Capillary Electrophoresis with Laser-Induced Fluorescence Detection. *J. Chromatogr.* **593** (1992) 265-274
[133] B. Nickerson and J.W. Jorgenson, High Sensitivity Laser Induced Fluorescence Detection in Capillary Zone Electrophoresis. *HRC&CC* **11** (1988) 878-881
[134] D.J. Rose and J.W. Jorgenson, Post-Capillary Fluorescence Detection in Capillary Zone Electrophoresis Using o-Phthaldialdehyde. *J. Chromatogr.* **447** (1988) 117-131
[135] D.J. Rose, Free-Solution Reactor for Post-Column Fluorescence Detection in Capillary Zone Electrophoresis. *J. Chromatogr.* **540** (1991) 343-353
[136] T. Tsuda, Y. Kobayashi, A. Hori, T. Matsumoto and O. Suzuki, Post-Column Detection for Capillary Zone Electrophoresis. *J. Chromatogr.* **456** (1988) 375-381
[137] X. Huang, J.A. Luckey, M.J. Gordon and R.N. Zare, Quantitative Analysis of Low Molecular Weight Carboxylic Acids by Capillary Zone Electrophoresis/Conductivity Detection. *Anal. Chem.* **61** (1989) 766-770
[138] X. Huang, M.J. Gordon and R.N. Zare, Effect of Electrolyte and Sample Concentration on the Relationship Between Sensitivity and Resolution in Capillary Zone Electrophoresis Using Conductivity Detection. *J. Chromatogr.* **480** (1989) 285-288
[139] X. Huang, T.-K. J. Pang, M.J. Gordon and R.N. Zare, On-Column Conductivity Detector for Capillary Zone Electrophoresis. *Anal. Chem.* **59** (1987) 2747-2749
[140] X. Huang, M.J. Gordon and R.N. Zare, Quantitation of Li^+ in Serum by Capillary Zone Electrophoresis with an On-Column Conductivity Detector. *J. Chromatogr.* **425** (1988) 385-390
[141] X. Huang, R.N. Zare, S. Sloss and A.G. Ewing, End-Column Detection for Capillary Zone Electrophoresis. *Anal. Chem.* **63** (1991) 189-192
[142] X. Huang and R.N. Zare, Improved End-Column Conductivity Detector for Capillary Zone Electrophoresis. *Anal. Chem.* **63** (1991) 2193-2196
[143] R.A. Wallingford and A.G. Ewing, Capillary Zone Electrophoresis with Electrochemical Detection in 12.7 µm Diameter Columns. *Anal. Chem.* **60** (1988) 1972-1975
[144] R.A. Wallingford and A.G. Ewing, Capillary Zone Electrophoresis with Electrochemical Detection. *Anal. Chem.* **59** (1987) 1762-1766
[145] X. Huang and R.N. Zare, Use of On-Column Frit in Capillary Zone Electrophoresis: Sample Collection. *Anal. Chem.* **62** (1990) 443-446
[146] T.J. O'Shea, R.D. Greenhagen, S.M. Lunte, C.E. Lunte, M.R. Smyth, D.M. Radzik and N. Watanabe, Capillary Electrophoresis with Electrochemical Detection Employing an On-Column Nafion Joint. *J. Chromatogr.* **593** (1992) 305-312
[147] R.A. Wallingford and A.G. Ewing, Amperometric Detection of Catechols in Capillary Zone Electrophoresis with Normal and Micellar Solutions, *Anal. Chem.* **60** (1988) 258-263
[148] T.W. Olefirowicz and A.G. Ewing, Capillary Electrophoresis in 2 and 5 µm Diameter Capillaries: Application to Cytoplasmic Analysis. *Anal. Chem.* **62** (1990) 1872-1876
[149] C.E. Engstrom-Silverman and A.G. Ewing, Copper Wire Amperometric Detector for Capillary Electrophoresis. *J. Microcol. Sep.* **3** (1991) 141-145
[150] T. Wang and R.A. Hartwick, Noise and Detection Limits of Indirect Absorption Detection in Capillary Zone Electrophoresis. *J. Chromatogr.* **607** (1992) 119-125
[151] F. Foret, S. Fanali, L. Ossicini and P. Bocek, Indirect Photometric Detection in Capillary Zone Electrophoresis. *J. Chromatogr.* **470** (1989) 299-308

[152] A.E. Vorndran, P.J. Oefner, H. Scherz and G.K. Bonn, Indirect UV Detection of Carbohydrates in Capillary Zone Electrophoresis. *Chromatographia* **33** (1992) 163-168

[153] S. Hjertén, K. Elenbring, K. Kilar, J. Liao, A. J. Chen, C. J. Sibert and M. D. Zhu, Carrier-Free Zone Electrophoresis, Displacement Electrophoresis and Isoelectric Focusing in a High-Performance Electrophoresis Apparatus. *J. Chromatogr.* **403** (1987) 47-61

[154] P. Jandik and W.R. Jones, Optimization of Detection Sensitivity in the Capillary Electrophoresis of Inorganic Anions. *J. Chromatogr.* **546** (1991) 431-443

[155] R. Kuhn, F. Erni, T. Bereuter and J. Häusler, Chiral Recognition and Enantiomeric Resolution Based on Host-Guest Complexation with Crown Ethers in Capillary Zone Electrophoresis. *Anal. Chem.* **64** (1992) 2815-2820

[156] W.R. Jones and P. Jandik, Controlled Changes of Selectivity in the Separation of Ions by Capillary Electrophoresis. *J. Chromatogr.* **546** (1991) 445-458

[157] F. Foret, S. Fanali, A. Nardi and P. Bocek, Capillary Zone Electrophoresis of Rare Earth Metals with Indirect UV Absorbance Detection. *Electrophoresis* **11** (1990) 780-783

[158] W. Beck and H. Engelhardt, Capillary Electrophoresis of Organic and Inorganic Cations with Indirect UV Detection. *Chromatographia* **33** (1992) 313-316

[159] P. Jandik, W.R. Jones, A. Weston and P.R. Brown, Electrophoretic Capillary Ion Analysis: Origins, Principles, and Applications. *LC-GC Intl.* **5** (1992) 20-27

[160] G.J.M. Bruin, A.C. van Asten, X. Xu and H. Poppe, Theoretical and Practical Aspects of Indirect Detection in Capillary Electrophoresis. *J. Chromatogr.* **608** (1992) 97-108

[161] M.W.F. Nielen, Quantitative Aspects of Indirect UV Detection in Capillary Zone Electrophoresis. *J. Chromatogr.* **588** (1991) 321-326

[162] W.G. Kuhr and E.S. Yeung, Indirect Fluorescence Detection of Native Amino Acids in Capillary Zone Electrophoresis. *Anal. Chem.* **60** (1988) 1832-1834

[163] W.G. Kuhr and E.S. Yeung, Optimization of Sensitivity and Separation in Capillary Zone Electrophoresis with Indirect Fluorescence Detection. *Anal. Chem.* **60** (1988) 2642-2646

[164] T.W. Garner and E.S. Yeung, Indirect Fluorescence Detection of Sugars Separated by Capillary Zone Electrophoresis with Visible Laser Excitation. *J. Chromatogr.* **515** (1990) 639-644

[165] L. Gross and E.S. Yeung, Indirect Fluorometric Detection of Cations in Capillary Zone Electrophoresis. *Anal. Chem.* **62** (1990) 427-431

[166] L.N. Amankwa and W.G. Kuhr, Indirect Fluorescence Detection in Micellar Electrokinetic Chromatography. *Anal. Chem.* **63** (1991) 1733-1737

[167] T.M Olefirowicz and A.G. Ewing, Capillary Electrophoresis with Indirect Amperometric Detection. *J. Chromatogr.* **499** (1990) 713-719

[168] C.-Y. Chen and M.D. Morris, On-Line Multichannel Raman Spectroscopic Detection System for Capillary Zone Electrophoresis. *J. Chromatogr.* **540** (1991) 355-363

[169] C.-Y. Chen and M.D. Morris, Raman Spectroscopic Detection System for Capillary Electrophoresis. *Appl. Spectrosc.* **42** (1988) 515-518

[170] J. Wu, T. Odake, T. Kitamori and T. Sawada, Ultrasensitive Detection for Capillary Zone Electrophoresis Using Laser-Induced Capillary Vibration. *Anal. Chem.* **63** (1991) 2216-2218

[171] A.E. Bruno, B. Krattiger, F. Maystre and H.M. Widmer, On-Column Laser-Based Refractive Index Detector for Capillary Electrophoresis. *Anal. Chem.* **63** (1991) 2689-2697
[172] C.-Y. Chen, T. Demana, S.-D. Huang and M.D. Morris, Capillary Zone Electrophoresis with Analyte Velocity Modulation. Application to Refractive Index Detection. *Anal. Chem.* **61** (1989) 1590-1593
[173] J. Pawliszyn, Nanoliter Volume Sequential Differential Concentration Gradient Detector. *Anal. Chem.* **60** (1988) 2796-2801
[174] J. Pawliszyn and J. Wu, Moving Boundary Capillary Electrophoresis with concentration Gradient Detection. *J. Chromatogr.* **559** (1991), 111-118
[175] J. Wu and J. Pawliszyn, Universal Detection for Capillary Isoelectric Focusing without Mobilization Using Concentration Gradient Imaging System. *Anal. Chem.* **64** (1992) 224-227
[176] J. Wu and J. Pawliszyn, Application of Capillary Isoelectric Focusing with Universal Concentration Gradient Detector to the Analysis of Protein Samples. *J. Chromatogr.* **608** (1992) 121-130
[177] K.C. Waldron and N.J. Dovichi, Sub-Femtomole Determination of Phenylthiohydantoin-Amino Acids: Capillary Electrophoresis and Thermooptical Detection. *Anal. Chem.* **64** (1992) 1396-1399
[178] C.W. Earle and N.J. Dovichi, Simultaneous Two-Color Thermo-Optical Absorbance Detector for Capillary Zone Electrophoresis. *J. Liq. Chromatogr.* **12** (1989) 2575-2585
[179] T. Demana, C.-Y. Chen and M.D. Morris, Non-Ideal Behavior in Analyte Velocity Modulation Capillary Electrophoresis. *HRC&CC* **13** (1992) 587-589
[180] S.L. Pentoney, R.N. Zare and J.F. Quint, On-Column Radioisotope Detection for Capillary Electrophoresis. *ACS Symp. Ser.* **434** (1990) 60-89
[181] S.L. Pentoney, R.N. Zare and J.F. Quint, Semiconductor Radioisotope Detector for Capillary Electrophoresis. *J. Chromatogr.* **480** (1989) 259-270
[182] S.L. Pentoney, R.N. Zare and J.F. Quint, On-Line Radioisotope Detection for Capillary Electrophoresis. *Anal. Chem.* **61** (1989) 1642-1647
[183] K.D. Altria, C.F. Simpson, A.K. Bharij and A.E. Theobald, A Gamma-Ray Detector for Capillary Zone Electrophoresis and its Use in the Analysis of Some Radiopharmaceuticals. *Electrophoresis* **11** (1990) 732-734
[184] H. Engelhardt, Lecture at "HPLC/CE '92 Bürgenstock Symposium" September 8-9, 1992, Bürgenstock, Switzerland
[185] R.M. McCormick and R.J. Zagursky, Polyimide Stripping Device for Producing Detection Windows on Fused-Silica Tubing Used in Capillary Electrophoresis. *Anal. Chem.* **63** (1991) 750-752
[186] J.A. Lux, U. Häusig and G. Schomburg, Production of Windows in Fused Silica Capillaries for In-Column Detection of UV-Absorbance or Fluorescence in Capillary Electrophoresis or HPLC. *HRC&CC* **13** (1990) 373-374
[187] Supelco SA International, New UV Transparent Column for Capillary Electrophoresis Allows Convenient On-Column Detection. *Supelco Biotext* **4** (1991) 12-14
[188] S. Wicar, M. Vilenchik, A. Belenkii, A.S. Cohen and B.L. Karger, Influence of Coiling on Performance in Capillary Electrophoresis Using Open Tubular and Polymer Network Columns. *J. Microcol. Sep.* **4** (1992) 339-348
[189] D.J. Rose and J.W. Jorgenson, Fraction Collector for Capillary Zone Electrophoresis. *J. Chromatogr.* **438** (1988) 23-34
[190] C. Fujimoto, T. Fulikawa and K. Jinno, Sample Collection by a Capillary Zone Electrophoretic System with an On-Column Fracture. *HRC&CC* **15** (1992) 201-203
[191] A. Guttman, A.S. Cohen, D.N. Heiger and B.L. Karger, Analytical and Micropreparative Ultrahigh Resolution of Oligonucleotides by Polyacrylamide Gel

High-Performance Capillary Electrophoresis. *Anal. Chem.* **62** (1990) 137-141

[192] Y.-F. Cheng, M. Fuchs, D. Andrews and W. Carson, Membrane Fraction Collection for Capillary Electrophoresis. *J. Chromatogr.* **608** (1992) 109-116

[193] D.M. Goodall, D.K. Lloyd and S.J. Williams, Current Trends in Capillary Electrophoresis. *LC-GC Intl.* **3** (1990) 29-38

[194] H. Wagner and E. Blasius (eds.), Praxis der elektrophoretischen Trennmethoden. Springer, Berlin-Heidelberg-New York-London-Paris-Tokyo (1989)

[195] S. Terabe, K. Otsuka, K. Ichikama, A. Tsuchiya and T. Ando, Electrokinetic Separations with Micellar Solutions and Open-Tubular Capillaries. Anal. Chem. **56** (1984) 111-113

[196] S. Hjertén, High Performance Electrophoresis: The Electrophoretic Counterpart of High-Performance Liquid Chromatography. *J. Chromatogr.* **270** (1983) 1-6

[197] A. S. Cohen and B. L. Karger, High-Performance Sodium Dodecyl Sulfate Polyacrylamide Gel Electrophoresis of Peptides and Proteins. *J. Chromatogr.* **397** (1987) 409-417

[198] A.S. Cohen, D.R. Najarian, A. Paulus, A. Guttman, J.A. Smith and B.L. Karger, Rapid Separation and Purification of Oligonucleotides by High-Performance Capillary Gel Electrophoresis. *Proc. Natl. Acad. Sci. USA* **85** (1988) 9660-9663

[199] M.V. Novotny, K.A. Cobb and J. Liu, Recent Advances in Capillary Electrophoresis of Proteins, Peptides and Amino Acids. *Electrophoresis* **11** (1990) 735-749

[200] K.A. Cobb, V. Dolnik and M. Novotny, Electrophoretic Separations of Proteins in Capillaries with Hydrolytically Stable Surface Structures. *Anal. Chem.* **62** (1990) 2478-2483

[201] J.K. Towns and F.R. Regnier, Capillary Electrophoretic Separations of Proteins Using Nonionic Surfactant Coatings. *Anal. Chem.* **63** (1991) 1126-1132

[202] J.K. Towns, J. Bao and F.R. Regnier, Synthesis and Evaluation of Epoxy Polymer Coatings for the Analysis of Proteins by Capillary Zone Electrophoresis. *J. Chromatogr.* **599** (1992) 227-237

[203] B.J. Herren, S.G. Shafer, J. Van Alstine, J.M. Harris and R.S. Snyder, Control of Electroosmosis in Coated Quartz Capillaries. *J. Colloid Interface Sci.* **115** (1987) 46-55

[204] G.J.M. Bruin, J.P. Chang, R.H. Kuhlman, K. Zegers, J.C. Kraak and H. Poppe, Capillary Zone Electrophoretic Separations of Proteins by Polyethylene Glycol-Modified Capillaries. *J. Chromatogr.* **471** (1989) 429-436

[205] S.A. Swedberg, Characterization of Protein Behavior in High-Performance Capillary Electrophoresis Using a Novel Capillary System. *Anal. Biochem.* **185** (1990) 51-56

[206] W. Nashabeh and Z. El Rassi, Capillary Zone Electrophoresis of Proteins with Hydrophilic Fused-Silica Capillaries. *J. Chromatogr.* **559** (1991), 367-384

[207] G.J.M. Bruin, R. Huisden, J.C. Kraak and H. Poppe, Performance of Carbohydrate-Modified Fused-Silica Capillaries for the Separation of Proteins by Zone Electrophoresis. *J. Chromatogr.* **480** (1989) 339-349

[208] J.K. Towns and F.E. Regnier, Polyethyleneimine-Bonded Phases in the Separation of Proteins by Capillary Electrophoresis. *J. Chromatogr.* **516** (1990) 69-78

[209] P.D. Grossman and D.S. Soane, Capillary Electrophoresis of DNA in entangled polymer solutions. *J. Chromatogr.* **559** (1991), 257-266

[210] J. Noolandi, Theory of DNA Gel Electrophoresis, in: A. Chrambach, M.J. Dunn and B.J. Radola (eds.), Advances in Electrophoresis, Volume 5. Verlag Chemie, Weinheim (1992)

[211] P.G. Righetti, S. Caglio, M. Saracchi and S. Quaroni, 'Laterally Aggregated' Polyacrylamide Gels for Electrophoresis. *Electrophoresis* **13** (1992) 587-595
[212] B.L. Karger and A.S. Cohen, Northeastern University; U.S. Patent 4,865,707, Sept. 12, 1989
[213] A. Paulus, E. Gassmann and M.J. Field, Calibration of Polyacrylamide Gel Columns for the Separation of Oligonucleotides by Capillary Electrophoresis. *Electrophoresis* **11** (1990) 702-708
[214] X.C. Huang, S.G. Stuart, P.F. Bente and T.M. Brennan, Capillary Gel Electrophoresis of Single-Stranded DNA Fragments with UV Detection. *J. Chromatogr.* **600** (1992) 289-295
[215] H.-F. Yin, J.A. Lux and G. Schomburg, Production of Polyacrylamide Gel Filled Capillaries for Capillary Gel Electrophoresis (CGE): Influence of Capillary Surface Pretreatment on Performance and Stability. *HRC&CC* **13** (1990) 624-627
[216] D.N. Heiger, A.S. Cohen and B.L. Karger, Separation of DNA Restriction Fragments by High Performance Capillary Electrophoresis with Low and Zero Crosslinked Polyacrylamide Using Continuous and Pulsed Electric Fields. *J. Chromatogr.* **516** (1990) 33-48
[217] J. Macek, U.R. Tjaden and J. van der Greef, Resolution and Concentration Detection Limit in Capillary Gel Electrophoresis. *J. Chromatogr.* **545** (1991) 177-182
[218] T. Wang, G.J. Bruin, J.C. Kraak and H. Poppe, Preparation of Polyacrylamide Gel-Filled Fused-Silica Capillaries by Photopolymerization with Riboflavin as the Initiator. *Anal. Chem.* **63** (1991) 2207-2208
[219] G.J.M. Bruin, T. Wang, X. Xu, J.C. Kraak and H. Poppe, Preparation of Polyacrylamide Gel-Filled Capillaries by Photopolymerization for Capillary Electrophoresis. *J. Microcol. Sep.* **4** (1992) 439-448
[220] J.A. Lux, H.-F. Yin and G. Schomburg, A Simple Method for the Production of Gel-Filled Capillaries for Capillary Gel Electrophoresis. *HRC&CC* **13** (1990) 436-437
[221] V. Dolnik, K.A. Cobb and M. Novotny, Preparation of Polyacrylamide Gel-Filled Capillaries for Capillary Electrophoresis. *J. Microcol. Sep.* **3** (1991) 155-159
[222] H. Drossman, J.A. Luckey, A.J. Kostichka, J. D'Cunha and L.M. Smith, High-Speed Separations of DNA Sequencing Reactions by Capillary Electrophoresis. *Anal. Chem.* **62** (1990) 900-903
[223] Y. Baba, T. Matsuura, K. Wakamoto, Y. Morita, Y. Nishitsu and M. Tsuhako, Preparation of Polyacrylamide Gel Filled Capillaries for Ultrahigh Resolution of Polynucleotides by Capillary Gel Electrophoresis. *Anal. Chem.* **64** (1992) 1221-1225
[224] S.R. Motsch, M.-H. Kleemiß and G. Schomburg, Production and Application of Capillaries Filled with Agarose Gel for Electrophoresis. *HRC&CC* **13** (1991) 629-632
[225] P. Bocek and A. Chrambach, Capillary Electrophoresis of DNA in Agarose Solutions at 40 °C. *Electrophoresis* **12** (1991) 1059-1061
[226] H.-J. Bode, The Use of Liquid Polyacrylamide in Electrophoresis. I. Mixed Gels Composed of Agar-Agar and Liquid Polyacrylamide *Anal. Biochem.* **83** (1977) 204-210
[227] H.-J. Bode, The Use of Liquid Polyacrylamide in Electrophoresis. II. Relationship Between Gel Viscosity and Molecular Sieving. *Anal. Biochem.* **83** (1977) 364-371

[228] M. Chiari, M. Nesi, M. Fazio and P.G. Righetti, Capillary Electrophoresis of Macromolecules in 'Syrupy' Solutions: Facts and Misfacts. *Electrophoresis* **13** (1992) 690-697
[229] K. Ganzler, K.S. Greve, A.S. Cohen and B.L. Karger, High-Performance Capillary Electrophoresis of SDS-Protein Complexes Using UV-Transparent Polymer Networks. *Anal. Chem.* **64** (1992) 2665-2671
[230] S. Terabe, K. Otsuka and T. Ando, Electrokinetic Chromatography with Micellar Solution and Open-Tubular Capillary. *Anal. Chem.* **57** (1985) 834-841
[231] K. Otsuka, S. Terabe and T. Ando, Electrokinetic Chromatography with Micellar Solutions - Retention Behavior and Separation of Chlorinated Phenols. *J. Chromatogr.* **348** (1985) 39-47
[232] U. Pfüller (ed.), Mizellen-Vesikel-Mikroemulsionen. Springer, Berlin-Heidelberg-New York-London-Paris-Tokyo (1986) 26
[233] M.G. Khaledi, S.C. Smith and J.K. Strasters, Micellar Electrokinetic Capillary Chromatography of Acidic Solutes: Migration Behavior and Optimization Strategies. *Anal. Chem.* **63** (1991) 1820-1830
[234] J.P. Foley, Optimization of Micellar Electrokinetic Chromatography. *Anal. Chem.* **62** (1990) 1302-1308
[235] D.E. Burton, M.J. Sepaniak and M.P. Mascarinec, Evaluation of the Use of Various Surfactants in Micellar Electrokinetic Capillary Chromatography. *J. Chromatogr. Sci.* **25** (1987) 514-518
[236] S.A. Swedberg, Use of Non-Ionic and Zwitterionic Surfactants to Enhance Selectivity in High-Performance Capillary Electrophoresis. *J. Chromatogr.* **503** (1990) 449-452
[237] R.O. Cole and M.J. Sepaniak, The Use of Bile Salt Surfactants in Micellar Electrokinetic Capillary Chromatography. *LC-GC* **10** (1991) 380-385
[238] H. Nishi, T. Fukuyama, M. Matsuo and S. Terabe, Separation and Determination of the Ingredients of a Cold Medicine by Micellar Electrokinetic Chromatography with Bile Salts. *J. Chromatogr.* **498** (1990) 313-323
[239] A.T. Balchunas and M.J. Sepaniak, Gradient Elution for Micellar Electrokinetic Capillary Chromatography. *Anal. Chem.* **60** (1988) 617-621
[240] O. Wu, H.A. Claessens and C.A. Cramers, The Separation of Herbicides by Micellar Electrokinetic Capillary Chromatography. *Chromatographia* **34** (1992) 25-30
[241] J. Gorse, A.T. Balchunas, D.F. Swaile and M.J. Sepaniak, Effects of Organic Mobile Phase Modifiers in Micellar Electrokinetic Capillary Chromatography. *HRC&CC* **11** (1988) 554-558
[242] G.M. Janini and H.I. Issaq, Micellar Electrokinetic Capillary Chromatography: Basic Considerations and Current Trends. *J. Liq. Chromatogr.* **15** (1992) 927-960)
[243] S. Terabe, Y. Ishihama, H. Nishi, F. Fukuyama and K. Otsuka, Effect of Urea Addition in Micellar Electrokinetic Chromatography. *J. Chromatogr.* **545** (1991) 359-368
[244] T. Kaneta, S. Tanaka, M. Taga and H. Yoshida, Effect of Addition of Glucose on Micellar Electrokinetic Capillary Chromatography with Sodium Dodecyl Sulfate. *J. Chromatogr.* **609** (1992) 369-374
[245] K. Otsuka, J. Kawahara, K. Tatekawa and S. Terabe, Chiral Separations by Micellar Electrokinetic Chromatography with Sodium N-Dodecanoyl-L-Valinate. *J. Chromatogr.* **559** (1991) 209-214
[246] R.A. Wallingford, P.D. Curry and A.G. Ewing, Retention of Catechols in Capillary Electrophoresis with Micellar and Mixed Micellar Solutions. *J. Microcol. Sep.* **1** (1989) 23-27

[247] H. Nishi, N. Tsumagari and S. Terabe, Effect of Tetraalkylammonium Salts on Micellar Electrokinetic Chromatography on Ionic Substances. *Anal. Chem.* **61** (1989) 2434-2439

[248] A.S. Cohen, S. Terabe, J.A. Smith and B.L. Karger, High-Performance Capillary Electrophoretic Separation of Bases, Nucleosides and Oligonucleotides: Retention Manipulation via Micellar Solutions and Metal Additives. *Anal. Chem.* **59** (1987) 1021-1027

[249] H. Nishi and M. Matsuo, Separation of Corticosteroids and Aromatic Hydrocarbons by Cyclodextrin-Modified Micellar Electrokinetic Chromatography. *J. Liq. Chromatogr.* **14** (1991) 973-986

[250] J. Liu, K.A. Cobb and M. Novotny, Capillary Electrophoretic Separations of Peptides Using Micelle-Forming Compounds and Cyclodextrins as Additives. *J. Chromatogr.* **519** (1990) 189-197

[251] J. Prunonosa, A.D. Gascon and L. Gouesclou, Chiral Separation of the Antihypertensive Drug Cicletanine by MECC Using P/ACE System 2000. *Beckman Instruments, Technical Information DS-798* (1991)

[252] Y. Miyashita and S. Terabe, Separation of Dansyl D,L-Amino Acids by Micellar Electrokinetic Capillary Chromatography with and without Cyclodextrins Using P/ACE System 2000. *Beckman Instruments, Technical Information DS-767* (1990)

[253] P. Bocek, M. Deml, P. Gebauer, V. Dolnik (eds.), Analytical Isotachophoresis. VCH, Weinheim, Germany, (1988)

[254] O. Vesterberg, Synthesis and Isoelectric Fractionation of Carrier Ampholytes. *Acta Chem. Scand.* **23** (1969) 2653-2666

[255] H. Svensson, Isoelectric Fractionation, Analysis and Characterization of Ampholytes in Natural pH Gradients II. *Acta Scand. Chem.* **16** (1962) 132-138

[256] O. Vesterberg and H. Svensson, Isoelectric Fractionation, Analysis and Characterization of Ampholytes in Natural pH Gradients IV. *Acta*

[257] S. Hjertén and M.-D. Zhu, Adaption of the Equipment of High-Performance Electrophoresis to Isoelectric Focusing. *J. Chromatogr.* **346** (1985) 265-270

[258] M. Zhu, R. Rodriguez and T. Wehr, Optimizing Separation Parameters in Capillary Isoelectric Focusing. *J. Chromatogr.* **559** (1991) 479-488

[259] J.R. Mazzeo and I.S. Krull, Capillary Isoelectric Focusing of Proteins in Uncoated Fused-Silica Capillaries Using Polymeric Additives. *Anal. Chem.* **63** (1991) 2852-2857

[260] T. Tsuda, Electrochromatography Using High Applied Voltage. *Anal. Chem.* **59** (1987) 521-523

[261] J.H. Knox and I.H. Grant, Miniaturization in Pressure and Electroosmotically Driven Liquid Chromatography: Some theoretical Considerations. *Chromatographia* **24** (1987) 135-143

[262] J.H. Knox and I.H. Grant, Electrochromatography in Packed Tubes Using 1.5 to 50 μm Silica Gels and ODS Bonded Silica Gels. *Chromatographia* **32** (1991) 317-328

[263] H. Yamamoto, J. Baumann and F. Erni, Electrokinetic Reversed Phase Chromatography with Packed Columns. *J. Chromatogr.* **593** (1992) 313-319

[264] T. Tsuda, Chromatographic Behavior in Electrochromatography. *Anal. Chem.* **60** (1988) 1677-1680

[265] T. Tsuda, High Performance Electrochromatography. *LC-GC Intl.* **5** (1992) 26-36

[266] J.A. Olivares, N.T. Nguyen, C.R. Yonker and R.D. Smith, On-Line Mass Spectrometric Detection for Capillary Electrophoresis. *Anal. Chem.* **59** (1987) 1230-1232

[267] R.D. Smith, C.J. Barinaga and H.R. Udseth, Improved Electrospray Ionization Interface for Capillary Zone Electrophoresis-Mass Spectrometry. *Anal. Chem.* **60** (1988) 1948-1952
[268] R.D. Smith, H.R. Udseth, C.J. Barinaga and C.G. Edmonts, Instrumentation for High-Performance Capillary Electrophoresis-Mass Spectrometry. *J. Chromatogr.* **559** (1991), 197-208
[269] E.D. Lee, W. Mück, J.D. Henion and T.R. Covey, Liquid Junction Coupling for Capillary Zone Electrophoresis/Ion Spray Mass Spectrometry. *Biomed. Envir. Mass Spectrom.* **18** (1989) 844-850
[270] W. Nichols, J. Zweigenbaum, F. Garcia, M. Johansson and J. Henion, CE-MS for Industrial Applications Using a Liquid Junction with Ion-Spray and CF-FAB Mass Spectrometry. *LC-GC* **10** (1991) 676-686
[271] R.M. Caprioli, W.T. Moore, M. Martin, B.B. DaGue, K. Wilson and S. Moring, Coupling Capillary Zone Electrophoresis and Continuous-Flow Fast Atom Bombardment Mass Spectrometry for the Analysis of Peptide Mixtures. *J. Chromatogr.* **480** (1989) 247-257
[272] M.A. Moseley, L.J. Deterging, K.B. Tomer and J.W. Jorgenson, Coupling of Capillary Zone Electrophoresis and Capillary Liquid Chromatography with Coaxial Continuous-Flow Fast Atom Bombardment Tandem Sector Mass Spectrometry. *J. Chromatogr.* **480** (1989) 197-209
[273] C.J. Giddings, Concepts and Comparisons in Multidimensional Separation. *HRC & CC* **10** (1987) 319-323
[274] M.M. Bushey and J.W. Jorgenson, Automated Instrumentation for Comprehensive Two-Dimensional High-Performance Liquid Chromatography/ Capillary Zone Electrophoresis. *Anal. Chem.* **62** (1990) 978-984
[275] D. Kaniansky and J. Marak, On-Line Coupling of Isotachophoresis with Capillary Zone Electrophoresis. *J. Chromatogr.* **498** (1990) 191-204
[276] V. Dolnik, K.A. Cobb and M. Novotny, Capillary Zone Electrophoresis of Dilute Samples with Isotachophoretic Preconcentration. *J. Microcol. Sep.* **2** (1990) 127-131
[277] F. Foret, V. Sustacek and P. Bocek, On-Line Isotachophoretic Sample Preconcentration for Enhancement of Zone Detectability in Capillary Zone Electrophoresis. *J. Microcol. Sep.* **2** (1990) 229-233
[278] D.S. Stegehuis, H. Irth, U.R. Tjaden and J. van der Greef, Isotachophoresis as an On-Line Concentration Pretreatment Technique in Capillary Electrophoresis. *J. Chromatogr.* **538** (1991) 393-402
[279] D.S. Stegehuis, U.R. Tjaden and J. van der Greef, Analyte Focusing in Capillary Electrophoresis Using On-Line Isotachophoresis. *J. Chromatogr.* **591** (1992) 341-349
[280] V. Horejsi and M. Ticha, Qualitative and Quantitative Applications of Affinity Electrophoresis for the Study of Protein-Ligand Interactions: A Review. *J. Chromatogr.* **376** (1986) 49-67
[281] K. Shimura, Review - Progress in Affinophoresis. *J. Chromatogr.* **510** (1990) 251-270
[282] Y.-H. Chu, L.Z. Avila, H.A. Biebuyck and G.M. Whitesides, Use of Affinity Capillary Electrophoresis to Measure Binding Constants of Ligands to Proteins. *J. Med. Chem.* **35** (1992) 2915-2917
[283] S. Honda, A. Taga, K. Suzuki, S. Suzuki and K. Kakehi, Determination of the Association Constant of Monovalent Mode Protein-Sugar Interaction by Capillary Zone Electrophoresis. *J. Chromatogr.* **597** (1992) 377-382
[284] Y.-H. Chu and G.M. Whitesides, Affinity Capillary Electrophoresis Can Simultaneously Measure Binding Constants of Multiple Peptides to Vancomycin. *J. org. Chem.* **57** (1992) 3524-3525

[285] J. L. Carpenter, P. Camilleri, D. Dhanak and G. Goodall, A Study of the Binding of Vancomycin to Dipeptides Using Capillary Electrophoresis. *J. Chem. Soc., Chem. Commun.* (1992) 804-806

[286] A. Guttman and N. Cooke, Capillary Gel Affinity Electrophoresis of DNA Fragments. *Anal. Chem.* **63** (1991) 2038-2042

[287] S.E. Moring, J.C. Colburn, P.D. Grossman and H.H. Lauer, Analytical Aspects of an Automated Capillary Electrophoresis System. *LC-GC Intl.* **3** (1990) 46-52

[288] R.-L. Chien and D.S. Burgi, On-Column Sample Concentration Using Field Amplification in CZE. *Anal. Chem.* **64** (1992) 489A-496A

[289] D.S. Burgi and R.-L. Chien, Optimization in Sample Stacking for High-Performance Capillary Electrophoresis. *Anal. Chem.* **63** (1991) 2042-2047

[290] D.S. Burgi and R.-L. Chien, Application of Sample Stacking to Gravity Injection in Capillary Electrophoresis. *J. Microcol. Sep.* **3** (1991) 199-202

[291] D.S. Burgi and R.-L. Chien, Improvement in the Method of Sample Stacking for Gravity Injection in Capillary Zone Electrophoresis. *Anal. Biochem.* **202** (1992) 306-309

[292] R.-L. Chien and D.S. Burgi, Field-Amplified Sample Injection in High-Performance Capillary Electrophoresis. *J. Chromatogr.* **559** (1991), 141-152

[293] R.-L. Chien and D.S. Burgi, Field-Amplified Polarity-Switching Sample Injection in High-Performance Capillary Electrophoresis. *J. Chromatogr.* **559** (1991), 153-162

[294] S.C. Smith, J.K. Strasters and M.G. Khaledi, Influence of Operating Parameters on Reproducibility in Capillary Electrophoresis. *J. Chromatogr.* **559** (1991) 57-68

[295] X. Huang, W.F. Coleman and R.N. Zare, Analysis of Factors Causing Peak Broadening in Capillary Zone Electrophoresis. *J. Chromatogr.* **480** (1989) 95-110

[296] E.V. Dose and G.A. Guiochon, Internal Standardization Technique for Capillary Zone Electrophoresis. *Anal. Chem.* **63** (1991) 1154-1158

[297] T.T. Lee and E.S. Yeung, Compensating for Instrumental and Sampling Biases Accompanying Electrokinetic Injection in Capillary Zone Electrophoresis. *Anal. Chem.* **64** (1992) 1226-1231

[298] S.E. Moring, J.C. Colburn, P.D. Grossman and H.H. Lauer, Analytical Aspects of an Automated Capillary Electrophoresis System. *LC-GC* **8** (1990) 34-46

[299] A. Nardi, S. Fanali and F. Foret, Capillary Zone Electrophoretic Separation of Cyclodextrins with Indirect UV Photometric Detection. *Electrophoresis* **11** (1990) 774-776

[300] A. Weston, P.R. Brown, P. Jandik, W.R. Jones and A.L. Heckenberg, Factors Affecting the Separation of Inorganic Metal Cations by Capillary Electrophoresis. *J. Chromatogr.* **593** (1992) 259-295

[301] A. Weston, P.R. Brown, P. Jandik, A.L. Heckenberg and W.R. Jones, Optimization of Detection Sensitivity in the Analysis of Inorganic Cations by Capillary Ion Electrophoresis Using Indirect Photometric Detection. *J. Chromatogr.* **608** (1992) 395-402

[302] K. Bächmann, K.-H. Steeg, T. Groh, I. Haumann, J. Boden and H. Holthues, Analysis of Cations and Anions in Small Volumes Using Micro-LC and CE. *J. Microcol. Sep.* **4** (1992) 431-438

[303] K. Bächmann, J. Boden and I. Haumann, Indirect Fluorimetric Detection of Alkali and Alkaline Earth Metal Ions in Capillary Zone Electrophoresis with Cerium(III) as Carrier Electrolyte. *J. Chromatogr.* **626** (1992) 259-265

[304] J. Romano, P. Jandik, W.R. Jones and P.E. Jackson, Optimization of Inorganic Capillary Electrophoresis for the Analysis of Anionic Solutes in Real Samples. *J. Chromatogr.* **546** (1991) 411-421
[305] T. Groh and K. Bächmann, Selective Influencing of the Migration Time of SO_4^{2-} by Forming Neutral Species with Pb^{2+} in Capillary Zone Electrophoresis. *Electrophoresis* **13** (1992) 458-461
[306] B.F. Kenney, Determination of Organic Acids in Food Samples by Capillary Electrophoresis. *J. Chromatogr.* **546** (1991) 423-430
[307] R.A. Carpio, R. Mariscal and J. Welch, Determination of Boron and Phosphorus in Borophosphosilicate Thin Films on Silicon Substrates by Capillary Electrophoresis. *Anal. Chem.* **64** (1992) 2123-2129
[308] T. Kaneta, S. Tanaka, M. Taga and H. Yoshida, Migration Behavior of Inorganic Anions in Micellar Electrokinetic Capillary Chromatography Using a Cationic Surfactant. *Anal. Chem.* **64** (1992) 798-801
[309] C.L. Ng, H.K. Lee and S.F.Y. Li, Analysis of Food Additives by Ion-Pairing Electrokinetic Chromatography. *J. Chromatogr. Sci.* **30** (1992) 167-170
[310] W. Buchberger and P.R. Haddad, Effects of Carrier Electrolyte Composition on Separation Selectivity in Capillary Zone Electrophoresis of Low-Molecular-Mass Anions. *J. Chromatogr.* **608** (1992) 59-64
[311] S. Fujiwara and S. Honda, Determination of Cinnamic Acid and its Analogues by Electrophoresis in a Fused Silica Capillary Tube. *Anal. Chem.* **58** (1986) 1811-1814
[312] C. Bjergegaard, S. Michaelsen and H. Sørensen, Determination of Phenolic Carboxylic Acids by Micellar Electrokinetic Capillary Chromatography and Evaluation of Factors Affecting the Method. *J. Chromatogr.* **608** (1992) 403-411
[313] H.J. Issaq, K.Delviks, G.M. Janini and G.M. Muschik, Capillary Zone Electrophoretic Separation of Homovanillic and Vanillylmandelic Acids. *HRC&CC* **15** (1992) 3193-3201
[314] P.A. Tienstra, J.A.M. van Riel, M.D. Mingorance and C. Olieman, Assessment of the Capabilities of Capillary Zone Electrophoresis for the Determination of Hippuric and Orotic Acid in Whey. *J. Chromatogr.* **608** (1992) 357-361
[315] T. Tsuda, K. Nomura and G. Nakagawa, Separation of Organic Ions by High-Voltage Capillary Electrophoresis. *J. Chromatogr.* **264** (1983) 385-392
[316] C.S. Weiss, J.S. Hazlett, M.H. Datta and M.H. Danzer, Determination of Quaternary Ammonium Compounds by Capillary Electrophoresis Using Direct and Indirect Detection. *J. Chromatogr.* **608** (1992) 325-332
[317] W.D. Pfeffer and E.S. Yeung, Electroosmotically Driven Electrochromatography of Anions Having Similar Electrophoretic Mobilities by Ion Pairing. *J. Chromatogr.* **557** (1991) 125-136
[318] P.L. Desbène, C. Rony, B. Desmazières and J.C. Jacquier, Analysis of Alkylaromatic Sulphonates by High-Performance Capillary Electrophoresis. *J. Chromatogr.* **608** (1992) 375-383
[319] J.A. Zweigenbaum, Analysis of Sodium Dodecylbenzenesulfonates by Capillary Electrophoresis with P/ACE System 2000,*Technical Information Bulletin No DS-805* (1990)
[320] J.B. Poli and M.R. Schure, Separation of Poly(styrenesulfonates) by Capillary Electrophoresis with Polymeric Additives. *Anal. Chem.* **64** (1992) 896-904
[321] M.E. Swartz, Method Development and Selectivity Control for Small Molecule Pharmaceutical Separations by Capillary Electrophoresis. *J. Liq. Chromatogr.* **14** (1991) 923-938

[322] H. Nishi and S. Terabe, Application of Micellar Electrokinetic Chromatography to Pharmaceutical Analysis. *Electrophoresis* **11** (1990) 691-701

[323] Y. Miyashita, S. Terabe and H. Nishi, Separation of Some Antibiotics and Corticosteroids by Micellar Electrokinetic Capillary Chromatography Using P/ACE System 2000. *Beckman Instruments, Appl. Brief DS-766* (1990)

[324] A. Wainwright, Capillary Electrophoresis Applied to the Analysis of Pharmaceutical Compounds. *J. Microcol. Sep.* **2** (1990) 166-175

[325] C.L. Ng, H.K. Lee and S.F.Y. Li, Systemic Optimization of Capillary Electrophoretic Separation of Sulfonamides. *J. Chromatogr.* **598** (1992) 133-138

[326] M.T. Ackermans, J.L. Beckers, F.M. Everaerts, H. Hoogland and M.J.H. Tomassen, Determination of Sulfonamides in Pork Meat Extracts by Capillary Zone Electrophoresis. *J. Chromatogr.* **596** (1992) 101-109

[327] I.J. Johansson, R.Pavelka and J.D. Henion, Determination of Small Drug Molecules by Capillary Electrophoresis-Atmospheric Pressure Ionization Mass Spectrometry. *J. Chromatogr.* **559** (1991) 515-528

[328] M.T. Ackermans, F.M. Everaerts and J.L. Beckers, Determination of Aminoglycoside Antibiotics in Pharmaceuticals by Capillary Zone Electrophoresis with Indirect UV Detection Coupled with Micellar Electrokinetic Capillary Chromatography. *J. Chromatogr.* **606** (1992) 229-235

[329] C.-X. Zhang, Z.-P. Sun, D.-K. Ling and Y.-J. Zhang, Separation of Tetracycline and its Degradation Products by Capillary Zone Electrophoresis. *J. Chromatogr.* **627** (1992) 281-286

[330] S. Fazio, R. Vivilecchia, L. Lesueur and J. Sheridan, Capillary Zone Electrophoresis: Some Promising Pharmaceutical Applications. *Am. Biotechnol. Lab.* **8** (1990) 10-17

[331] W. Thorman, A. Minger, S. Molteni, J. Caslavska and P. Gebauer, Determination of Substituted Purines in Body Fluids be Micellar Electrokinetic Capillary Chromatography with Direct Sample Injection. *J. Chromatogr.* **593** (1992) 275-288

[332] K.-J. Lee, G. S. Heo, N.J. Kim and D.-C. Moon, Separation of Theophylline and its Analogeus by Micellar Electrolkinetic Chromatography: Application to the Determination of Theophylline in Human Plasma. *J. Chromatogr.* **577** (1992) 135-141

[333] R.R. Kerr and L. Jung, Capillary Electrophoresis of Opiates with the P/ACE System 2000. *Beckman Instruments, Appl. Brief DS-783* (1990)

[334] J. Vindevogel and P. Sandra, Resolution Optimization in Micellar Electrokinetic Chromatography: Use of Plackett-Burman Statistical Design for the Analysis of Testosterone Esters. *Anal. Chem.* **63** (1991) 1530-1536

[335] E.W. Tsai, M.M. Singh, H.H. Lu, D.P. Ip and M.A. Brooks, Application of Capillary Electrophoresis to Pharmaceutical Analysis - Determination of Alendronate in Dosage Forms. *J. Chromatogr.* **626** (1992) 245-250

[336] N.J. Reinhoud, U.R. Tjaden, H. Irth and J. van der Greef, Bioanalysis of Some Anthracyclines in Human Plasma by Capillary Electrophoresis with Laser-Induced Fluorescence Detection. *J. Chromatogr.* **574** (1992) 327-334

[337] K.-L. Lee and G.S. Heo, Analysis of Antiepileptic Drugs in Human Plasma Using Micellar Electrokinetic Capillary Chromatography. *J. Chromatogr.* **608** (1992) 243-250

[338] C.P. Ong, C.L. Ng, H.K. Lee and S.F.Y. Li, Determination of Antihistamines in Pharmaceuticals by Capillary Electrophoresis. *J. Chromatogr.* **588** (1991) 335-339

[339] W. Thormann, P. Meier, C. Marcolli and F. Binder, Analysis of Barbiturates in Human Serum and Urine by High-performance Capillary Electrophoresis-Micellar Electrokinetic Capillary Chromatography with On-Column Multi-Wavelength Detection. *J. Chromatogr.* **545** (1991) 445-460

[340] H. Soini, T. Tsuda and M. Novotny, Electrochromatographic Solid-Phase Extraction for Determination of Cimetidine in Serum by Micellar Electrokinetic Capillary Chromatography. *J. Chromatogr.* **559** (1991) 547-558

[341] X.-Z. Qin, D.P. Ip and E.W. Tsai, Determination and Rotamer Separation of Enalapril Maleate by Capillary Electrophoresis. *J. Chromatogr.* **629** (1992) 251-258

[342] P. Lukkari, J. Jumppanen, K. Jinno, H. Elo and M.-L. Riekkola, Determination of Bis(amidinohydrazones) by Micellar Electrokinetic Capillary Chromatography. *J. Pharm. & Biomed. Anal.* **10** (1992) 561-575

[343] P. Lukkari, J. Jumppanen, T. Holma, H. Sirén, K. Jinno, H. Elo and M.-L. Riekkola, Effect of Buffer Solution on the Elution Order and Separation of Bis(amidinohydrazones) by Micellar Electrokinetic Capillary Chromatography. *J. Chromatogr.* **608** (1992) 317-324

[344] U. Seitz, P.J. Oefner, S. Nathakarnkitkool, M. Popp and G.K. Bonn, Capillary Electrophoretic Analysis of Flavonoids. *Electrophoresis* **13** (1992) 35-38

[345] C.L. Ng., C.P. Ong, H.K. Lee and S.F.Y. Li, Systematic Optimization of Micellar Electrokinetic Chromatographic Separation of Flavonoids. *Chromatographia* **34** (1992) 166-172

[346] C.L. Ng., C.P. Ong, H.K. Lee and S.F.Y. Li, Analysis of Flavonoids by Micellar Electrokinetic Chromatography with P/ACE System 2000. *Beckman Instruments, Technical Information DS-821* (1992)

[347] P. Morin, F. Villard and M. Dreux, Borate Complexation of Flavonoid-O-Glycosides in Capillary Electrophoresis. I. Separation of Flavonoid-7-O-Glycosides Differing in Their Flavonoid Aglycone. *J. Chromatogr.* **628** (1993) 153-160

[348] P. Morin, F. Villard and M. Dreux, Borate Complexation of Flavonoid-O-Glycosides in Capillary Electrophoresis. II. Separation of Flavonoid-3-O-Glycosides Differing in Their Sugar Moiety. *J. Chromatogr.* **628** (1993) 161-169

[349] R.A. Wallingford and A.G. Ewing, Retention of Ionic and Non-Ionic Catechols in Capillary Zone Electrophoresis with Micellar Solutions. *J. Chromatogr.* **441** (1988) 299-309

[350] T. Kaneta, S. Tanaka and H. Yoshida, Improvement of Resolution in the Capillary Electrophoretic Separation of Catecholamines by Complex Formation with Boric Acid and Control of Electroosmosis with a Cationic Surfactant. *J. Chromatogr.* **538** (1991) 385-391

[351] C.P Ong, S.F. Pang, S.P. Low, H.K. Lee and S.F.Y. Li, Migration Behaviour of Catechols and Catecholamines in Capillary Electrophoresis. *J. Chromatogr.* **559** (1991), 529-536

[352] T. Tsuda, Y. Kobayashi, A. Hori, T. Matsumoto and O. Suzuki, Separation of Polyamines in Rat Tissues by Capillary Electrophoresis. *J. Microcol. Sep.* **2** (1990) 21-25

[353] Y. Ma, R. Zhang and C.L. Cooper, Indirect Photometric Detection of Polyamines in Biological Samples Separated by High-Performance Capillary Electrophoresis. *J. Chromatogr.* **608** (1992) 93-96

[354] J.A. Zweigenbaum, Analysis of Water-Soluble Vitamins by Capillary Electrophoresis with P/ACE System 2000. *Beckman Instruments, Technical Information DS-804* (1991)

[355] D.F. Swaile, D.E. Burton, A.T. Balchunas and M.J. Sepaniak, Pharmaceutical Analysis Using Micellar Electrokinetic Capillary Chromatography. *J. Chromatogr. Sci.* **26** (1988) 406-409

[356] D. Lambert, C. Adjalla, F. Felden, S. Benhayoun, J.P. Nicolas and J.L. Guéant, Identification of Vitamin B_{12} and Analogues by High-Performance Capillary Electrophoresis and Comparison with High-Performance Liquid Chromatography. *J. Chromatogr.* **608** (1992) 311-316

[357] S. Fujiwara, S. Iwase and S. Honda, Analysis of Water-Soluble Vitamins by Micellar Electrokinetic Capillary Chromatography. *J. Chromatogr.* **447** (1988) 133-140

[358] Y.F. Yik, H.K. Lee, S.F.Y. Li and S.B. Khoo, Micellar Electrokinetic Capillary Chromatography of Vitamin B_6 with Electrochemical Detection. *J. Chromatogr.* **585** (1991) 139-144

[359] Y.F. Yik, H.K. Lee and S.F.Y. Li, Separation of Nitroaromatics by Micellar Electrokinetic Capillary Chromatography. *HRC&CC* **15** (1992) 198-200

[360] C.P. Palmer, M.Y. Khaled and H.M. McNair, A Monomolecular Pseudostationary Phase for Micellar Electrokinetic Capillary Chromatography. *HRC&CC* **15** (1992) 756-762

[361] C.D. Gaitonde and P.V. Pathak, Capillary Zone Electrophoretic Separation of Chlorophenols in Industrial Waste Water with On-Column Electrochemical Detection. *J. Chromatogr.* **514** (1990) 389-393

[362] C. Desiderio and S. Fanali, Atrazine and Simazine Determination in River Water Samples by Micellar Electrokinetic Capillary Chromatography. *Electrophoresis* **13** (1992) 698-700

[363] J. Cai and Z. El Rassi, On-Line Preconcentration of Triazine Herbicides with Tandem Octadecyl Capillaries - Capillary Zone Electrophoresis. *J. Liq. Chromatogr.* **15** (1992) 1179-1192

[364] J. Cai and Z. El Rassi, Capillary Zone Electrophoresis of Two Cationic Herbicides, Paraquat and Diquat. *J. Liq. Chromatogr.* **15** (1992) 1193-1200

[365] J. Cai and Z. El Rassi, Micellar Electrokinetic Capillary Chromatography of Neutral Solutes with Micelles of Adjustable Surface Charge Density. *J. Chromatogr.* **608** (1992) 31-45

[366] C.P. Ong, C.L. Ng, H.K. Lee and S.F.Y. Li, Separation of DNS-Amino Acids and Vitamins by Micellar Electrokinetic Chromatography. *J. Chromatogr.* **559** (1991) 537-546

[367] B.J. Wanders and A.A.A.M. van de Goor, Derivatization of Amino Acids with CBQCA for Ultrasensitive Detection by P/ACE Capillary Electrophoresis - Laser Induced Fluorescence Detection. *Beckman Instruments, Technical Information DS-825* (1992)

[368] J.Y. Zhao, K.C. Waldron, J. Miller, J.Z. Zhang, H. Harke and N.J. Dovichi, Attachment of a Single Fluorescent Label to Peptides for Determination by Capillary Zone Electrophoresis. *J. Chromatogr.* **608** (1992) 239-242

[369] P.D. Grossman, K.J. Wilson, G. Petrie and H.H. Lauer, Effect of Buffer pH and Peptide Composition on the Selectivity of Peptide Separations by Capillary Zone Electrophoresis. *Anal. Biochem.* **173** (1988) 265-270

[370] MECC of Neutral and Charged Peptides. *Applied Biosystems, Capillary Electrophoresis Appl. Note 4* (1990)

[371] Peptide Hormones: Peak Reproducibility. *Applied Biosystems, Capillary Electrophoresis Appl. Note 16* (1990)

[372] T.A.A.M. van de Goor, P.S.L. Janssen, J.W. van Nispen, M.J.M. van Zeeland and F.M. Everaerts, Capillary Electrophoresis of Peptides - Analysis of Adrenocorticotropic Hormone-Related Fragments. *J. Chromatogr.* **545** (1991) 379-389

[373] T. Sakow, A. Machida, K. Funakushi and R. Palmieri, Effects of the Sample Matrix on the Separation of Peptides by High Performance Capillary Electrophoresis. *HRC&CC* **14** (1991) 276-279

[374] C.-T. Wu, T. LOpes, B. Patel and C.S. Lee, Effect of Direct Control of Electroosmosis on Peptide and Protein Separations in Capillary Electrophoresis. *Anal. Chem.* **64** (1992) 886-891

[375] M. Zhu, D. Hansen, S. Burd, V. Huebner, P. Balasubramanian and A.J.C. Chen, Use of High Performance Electrophoresis in Monitoring Peptide Synthesis. *Bio-Rad Bulletin 1482* (1989)

[376] A. Pessi, E. Bianchi, L. Chappinelli, A. Nardi and S. Fanali, Application of Capillary Zone Electrophoresis to the Characterization of Multiple Antigen Peptides. *J. Chromatogr.* **557** (1991) 307-313

[377] M.G. Quaglia, S. Fanali, A. Nardi, C. Rossi and M. Ricci, Separation of Leucinostatins by Capillary Zone Electrophoresis. *J. Chromatogr.* **593** (1992) 259-263

[378] Z. Deyl, V. Rohlicek and R. Struzinsky, Some Rules Applicable to Capillary Zone Electrophoresis of Peptides and Proteins. *J. Liq. Chromatogr.* **12** (1989) 2515-2526

[379] P.D. Grossman, J.C. Colburn, H.H. Lauer, R.G. Nielsen, R.M. Riggin, G.S. Sittampalam and E.C. Rickard, Application of Free-Solution Capillary Electrophoresis to the Analytical Scale Separation of Proteins and Peptides. *Anal. Chem.* **61** (1989) 1186-1194

[380] P.D. Grossman, J.C. Colburn and H.H. Lauer, A Semiempirical Model for the Electrophoretic Mobilities of Peptides in Free-Solution Capillary Electrophoresis. *Anal. Biochem.* **179** (1989) 28-33

[381] Separation of Endorphin Fragments by Capillary Electrophoresis. *Applied Biosystems, Capillary Electrophoresis Appl. Note 23*

[382] R.J. Florance, Z.D. Konteatis, M.J. Macielag, R.A. Lessor and A. Galdes, Capillary Zone Electrophoresis Studies of Motilin Peptides. Effects of Charge, Hydrophobicity, Secondary Structure and Length. *J. Chromatogr.* **559** (1991), 391-400

[383] Polypeptide Hormones: Analysis of Calcitonin Variants. *Applied Biosystems, Capillary Electrophoresis Appl. Note 17*

[384] J. Soucheleau and L. Denoroy, Determination of Vasoactive Intestinal Peptide in Rat Brain by High-Performance Capillary Electrophoresis. *J. Chromatogr.* **608** (1992) 181-188

[385] E. Yildiz, G. Grübler, S. Hörger, H. Zimmermann, H. Echner, S. Stoeva and W. Voelter, Application of Capillary Electrophoresis in Peptide Research. *Electrophoresis* **13** (1992) 683-686

[386] M. Idei, I. Mezo, Zs. Vadasz, A. Horvath, I. Teplan and Gy. Keri, Capillary Electrophoretic Analysis of Somatostatin Analog Peptides. Effect of Organic Solvents as Buffer Modifiers

[387] A.J.C. Chen, M. Zhu, D. Hansen and S. Burd, Use of High Performance Electrophoresis in Peptide Analysis and Peptide Mapping. *Bio-Rad Bulletin 1480* (1989)

[388] J. Frenz, S.-L. Wu and W.S. Hancock, Characterization of Human Growth Hormone by Capillary Electrophoresis. *J. Chromatogr.* **480** (1989) 379-391

[389] H. Lüdi, E. Gassmann, H. Grossenbacher and W. Märki, Capillary Zone Electrophoresis for the Analysis of Peptides Synthesized by Recombinant DNA Technology. *Anal. Chim. Acta* **213** (1988) 215-219

[390] N.A. Guzman, J. Moschera, C.A. VBailey, K. Iqbal and A.A. Malick, Assay of Protein Drug Substances Present in Solution Mixtures by Fluorescamine Derivatization and Capillary Electrophoresis. *J. Chromatogr.* **598** (1992) 123-131

[391] N.A. Guzman. H. Ali, J, Moschera, K. Iqbal and A.W. Malick, Assessment of Capillary Electrophoresis in Pharmaceutical Applications. Analysis and Quantification of a Recombinant Cytokine in an Injectable Dosage Form. *J. Chromatogr.* **559** (1991), 307-316

[392] K.W. Yim, Fractionation of the Human Recombinant Tissue Plasminogen Activator (rtPA) Glycoforms by High-Performance Capillary Zone Electrophoresis. *J. Chromatogr.* **559** (1991), 401-410

[393] R.G. Nielsen and E.C. Rickard, Method Optimization in Capillary Zone Electrophoretic Analysis of hGH Tryptic Digest Fragments. *J. Chromatogr.* **516** (1990) 99-114

[394] Protein Digests: tryptic Map of ß-Lactoglobulin A. *Applied Biosystems, Capillary Electrophoresis Appl. Note 19* (1990)

[395] A.A.A.M. van de Goor, B.J. Wanders and W.J. Henzel, High-Sensitivity Peptide Mapping with Capillary Electrophoresis Using Laser-Induced Fluorescence Detection on a P/ACE System. *Beckman Instruments, Appl. Brief DS-824* (1992)

[396] R. Palmieri, Peptide Mapping with Capillary Electrophoresis Using the P/ACE System 2000. *Beckman Instruments, Appl. Brief DS-774* (1990)

[397] K.A. Cobb and M.V. Novotny, Peptide Mapping of Complex Proteins at the Low-Picomole Level with Capillary Electrophoretic Separations. *Anal. Chem.* **64** (1992) 879-886

[398] Z. Zhao, A. Malik and M.L. Lee, Separations of Proteins and Proteolytic Digests of Proteins by Capillary Electrophoresis on Superox-Coated Open Tubular Columns. *J. Microcol. Sep.* **4** (1992) 411-417

[399] R. Nelson, A.S. Cohen, R.S. Rush and B.L. Karger, Separation of Proteins by Isoelectric-Focusing Capillary Electrophoresis in P/ACE System 2000. *Beckman Instruments, Appl. Data DS-750* (1989)

[400] Coatings: The Key to Success in Capillary Electrophoresis of Proteins. *Bio-Rad Bulletin 1641* (1990)

[401] Å. Emmer, M. Jansson and J. Roeraade, Improved Capillary Zone Electrophoretic Separation of Basic Proteins, Using a Fluorosurfactant Buffer additive. *J. Chromatogr.* **547** (1991) 544-550

[402] M.J. Gordon, K.-J. Lee, A.A. Arias and R.N. Zare, Protocol for Resolving Protein Mixtures in Capillary Zone Electrophoresis. *Anal. Chem.* **63** (1991) 69-72

[403] K.-J. Lee and G.S. Heo, Free-Solution Capillary Electrophoresis of Proteins Using Untreated Fused-Silica Capillaries. *J. Chromatogr.* **559** (1991), 317-324

[404] F.-T.A. Chen, Rapid Protein Analysis by Capillary Electrophoresis. *J. Chromatogr.* **559** (1991), 445-454

[405] F.-T.A. Chen, Protein Separations by P/ACE System 2000 Capillary Electrophoresis on Fused-Silica Columns. *Beckman Instruments, Technical Information DS-818*

[406] M. Gilges, H. Husmann, M.-H. Kleemiß, S.R. Motsch and G. Schomburg, CZE Separations of Basic Proteins at Low pH in Fused Silica Capillaries with Surfaces Modified by Silane Derivatization and/or Adsorption of Polar Polymers. *HRC&CC* **15** (1992) 452-457

[407] M. Huang and M.L. Lee, Hydrogel Polymer Coatings for Capillary Electrophoretic Separation of Proteins. *J. Microcol. Sep.* **4** (1992) 491-496

[408] D. Wu and F.E. Regnier, Sodium Dodecyl Sulfate - Capillary Gel Electrophoresis of Proteins Using Non-Cross-Linked Polyacrylamide. *J. Chromatogr.* **608** (1992) 349-356

[409] K. Tsuji, High-Performance Capillary Electrophoresis of Proteins. Sodium Dodecyl Sulphate - Polyacrylamide Gel-Filled Capillary Column for the Determination of Recombinant Biotechnology-Derived Proteins. *J. Chromatogr.* **550** (1991) 823-830

[410] A. Widhalm, C. Schwer, D. Blaas and E. Kenndler, Capillary Zone Electrophoresis with a Linear, Non-Cross-Linked Polyacrylamide Gel: Separation of Proteins According to Molecular Mass. *J. Chromatogr.* **549** (1991) 446-451

[411] A. Guttman, P. Shieh and N. Cooke, P/ACE SDS - Capillary Gel Electrophoresis of Proteins. *Beckman Instruments, Technical Information DS-827* (1992)

[412] F. Kilar and S. Hjertén, Separation of the Human Transferrin Isoforms by Carrier-Free High-Performance Zone Electrophoresis and Isoelectric Focusing. *J. Chromatogr.* **480** (1989) 351-257

[413] S.J. Harrington, R. Varro and T.M. Li, High-Performance Capillary Electrophoresis as a Fast In-Process Control Method for Enzyme-Labeled Monoclonal Antibody Conjugates. *J. Chromatogr.* **559** (1991), 385-390

[414] HPE 100 Capillary Electrophoresis System: Monoclonal Antibodies. *Bio-Rad Appl. Note 21* (1991)

[415] M.A. Costello, C. Woititz, J. de Feo, D. Stremlo, F.-F.L. Wen, D.J. Palling, K. Iqbal and N.A. Guzman, Characterization of Humanized Anti-TAC Monoclonal Antibody by Traditional Separation Techniques and Capillary Electrophoresis. *J. Liq. Chromatogr.* **15** (1992) 1081-1097

[416] B.J. Compton, Electrophoretic Mobility Modeling of Proteins in Free Zone Capillary Electrophoresis and its Application to Monoclonal Antibody Microheterogeneity Analysis. *J. Chromatogr.* **559** (1991), 357- 366

[417] P.G. Pande, R.V. Nellore and H.R. Bhagat, Optimization and Validation of Analytical Conditions for Bovine Serum Albumin Using Capillary Electrophoresis. *Anal. Biochem.* **204** (1992) 103-106

[418] C.-N. Ong, L.S. Liau and H.Y. Ong, Separation of Globins Using Free Zone Capillary Electrophoresis. *J. Chromatogr.* **576** (1992) 346-350

[419] M. Zhu, R. Rodriguez, T. Wehr and C. Siebert, Capillary Electrophoresis of Hemoglobins and Globin Chains. *J. Chromatogr.* **608** (1992) 225-237

[420] J.E. Wiktorowicz and J.C. Colburn, Separation of Cationic Proteins *via* Charge Reversal in Capillary Electrophoresis. *Electrophoresis* **11** (1990) 769-773

[421] L.R. Gurley, J.E. London and J.G. Valdez, High-Performance Capillary Electrophoresis of Histones. *J. Chromatogr.* **559** (1991) 431-444

[422] H. Lindner, W. Helliger, A. Dirschlmayer, H. Talasz, M. Wurm, B. Sarg, M. Jaqumar and B. Puschendorf, Separation of Phosphorylated Histone H1 Variants by High-Performance Capillary Electrophoresis. *J. Chromatogr.* **608** (1992) 211-215

[423] W. Schütz, Separation of Myoglobine-Subunits by High Performance Isoelectric Focusing Using P/ACE System 2000 with Polyacrylamide-Coated Capillaries. *Beckman Instruments, Appl. Data* (1990)

[424] R.I. Hecht, S.C. Howard and G.S. Jacob, The Analysis of a Model Glycoprotein, Ribonuclease B, by Capillary Electrophoresis. *Beckman Instruments, Technical Information DS-822* (1992)

[425] A.D. Tran, T. Blanc, S.Park and P.J. Lisi, Effect of Phosphate Ions in the Analysis of Glycoproteins at Low pH by Capillary Electrophoresis Using P/ACE System 2000. *Beckman Instruments, Technical Information DS-771* (1990)

[426] Structural Analysis of RNAse T1 Native and Site-Directed Mutants. *Applied Biosystems, Capillary Electrophoresis Appl. Note 27* (1990)

[427] P. Pascual, E. Martinez-Lara, J.A. Barcena, J. Lopez-Barea and F. Toribio, Direct Assay of Glutathione Peroxidase Activity Using High-Performance Capillary Electrophoresis. *J. Chromatogr.* **581** (1992) 49-56

[428] J.P. Landers, R.P. Oda, B.J. Madden and T.C. Spelsberg, High-Performance Capillary Electrophoresis of Glycoproteins: The Use of Modifiers of Electroosmotic Flow for Analysis of Microheterogeneity. *Anal. Biochem.* **205** (1992) 115-124

[429] H. Kajiwara, Application of High-Performance Capillary Electrophoresis to the Analysis of Conformation and Interaction of Metal-Binding Proteins. *J. Chromatogr.* **559** (1991), 345-356
[430] S. Honda, S. Iwase, A. Makino and S. Fujiwara, Simultaneous Determination of Reducing Monosaccharides by Capillary Zone Electrophoresis as the Borate Complexes of *N*-2-Pyridylglycamines. *Anal. Biochem.* **174** (1989) 72-77
[431] W. Nashabeh and Z. El Rassi, Capillary Zone Electrophoresis of Pyridylamino Derivatives of Maltooligosaccharides. *J. Chromatogr.* **514** (1990) 57-64
[432] S. Honda, S. Suzuki, A. Nose, K. Yamamoto and K. Kakehi, Capillary Zone Electrophoresis of Reducing Mono- and Oligo-Saccharides as the Borate Complexes of Their 3-Methyl-1-Phenyl-2-Pyrazolin-5-One Derivatives. *Carbohydrate Res.* **215** (1991) 193-198
[433] J. Liu, O. Shirota and M. Novotny, Separation of Fluorescent Oligosaccharide Derivatives by Microcolumn Techniques Based on Electrophoresis and Liquid Chromatography. *J. Chromatogr.* **559** (1991), 223-236
[434] S. Honda, K. Yamamoto, S. Suzuki, M. Ueda and K. Kakehi, High-Performance Capillary Zone Electrophoresis of Carbohydrates in the Presence of Alkaline Earth Metal Ions. *J. Chromatogr.* **588** (1992), 327-333
[435] A.E. Vorndran, E. Grill, C. Huber, P.J. Oefner and G.K. Bonn, Capillary Zone Electrophoresis of Aldoses, Ketoses and Uronic Acids Derivatized with Ethyl *p*-Aminobenzoate. *Chromatographia* **34** (1992) 109-114
[436] P.J. Oefner, A.E. Vorndran, E. Grill, C. Huber and G.K. Bonn, Capillary Zone Electrophoretic Analysis of Carbohydrates by Direct in Indirect UV Detection. *Chromatographia* **34** (1992) 308-316
[437] W. Nashabeh and Z. El Rassi, Capillary Zone Electrophoresis of Linear and Branched Oligosaccharides. *J. Chromatogr.* **600** (1992) 279-287
[438] S. Honda, A. Makino, K. Suzuki and K. Kakehi, Analysis of the Oligosaccharides of Ovalbumin by High-Performance Capillary Electrophoresis. *Anal. Biochem.* **191** (1990) 228-234
[439] W. Nashabeh and Z. El Rassi, Capillary Zone Electrophoresis of α_1-Acid Glycoprotein Fragments from Trypsin and Endoglycosidase Digestions. *J. Chromatogr.* **536** (1991) 31-42
[440] Y. Liu and K.-F.J. Chan, High-Performance Capillary Electrophoresis of Gangliosides. *Electrophoresis* **12** (1991) 402-408
[441] A. Al-Hakim and R.J. Linhardt, Capillary Electrophoresis for the Analysis of Chondroitin Sulfate- and Dermatan Sulfate-Derived Disaccharides. *Anal. Biochem.* **195** (1991) 68-73
[442] S.A. Ampofo, H.M. Wang and R.J. Linhardt, Disaccharide Compositional Analysis of Heparin and Heparan Sulfate Using Capillary Zone Electrophoresis. *Anal. Biochem.* **199** (1991) 249-255
[443] J.B.L. Damm, G.T. Overklift, B.W.M. Vermeulen, C.F. Fluitsma and G.W.K. van Dedem, Separation of Natural and Synthetic Heparin Fragments by High-Performance Capillary Electrophoresis. *J. Chromatogr.* **608** (1992) 297-309
[444] S. Honda, T. Ueda and K. Kakehi, High-Performance Capillary Electrophoresis of Unsaturated Oligosaccharides Derived from Glycosaminoglycans by Digestion with Chondroitinase ABC as 1-Phenyl-3-Methyl-5-Pyrazolone Derivatives. *J. Chromatogr.* **608** (1992) 289-295
[445] S. Michaelsen, P. Møller and H. Sørensen, Factors Influencing the Separation and Quantitation of Intact Glucosinolates and Desulphoglucosinolates by Micellar Electrokinetic Capillary Chromatography. *J. Chromatogr.* **608** (1992) 363-374
[446] A. Henshall, M.P. Harrold and J.M.Y. Tso, Separation of Inositol Phosphates by Capillary Electrophoresis. *J. Chromatogr.* **608** (1992) 413-419

[447] J. Liu, F. Banks and M. Novotny, High-Speed Micellar Electrokinetic Capillary Chromatography of the Common Phosphorylated Nucleosides. *J. Microcol. Sep.* **3** (1989) 136-141

[448] R. Takigiku and R.E. Schneider, Reproducibility and Quantitation of Separation for Ribonucleoside Triphosphates and Deoxyribonucleoside Triphosphates by Capillary Zone Electrophoresis. *J. Chromatogr.* **559** (1991), 247-256

[449] W. Li, A. Moussa and R.W. Giese, Capillary Electrophoresis of Fluorescein-Ethylenediamine-5'-deoxynucleotides. *J. Chromatogr.* **608** (1992) 171-174

[450] M. Ng, T.F. Blaschke, A.A. Arias and R.N. Zare, Analysis of Free Intracellular Nucleotides Using High-Performance Capillary Electrophoresis. *Anal. Chem.* **64** (1992) 1682-1684

[451] L. Hernandez, B.G. Hoebel and N.A. Guzman, Analysis of Cyclic Nucleotides by Capillary Electrophoresis Using Ultraviolet Detection. *ACS Symp. Ser.* **434** (1990) 50-59

[452] A.S. Cohen, A. Paulus and B.L. Karger, High-Performance Capillary Electrophoresis Using Open Tubes and Gels. *Chromatographia* **24** (1987) 15-24

[453] A.S. Cohen, D. Najarian, J.A. Smith and B.L. Karger, Rapid Separation of DNA Restriction Fragments Using Capillary Electrophoresis. *J. Chromatogr.* **458** (1988) 323-333

[454] A.J.C. Chen, M. Zhu, D. Hansen and S. Burd, Application of High Performance Electrophoresis to DNA Restriction Fragment Analysis. *Bio-Rad Bulletin 1479* (1989)

[455] V. Dolnik, J. Liu, F. Banks, M. Novotny and P. Bocek, Capillary Zone Electrophoresis of Oligonucleotides. Factors Affecting Separation. *J. Chromatogr.* **480** (1989) 321-330

[456] A.S. Cohen, D.R. Najarian and B.L. Karger, Separation and Analysis of DNA Sequence Reaction Products by Capillary Gel Electrophoresis. *J. Chromatogr.* **516** (1990) 49-60

[457] A. Paulus and J.I. Ohms, Analysis of Oligonucleotides by Capillary Gel Electrophoresis. *J. Chromatogr.* **507** (1990) 113-123

[458] D.Y. Chen, H.P. Swerdlow, H.R. Harke, J.Z. Zhang and N.J. Dovichi, Low-Cost, High-Sensitivity Laser-Induced Fluorescence Detection for DNA Sequencing by Capillary Electrophoresis. *J. Chromatogr.* **559** (1991), 237-246

[459] H.E. Schwartz, K. Ulfelder, F.J. Sunzeri, M.P. Busch and R.G. Brownlee, Analysis of DNA Restriction Fragments and Polymerase Chain Reaction Products Towards Detection of the AIDS (HIV-1) Virus in Blood. *J. Chromatogr.* **559** (1991), 267-284

[460] V. Dolnik and M. Novotny, Capillary Electrophoresis of DNA Fragments in En-tangled Polymer Solutions: A Study of Separation Variables. *J. Microcol. Sep.* **4** (1992) 515-519

[461] A. Guttman, B. Wanders and N. Cooke, Enhanced Separation of DNA Restriction Fragments by Capillary Gel Electrophoresis Using Field Strength Gradients. *Anal. Chem.* **64** (1992) 2348-2351

[462] P.J. Oefner, G.K. Bonn, C.G. Huber and S. Nathakarnkitkool, Comparative Study of Capillary Zone Electrophoresis and High-Performance Liquid Chromatography in the Analysis of Oligonucleotides and DNA. *J. Chromatogr.* **625** (1992) 331-340

[463] Counter-Migration Capillary Electrophoresis (CMCE) in DNA Restriction Fragment Analysis. *Applied Biosystems, Capillary Electrophoresis Appl. Note 21*

[464] H.E. Schwartz and K.J. Ulfelder, Capillary Electrophoresis with Laser-Induced Fluorescence Detection of PCR Fragments Using Thiazole Orange. *Anal. Chem.* **64** (1992) 1737-1740

[465] H.R. Harke, S. Bay, J.Z. Zhang, M.J. Rocheleau and N.J. Dovichi, Effect of Total Percent Polyacrylamide in Capillary Gel Electrophoresis for DNA Sequencing of Short Fragments. A Phenomenological Model. *J. Chromatogr.* **608** (1992) 143-150

[466] J.W. Chen, A.S. Cohen and B.L. Karger, Identification of DNA Molecules by Pre-Column Hybridization Using Capillary Electrophoresis. *J. Chromatogr.* **559** (1991), 295-306

[467] M.-J. Rocheleau and N.J. Dovichi, Separation of DNA Sequencing Fragments at 53 Bases/Minute by Capillary Gel Electrophoresis. *J. Microcol. Sep.* **4** (1992) 449-453

[468] H. Swerdlow and R. Gesteland, Capillary Gel Electrophoresis for Rapid, High Resolution DNA Sequencing. *Nucl. Acids Res.* **18** (1990) 1415-1419

[469] C.E. Dalgliesh, The Optical Resolution of Aromatic Amino-Acids on Paper Chromatograms. *J. Chem. Soc.*, **137**, (1952) 3940-3943

[470] J. Snopek, I. Jelinek and E. Smolkova-Keulemansova, Chiral Separation by Analytical Electromigration Methods. *J. Chromatogr.* **609** (1992) 1-17

[471] S. Fanali and P. Bocek, Enantiomer resolution by Using Capillary Zone Electrophoresis: Resolution of Racemic Tryptophan and Determination of the Enantiomer Composition of Commercial Pharmaceutical Epinephrine. *Electrophoresis* **11** (1990) 757-760

[472] S. Fanali, Separation of Optical Isomers by Capillary Zone Electrophoresis Based on Host-Guest Complexation with Cyclodextrins. *J. Chromatogr.* **474** (1989) 441-446

[473] T.E. Peterson and D. Trowbridge, Quantitation of *l*-Epinephrine and Determination of *d*-/*l*-Epinephrine Enantiomer Ratio in a Pharmaceutical Formulation by Capillay Electrophoresis. *J. Chromatogr.* **603** (1992) 298-301

[474] H. Nishi, T. Fukuyama and S. Terabe, Chiral Separation by Cyclodextrin-Modified Micellar Electrokinetic Chromatography. *J. Chromatogr.* **553** (1991) 503-516

[475] S. Fanali, M. Flieger, N. Steinerova, A. Nardi, Use of Cyclodextrins for Enantioselective Separation of Ergot Alkaloids by Capillary Zone Electrophoresis. *Electrophoresis* **13** (1992) 39-43

[476] S. Fanali, Use of Cyclodextrins in Capillary Zone Electrophoresis, Resolution of Terbutaline and Propranolol Enantiomers. *J. Chromatogr.* **545** (1991) 437-444

[477] K.D. Altria, D.M. Goodall and M.M. Rogan, Chiral Separation of β-Amino Alcohols by Capillary Electrophoresis Using Cyclodextrins as Buffer Additives. I. Effect of Varying Operating Parameters. *Chromatographia* **34** (1992) 19-24

[478] H. Soini, M.-L. Riekkola and M.V. Novotny, Chiral Separations of Basic Drugs and Quantitation of Bupivacaine Enantiomers in Serum by Capillary Electrophoresis with Modified Cyclodextrin Buffers. *J. Chromatogr.* **608** (1992) 265-274

[479] J. Snopek, H. Soini, M. Novotny, E. Smolkova-Keulemansova and I. Jelinek, Selected Applications of Cyclodextrin Selectors in Capillary Electrophoresis. *J. Chromatogr.* **559** (1991) 215-222

[480] R. Kuhn, F. Stoeklin and F. Erni, Chiral Separations by Host-Guest Complexation with Cyclodextrin and Crown Ether in Capillary Zone Electrophoresis. *Chromatographia* **33** (1992) 32-36

[481] H. Nishi, T. Fukuyama, M. Matsuo and S. Terabe, Chiral Separation of Diltiazem, Trimetoquinol and Related Compounds by Micellar Electrokinetic Chromatography with Bile Salts. *J. Chromatogr.* **515** (1990) 233-243

[482] H. Nishi, T. Fukuyama, M. Matsuo and S. Terabe, Chiral Separation of Optical Isomeric Drugs Using Micellar Electrokinetic Chromatography and Bile Salts. *J. Microcol. Sep.* **1** (1989) 234-241
[483] H. Nishi, T. Fukuyama, M. Matsuo and S. Terabe, Chiral Separation of Trimetoquinol Hydrochloride and Related Compounds by Micellar Electrokinetic Chromatography Using Sodium Taurodeoxycholate Solutions and Application to Optical Purity Determination. *Anal. Chim. Acta* **236** (1990) 281-286
[485] A. D'Hulst and N.Verbeke, Chiral Separation by Capillary Electrophoresis with Oligosaccharides. *J. Chromatogr.* **608** (1992) 275-287
[486] A. Nardi, L. Ossicini and S. Fanali, Use of Cyclodextrins in Capillary Zone Electrophoresis for the Separation of Optical Isomers: Resolution of Racemic Tryptophan Derivatives. *Chirality* **4** (1992) 56-61
[487] S. Terabe, M. Shibata and Y. Miyashita, Chiral Separation by Electrokinetic Chromatography with Bile Salt Micelles. *J. Chromatogr.* **480** (1989) 403-411.
[488] A. Guttman, A. Paulus, S. Cohen and N. Grinberg, B.L. Karger, Use of Complexing Agents for Selective Separation in High-Performance Capillary Electrophoresis: Chiral Resolution via Cyclodextrins Incorporated Within Polyacrylamide Gel Columns*J. Chromatogr.* **448** (1988) 41-53.
[489] E. Gassmann, J.E. Kuo and R.N. Zare, Electrokinetic Separation of Chiral Compounds.*Science* **230** (1985) 813-814.
[490] P. Gozel, E. Gassmann, H. Michelson and R.N. Zare, Electrokinetic Resolution of Amino Acid Enantiomers with Copper(II)-Aspartame Support Electrolyte. *Anal. Chem.* **59** (1987) 44-49.
[491] T. Ueda, F. Kitamura, R. Mitchell, T. Metcalf, T. Kuwana and A. Nakamoto, Chiral Separation of Naphthalene-2,3-dicarboxaldehyde-Labeled Amino Acid Enantiomers by Cyclodextrin-Modified Micellar Electrokinetic Chromatography with Laser-Induced Fluorescence Detection. *Anal. Chem.* **63** (1991) 2979-2981.
[492] K. Otsuka and S. Terabe, Enantiomeric Resolution by Micellar Electrokinetic Chromatography with Chiral Surfactants. *J. Chromatogr.* **515** (1990) 221-226.
[493] S. Mayer and V. Schurig, Enantiomer Separation by Electrochromatography on Capillaries Coated with Chirasil-Dex *HRC&CC* **15** (1992) 129-131.
[494] R.O. Cole, M.J. Sepaniak, W.L. Hinze, Optimization of Binaphthyl Enantiomer Separations by Capillary Zone Electrophoresis Using Mobile Phases Containing Bile Salts and Organic Solvent. *HRC&CC* **13** (1990) 579-582
[495] E. Höhne, G.-J. Krauss and G. Gübitz, Capillary Zone Electrophoresis of the Enantiomers of Aminoalcohols Based on Host-Guest Complexation with a Chiral Crown Ether. *HRC&CC* **15** (1992) 698-700
[496] N. Banke, K. Hansen and I. Diers, Detection of Enzyme Activity in Fractions Collected From Free Solution Capillary Electrophoresis of Complex Samples. *J. Chromatogr.* **559** (1991), 325-336
[497] E. Jellum, A.K. Thorsrud and E. Time, Capillary Electrophoresis for Diagnosis and Studies of Human Disease, Particularly Metabolic Disorders. *J. Chromatogr.* **559** (1991), 455-466
[498] W.M. Hurni and W.J. Miller, Analysis of Vaccine Purification Process by Capillary Electrophoresis. *J. Chromatogr.* **559** (1991), 337-344
[499] Effect of Low Temperature Storage on Collected Urine Specimens) in the Identification of Urinary Metabolites by Capillary Electrophoresis. *J. Liq. Chromatogr.* **12** (1989) 2563-2574
[500] Is a Brew Really Just a Brew? *Applied Biosystems, Capillary Electrophoresis Appl. Note 30* (1990)

[501] A. Vinther, J. Petersen and H. Søeberg, Capillary Electrophoretic Determination of the Protease Savinase in Cultivation Broth. *J. Chromatogr.* **608** (1992) 205-210

Subject Index

MIX
Papier aus verantwortungsvollen Quellen
Paper from responsible sources
FSC® C105338

If you have any concerns about our products,
you can contact us on
ProductSafety@springernature.com

In case Publisher is established outside the EU,
the EU authorized representative is:
Springer Nature Customer Service Center GmbH
Europaplatz 3, 69115 Heidelberg, Germany

Printed by Libri Plureos GmbH
in Hamburg, Germany